P9-DJL-030

The Best American Science Writing 2002

THE BEST AMERICAN SCIENCE WRITING

EDITORS

2000: James Gleick
2001: Timothy Ferris

FORTHCOMING

2003: Oliver Sacks
2004: Dava Sobel
2005: Alan Lightman

The Best American SCIENCE WRITING 2002

EDITOR: MATT RIDLEY

Series Editor: Jesse Cohen

An Imprint of HarperCollins*Publishers*

THE BEST AMERICAN SCIENCE WRITING 2002

Permissions appear following page 324.

HarperCollins books may be purchased for educational, business, or sales promotional use. For information please write: Special Markets Department, HarperCollins Publishers, Inc. 10 East 53rd Street, New York, NY 10022.

FIRST EDITION

Designed by Cassandra J. Pappas

Library of Congress Cataloging-in-Publication Data

ISBN 0-06-621162-X HARDCOVER

ISBN 0-06-093650-9 TRADE PAPERBACK

02 03 04 05 06 BVG/RRD 10 9 8 7 6 5 4 3 2 1

Contents

Introduction by Matt Ridley ix

LAUREN SLATER | *Dr. Daedalus* 1

ATUL GAWANDE | *Crimson Tide* 21

LISA BELKIN | *The Made-to-Order Savior* 34

MARGARET TALBOT | *A Desire to Duplicate* 54

SALLY SATEL | *Medicine's Race Problem* 72

JEROME GROOPMAN | *The Thirty Years' War* 83

GARY TAUBES | *The Soft Science of Dietary Fat* 104

JOSEPH D'AGNESE | *Brothers with Heart* 124

CHRISTOPHER DICKEY | *I Love My Glow Bunny* 136

MICHAEL SPECTER | *Rethinking the Brain* 151

MARY ROGAN | *Penninger* 171

SARAH BLAFFER HRDY | *Mothers and Others* 189

NATALIE ANGIER | *Of Altruism, Heroism and Nature's Gifts in the Face of Terror* 202

JULIAN DIBBELL | *Pirate Utopia* 207

CAROLYN MEINEL | *Code Red for the Web* 216

DAVID BERLINSKI | *What Brings a World into Being?* 225

TIM FOLGER | *Quantum Shmantum* 239

OLIVER MORTON | *Shadow Science* 247

STEVEN WEINBERG | *Can Science Explain Everything? Anything?* 259

NICHOLAS WADE | *The Eco-Optimist* 273

DARCY FREY | *George Divoky's Planet* 279

About the Contributors 307

Introduction by Matt Ridley

In Douglas Adams's book *The Hitchhiker's Guide to the Galaxy,* a computer called Deep Thought is set the task of computing the key to life, the universe, and everything. After much cogitation, it comes up with the answer: 42. Adams, who died suddenly last year in California, would have guffawed at being called a philosopher. But in that little story he extracted a profound truth about the nature of knowledge: that science is just the search for new forms of ignorance.

Is "I don't know" the most under-used phrase in the English language? It is a phrase we need more now than ever, as the start of a new millennium pitches us queasily out of the present and into the future. Perhaps because of that millennial cusp, or perhaps because of September 11th, editing this collection of essays has brought home to me just how much I do not know. I don't know whether bio-terrorism is going to imperil the world, whether human genetic engineering will be in demand, whether cloning can be safe, whether there are many universes, or how much the climate will change during the coming century.

This may alarm those who prefer certainty, but it is meat and drink to science writers. The fuel on which science runs is ignorance and mystery. Go into a modern laboratory, ask the men and women in white coats what excites them and you will be given a list of enigmas and mysteries, not a catalogue of facts. They are as bored as the next person by what is already known. A known fact? Stack it on the shelf and feed it to the students.

The achievements of Newton lay not in the answers he gave but the questions he dared to ask; the thrill of the Einsteinian revolution lay in the realization that physics was not complete. What took Jim Watson to the Cavendish Laboratory in Cambridge fifty years ago last year was the enigma of the gene, not the things he already knew. Inspired by Erwin Schrodinger's little book *What Is Life?* he thought the answer to the secret of life must lie in the properties or structure of a molecule. He found the answer and it posed only more questions. The excitement of reading the human genome lies in the mighty mother lode of mystery and surprise that it undoubtedly contains. Somewhere in that digital gigabyte of DNA lies a real world-changer of a new idea.

This is a truth that escaped many science writers until quite recently. They thought their job was to take the reader through a catalog of the known facts. But that's called education and nobody does it for fun. Nowadays—since Dawkins, Hawking, Gleick, and Pinker—science writing is different; it takes the reader to the brink of the abyss of ignorance, waves its hand in the air, and cries: "One day, all this could be yours."

In this narrow sense, good science writing almost has more in common with books about telepathy and the pyramids than with student textbooks. It is about the occult—the hidden. But unlike pseudoscience, or science fiction for that matter, these are new mysteries, undreamt of in our philosophy. What mysteries has fiction produced to rival mind bogglers like deep geological time, a boundless universe, the big bang, relativity, quantum mechanics, the double helix, natural selection, mass extinction, the language instinct, and chaos theory? I can think of one—time travel—but not more.

In choosing the best American science writing, I have given full rein to my own prejudices. First among these is a bias in favor of mystery and ignorance. It is in this spirit that I have chosen "Dr. Daedalus" by Lauren Slater—an article about a man who refuses to be constrained by what is possible, by a writer who refuses to be constrained by what is normal. Slater pushes the envelope of science writing as much as her subject, Joe Rosen, plans to push the envelope of cosmetic surgery. I hope a shiver runs up your spine when you read the second paragraph and realize that it was written several months before September 11th.

Atul Gawande's article on blushing ("Crimson Tide") is about cosmetic surgery, too. But it is also an elegant illustration of the indivisibility of body and mind, told through a moving personal story.

In the last century, medical science was all about cure. Now, suddenly, it seems to be about creation. Lisa Belkin's "The Made-to-Order Savior" captures the ethical and medical journey of two desperate families searching for an unprecedented cure by means of the creation of a genetically matched child. This case was one of the landmarks of the year 2001 in science.

But it is ethical chicken feed compared to what is coming next. Margaret Talbot's "A Desire to Duplicate" reveals the real possibility that the cloning genie will soon be out of the bottle before it is known to be safe. "If someone can be replaced," she writes, "then he is no longer singular, which is to say priceless."

Another of my biases is in favor of the idea that good science writing is often contrarian. Anybody can recite the conventional wisdom. It takes courage to challenge it. A good writer should not be able to see a nostrum to which everybody subscribes to without wanting to poke it and see if it explodes. The next three pieces are all boldly contrarian. Sally Satel's "Medicine's Race Problem" worries that with the best intentions, medical researchers are ignoring race; in doing so they may miss vital clinical clues to an individual's health.

Jerome Groopman's frank confession that the war on cancer has largely failed ("The Thirty Years' War") is rare in its honesty. Political *dirigisme* is as bad at picking winners in science as in industry. In a similar vein, Gary Taubes's "The Soft Science of Dietary Fat" shows just how tall you can build a scientific house of cards: There is no evidence that eating less fat saves lives, yet it has acquired the status of dogma.

After such gloom, you will need cheering up, and there is no better way to do so than to read Joseph D'Agnese's profile of four open-minded, optimistic brothers who had been taught as children that "no theory, no question, no design of theirs was too ridiculous to explore" ("Brothers with Heart"). Their curiosity might save your life one day. Scientists are far too good at telling each other what is wrong with new ideas, and not nearly good enough at imagining what might be right.

I have a prejudice toward humor: who doesn't? "I Love My Glow Bunny" by Christopher Dickey brings genetic engineering right to the heart of culture, watches it collide with conceptual art, and steps back to enjoy the fireworks. Rarely has anybody captured the contrast between scientific and artistic creativity so well—let alone with such wit.

Who can resist the story of any underdog taking on the establishment? Medical Specter's account of Elizabeth Gould's long battle to upset the conventional wisdom about neural plasticity in the brain ("Rethinking the Brain") is a landmark piece about a crucial moment in neuroscience. Mary Rogan's hero, Joseph Penninger, has had less of a struggle to win recognition for his work. What appeals about this story is the combination of good biography and good biology—and a Gonzo style worthy of Hunter S. Thompson.

Good science writing does not have to be secondhand reporting. It can put forward original hypotheses. Some of the great ideas in science have been developed not in obscure journals, but in popular writing: natural selection, for instance, or selfish genes. This is why I include Sarah Hrdy's hypothesis that human beings evolved with cooperative rearing ("Mothers and Others"). It is a new idea, persuasively argued in her typically elegant style.

The same science of evolutionary psychology gives Natalie Angier a chance to reflect on the awful events of September 11th ("Of Altruism, Heroism and Nature's Gifts in the Face of Terror"). With great feeling she extracts insights into human heroism and self-sacrifice from the carnage of that dreadful day. Sympathy and altruism run too deep in human behavior to be mere artifacts of culture. They are part of our evolved nature.

Julian Dibbell's piece on steganography ("Pirate Utopia") begins with Osama bin Laden's deeds, too, but it was written more than six months before September 11th—another reminder of how the shade of the future haunted so many of these essays for good and ill. That is not what earned it its place so much as its almost nostalgic evocation of a distant past, of an age, now so long ago—1992—when the Internet was young, before the "anarchovisionary energy of the Cypherpunk movement [was] siphoned off into the mundane importance of making the world safe for online credit-card transactions."

There is a maturity about the Internet these days that contrasts with the genome's newness. But in one vital respect the Web can learn from genes: viruses. Another of my prejudices is an obsession with analogy and metaphor—the lifeblood (*sic*) of good science prose. But metaphor can be more than a trick; it can reveal the truth. In the case of computer viruses—so well analyzed by Carolyn Meinel in "Code Red for the Web"—it is increasingly clear that the analogy with infection is profound, because the genome, too, is digital. We are alarmingly vulnerable to both carbon and silicon pathogens.

That biology is at root a digital technology is to me the great discovery of the twentieth century, so I expected to like "What Brings a World into Being?" David Berlinski's contrarian answer, that information seems more magic than a valid scientific theory, is nonetheless fascinating. He decries rather than glories in the mystery within that answer, but that did not shake my enjoyment of his piece.

Which brings us neatly to quantum theory. Richard Feynman's famous quip, that if you think you understand quantum theory then you've missed the point of it, captures this vision of science as the worship of mystery. I find my mind well boggled by David Deutsch's multiple universes ("Quantum Shmantum" by Tim Folger)—which is a good sign. I cannot begin to tell you if Deutsch is right, but I can say that the world is a better place for having such eccentrics in it who think such thoughts.

Now comes the time for admission. Despite editing this collection of American writing, I am not American. Nor is the next author, Oliver Morton. But both of us have married Americans, and we share a rather un-European enthusiasm for new ideas and possibilities, so given that Morton's article ("Shadow Science") appeared in an American magazine, I have been graciously allowed to include it. It is about the one subject that could—and one day probably will—change the world completely, reducing the genome and September 11th to historical footnotes. For if we can find lots of other solar systems, then we surely will find other forms of digital replicators on one of their planets, and perhaps some of them will be intelligent.

Steven Weinberg's essay on the nature of explanation ("Can Science Explain Everything? Anything?") is written with his usual economy and clarity. He finds a mystery at the heart of the scientific method itself—that explanation is itself a concept that philosophers cannot pin down. And he will not let them get away with redefining the terms.

One of the most significant and controversial books of the year was Bjørn Lomborg's *The Skeptical Environmentalist.* Lomborg argues that the "litany" of the environmental movement—that things are getting worse—is largely wrong and that a dispassionate view of the statistics shows that on most measures the environment is improving, because, beyond a certain level of prosperity, more economic growth leads to less environmental damage. This appeals to two of my prejudices: first it is contrarian, and second it is optimistic. Lomborg's argument and his facts stood up well to the inevitable assault from the environmental movement, as Nicholas Wade's article shows ("The Eco-Optimist").

I doubt that George Divoky is a fan of Lomborg. As a pessimist about global warming, he expresses exasperation with those who deny it is happening at all (not that Lomborg goes that far). His own work on black guillemots off Alaska shows indubitably that the spring thaw is coming earlier every year. But what catches my attention in Darcy Frey's extraordinary profile of Divoky is the astonishing dedication of the individual scientist. For more than 25 years, Divoky has marooned himself on a bleak Arctic island each summer to study the guillemots. The length of the study makes it a valuable test of climate change, but that was not what made Divoky do it. He did it for love of the birds and the mystery, and at some personal cost. The article reminds us forcefully of the unsung heroes of science: the many men and women who put in long hours, fill in small details, and take great trouble. It is on their data that great breakthroughs stand.

I liked this article for another reason. My first scientific paper was also on the breeding behavior of a bird on a remote Arctic island: the red phalarope on Spitsbergen. Like Divoky, I find the Arctic and its ecosystem eternally fascinating. Like Divoky, I woke one day to find a polar bear right outside my tent. I went back north twice, but unlike Divoky, I could never have gone back year after year for a quarter of a century.

In 2001, I increased my own personal store of ignorance. So did humankind. The writers represented in this volume have mined some magnificent nuggets of the stuff. There is now a whole set of new questions to which the only honest answer is "I don't know."

The Best American Science Writing 2002

LAUREN SLATER

Dr. Daedalus

FROM *HARPER'S MAGAZINE*

Whether it's a skin graft for a burn victim or a face-lift for an aging movie star, plastic surgery works in the realm that most affects our identity: our appearance. The writer and psychologist Lauren Slater profiles a plastic surgeon in Massachusetts whose radical ideas go well beyond the cosmetic—and confront our basic sense of what it means to be human.

Part I: Beautiful People

Joe Rosen, plastic surgeon at the renowned Dartmouth-Hitchcock Medical Center, and by any account an odd man, has a cold. But then again, he isn't sure it's a cold. "It could be anthrax," he says as he hurries to the car, beeper beeping, sleet sleeting, for it's a freezing New England midwinter day when all the world is white. Joe Rosen's nose is running, his throat is raw, and he's being called into the ER because some guy made meat out of his forefinger and a beautiful teenager split her fine forehead open on the windshield of her SUV. It seems unfair, he says, all these calls coming in on a Sunday, especially because he's sick and he isn't sure whether it's the flu or the first subtle signs of a biological attack. "Are you serious?" I say to him. Joe Rosen is smart. He graduated cum laude from Cornell and got a medical degree from Stanford in 1978. And we're in his car now, speeding toward the hospital where he reconstructs faces, appends limbs, puffs and preens the female form. "You really wonder," I say, "if your cold is a sign of a terrorist attack?"

Joe Rosen, a respected and controversial plastic surgeon, wonders a lot of things, some of them directly related to his field, others not. Joe Rosen wonders, for instance, whether Osama bin Laden introduced the West Nile virus to this country. Joe Rosen wonders how much bandwidth it would take to make virtual-reality contact lenses available for all. Joe Rosen wonders why both his ex-wife and his current wife are artists, and what that says about his deeper interests. Joe Rosen also wonders why we insist on the kinds of conservative medical restraints that prevent him from deploying some of his most creative visions: wings for human beings; cochlear implants to enhance hearing, beefing up our boring ears and giving us the range of an owl; super-duper delicate rods to jazz up our vision—binocular, beautiful—so that we could see for many miles and into depths as well. Joe Rosen has ideas: implants for this, implants for that, gadgets, gears, discs, buttons, sculpting soft cartilage that would enable us, as humans, to cross the frontiers of our own flesh and emerge as something altogether . . . what? Something other.

And we're in the car now, speeding on slick roads toward the hospital, beeper beeping, sleet sleeting, passing cute country houses with gingerbread trim, dollops of smoke hanging above bright brick chimneys; his New Hampshire town looks so sweet. We pull into the medical center. Even this has a slight country flair to it, with gingham curtains hanging in the rows of windows. We skid. Rosen says, "One time I was in my Ford Explorer with my daughter, Sam. We rolled, and the next thing I knew we were on the side of the highway, hanging upside down like bats." He laughs.

We go in. I am excited, nervous, running by his bulky side with my tape recorder to his mouth. A resident in paper boots comes up to us. He eyes the tape recorder, and Rosen beams. Rosen is a man who enjoys attention, credentials. A few days ago he boasted to me, "You shouldn't have any trouble with the PR people in this hospital. I've had three documentaries made of me here already."

"Can I see them?" I asked.

"I don't know," Rosen answered, suddenly scratching his nose very fast. "I guess I'm not sure where I put them," and something about his voice, or his nose, made me wonder whether the documentaries were just a tall tale.

Now the resident rushes up to us, peers at the tape recorder, peers at me. "They're doing a story on me," Rosen says. "For *Harper's*."

"Joe is a crazy man, a nutcase," the resident announces, but there's affection in his voice.

"Why the beeps?" Rosen asks.

"This guy, he was working in his shop, got his finger caught in an electric planer . . . The finger's hamburger," the resident says. "It's just hamburger."

We go to the carpenter's cubicle. He's a man with a burly beard and sawdust-caked boots. He lies too big for the ER bed, his dripping finger held high in the air and splinted. It does look like hamburger.

I watch Rosen approach the bed, the wound. Rosen is a largish man, with a curly head of hair, wearing a Nordstrom wool coat and a cashmere scarf. As a plastic surgeon, he thinks grand thoughts but traffics mostly in the mundane. He has had over thirty papers published, most of them with titles like "Reconstructive Flap Surgery" or "Rhinoplasty for the Adolescent." He is known among his colleagues only secondarily for his epic ideas; his respect in the field is rooted largely in his impeccable surgical skill with all the toughest cases: shotgunned faces, smashed hands.

"How ya doin'?" Rosen says now to the carpenter. The carpenter doesn't answer. He just stares at his mashed finger, held high in the splint.

Rosen speaks softly, gently. He puts his hand on the woodworker's dusty shoulder. "Looks bad," he says, and he says this with a kind of simplicity—or is it empathy?—that makes me listen. The patient nods. "I need my finger," he says, and his voice sounds tight with tears. "I need it for the work I do."

Rosen nods. His tipsiness, his grandiosity, seem to just go away. He stands close to the man. "Look," he says, "I'm not going to do anything fancy right now, okay? I'll just have my guys sew it up, and we'll try to let nature take its course. I think that's the best thing, right now. To let nature take its course."

The carpenter nods. Rosen has said nothing really reassuring, but his tone is soothing, his voice rhythmic, a series of stitches that promises to knit the broken together.

We leave the carpenter. Down the hall, the teenage beauty lies in still more serious condition, the rent in her forehead so deep we can see, it seems, the barest haze of her brain.

"God," whispers Rosen as we enter the room. "I dislike foreheads. They get infected so easily."

He touches the girl. "You'll be fine," he says. "We're not going to do anything fancy here. Just sew you up and let nature take its course."

I think these are odd, certainly unexpected words coming from a man who seems so relentlessly anti-nature, so visionary and futuristic in his interests. But then again, Rosen himself is odd, a series of swerves, a topsy-turvy, upside-down, smoke-and-mirrors sort of surgeon, hanging in his curious cave, a black bat.

"I like this hospital," Rosen announces to me as we leave the girl's room. "I like its MRI machines." He pauses.

"I should show you a real marvel," he suddenly says. He looks around him. A nurse rushes by, little dots of blood on her snowy smock. "Come," Rosen says.

We ride the elevator up. The doors whisper open. Outside, the sleet has turned to snow, falling fast and furious. The floor we're on is ominously quiet, as though there are no patients here, or as though we're in a morgue. Rosen is ghoulish and I am suddenly scared. I don't know him really. I met him at a medical-ethics convention at which he discussed teaching *Frankenstein* to his residents and elaborated, with a little light in his eye, on the inherent beauty in hybrids and chimeras, if only we could learn to see them that way. "Why do we only value the average?" he'd asked the audience. "Why are plastic surgeons dedicated only to restoring our current notions of the conventional, as opposed to letting people explore, if they want, what the possibilities are?"

Rosen went on to explain other things at that conference. It was hard for me to follow his train of thought. He vacillates between speaking clearly, almost epically, to mumbling and zigzagging and scratching his nose. At this conference he kangaroo-leapt from subject to subject: the army, biowarfare, chefs with motorized fingers that could whip eggs, noses that doubled as flashlights, soldiers with sonar, the ocean, the monsters, the marvels. He is a man of breadth but not necessarily depth. "According to medieval man," Rosen said to the convention, finally coming clear, "a monster is someone born with congenital deformities. A marvel," he explained, "is a person with animal parts—say, a tail or wings." He went on to show us pictures, a turn-of-the-century newborn hand with syphilitic sores all over it, the fingers webbed in a way that might have been beautiful but not to me, the pearly skin stretched to nylon netting in the crotch of each crooked digit.

And the floor we're on now is ominously quiet, except for a hiss somewhere, maybe some snake somewhere, with a human head. We walk for what seems a long time. My tape recorder sucks up the silence.

Rosen turns, suddenly, and with a flourish parts the curtains of a cubicle. Before me, standing as though he were waiting for our arrival, is a man, a real man, with a face beyond description. "Sweeny,"* Rosen says, gesturing toward

*Not his real name.

the man, "has cancer of the face. It ate through his sinus cavities, so I scraped off his face, took off his tummy fat, and made a kind of, well, a new face for him out of the stomach. Sweeny, you look good!" Rosen says.

Sweeny, his new face, or his old stomach, oozing and swollen from this recent, radical surgery, nods. He looks miserable. The belly-face sags, the lips wizened and puckered like an anus, the eyes in their hills of fat darting fast and frightened.

"What about my nose?" Sweeny says, and then I notice: Sweeny has no nose. The cancer ate that along with the cheeks, etc. This is just awful. "That comes next. We'll use what's left of your forehead." A minute later, Rosen turns to me and observes that pretty soon women will be able to use their buttocks for breast implants. "Where there's fat," Rosen says, "there are possibilities."

THE COFFEE IS HOT and good. We drink it in the hospital cafeteria while we wait for the weather to clear. "You know," Rosen says, "I'm really proud of that face. I didn't follow any protocol. There's no textbook to tell you how to fashion a face eaten away by cancer. Plastic surgery is the intersection of art and science. It's the intersection of the surgeon's imagination with human flesh. And human flesh," Rosen says, "is infinitely malleable. People say cosmetic surgery is frivolous—boobs and noses. But it's so much more than that! The body is a conduit for the soul, at least historically speaking. When you change what you look like, you change who you are."

I nod. The coffee, actually, is too damn hot. The delicate lining of skin inside my mouth starts to shred. The burn-pain distracts me. I have temporarily altered my body, and thus my mind. For just one moment, I am a burned-girl, not a writer-girl. Rosen may be correct. With my tongue I flick the loose skin, picture it, pink and silky, on fire.

NO, PLASTIC SURGERY is not just boobs and noses. Its textbooks are tomes—thick, dusty, or slick, no matter—that all open up to images of striated muscle excised from its moorings, bones—white, calcium-rich—elongated by the doctor's finest tools. Plastic surgery, as a medical specialty, is very confusing. It aims, on the one hand, to restore deformities and, on the other hand, to alter the normal. Therefore, the patients are a motley crew. There is the gorgeous blonde with the high sprayed helmet of hair who wants a little tummy tuck, even though she's thin, and then there is the Apert

Syndrome child, the jaw so foreshortened the teeth cannot root in their sockets. Plastic surgery—like Rosen, its premier practitioner—is flexible, high-minded, and wide-ranging, managing to be at once utterly necessary and ridiculously frivolous, all in the same breath, all in the same scalpel.

According to the American Society of Plastic Surgeons, last year more than 1.3 million people had cosmetic surgery performed by board-certified plastic surgeons, an increase of 227 percent since 1992. (These numbers do not include medically necessary or reconstructive surgeries.) The five most popular procedures were liposuction (229,588), breast augmentation (187,755), eyelid surgery (172,244), the just available Botox injections (118,452), and face lifts (70,882). Most cosmetic surgeries are performed on women, but men are catching up: the number of men receiving nose jobs—their most popular procedure—has increased 141 percent since 1997. The vast majority of patients are white, but not necessarily wealthy. A 1994 study found that 65 percent of cosmetic-surgery patients had a family income of less than $50,000, even though neither state nor private health insurance covers the cost of cosmetic surgeries. These figures alone point to the tremendous popularity and increasing acceptance of body alteration, and suggest that the slippery slope from something as bizarre as eyelid tucks to something still more bizarre, like wings, may be shorter than we think.

This medical specialty is ancient, dating back to 800 B.C., when hieroglyphics describe crude skin grafts. Rosen once explained to me that plastic surgery started as a means to blur racial differences. "A long time ago," he'd said, "Jewish slaves had clefts in their ears. And some of the first plastic-surgery operations were to remove those signs of stigma."

One history book mentions the story of a doctor named Joseph Dieffenbach and a man with grave facial problems. This man had the sunken nose of syphilis, a disease widely associated with immorality. Dieffenbach, one of the fathers of plastic surgery, so the story goes, devised a gold rhinoplasty bridge for this marginal man, thus giving him, literally, a Midas nose and proving, indeed, that medicine can make criminals kings.

As a field, plastic surgery is troubled, insecure. It is a lot like psychiatry, or dentistry, in its inferior status as a subspecialty of medicine. In fact, the first plastic-surgery association, started in 1921, was an offshoot of oral practitioners. Read: teeth people. Not to digress, but the other day I woke up with a terrible toothache and rushed in to see a dentist. I said to him, just to be friendly, "What sort of training do you need for your profession?" He said, "You need A LOT of training, believe me. I trained with the same guys who cure your cancer, but I don't get the same respect."

I wonder if Rosen ever feels like my dentist, and if that's why he's so grandiose, like the little boy who is a bully. Sander Gilman, a cultural critic of plastic surgery, writes that, in this group of doctors, there are a lot of big words thrown around in an effort to cover up the sneaking suspicion that their interventions are not important. One is not ever supposed to say "nose job"; it's called rhinoplasty. Gilman writes, "The lower the perceived status of a field . . . the more complex and 'scientific' the discourse of the field becomes."

Of course, I rarely meet a doctor who doesn't like jargon and doesn't like power. Rosen may be different only in intensity. "I'm not a cosmetic surgeon," Rosen keeps repeating to me.

He says, "Really, there's no such thing as just cosmetic surgery. The skin and the soul are one." On paper, maybe, this comment seems a little overblown, but delivered orally, in a New England town when all the world is white, it has its lyrical appeal.

When Rosen cries out that he's not "just a cosmetic surgeon," he's put his finger on a real conflict in his field. Where does necessary reconstruction end and frivolous interventions begin? Are those interventions really frivolous, or are they emblematic of the huge and sometimes majestic human desire to alter, to transcend? If medicine is predicated upon the notion of making the sick well, and a plastic surgeon operates on someone who is not sick, then can the patient truly be called a patient, and the doctor a doctor? Who pays for this stuff, when, where, and how? These are the swirling questions. Over a hundred years ago Jacques Joseph, another of plastic surgery's founding fathers, wrote that beauty was a medical necessity because a person's looks can create social and economic barriers. Repairing the deformity, therefore, allows the man to function in a fully healthy way in society. Voilà. Function and form, utilitarianism and aestheticism, joined at the hip, grafted together: skin tight.

PERHAPS WE CAN accept Joseph's formulation. Okay, we say. Calm down. We say this to all the hopping, hooting cosmetic surgeons who want to stake out their significance. Okay, we respect you. I'd like to say this to Rosen, but I can't. Rosen's ideas and aspirations, not to mention his anthrax concerns, go beyond what I am comfortable with, though I can't quite unearth the architecture of my concerns. After all, he doesn't want to hurt anyone. Maybe it's because Rosen isn't just talking about everyday beauty and its utilitarian aspects. He is talking EXTREMES. When Rosen thinks of beauty, he thinks of the human form stretched on the red-hot rack of his imagination, which is mired in medieval texts and books on trumpeter swans. At its outermost limits, beauty becomes fantastical, perhaps absurd. Here is where Rosen rests. He dreams of making wings for human beings. He has shown me blue-prints, sketches of the scalpel scissoring into skin, stretching flaps of torso fat to fashion gliders piped with rib bone. When the arm stretches, the gliders unfold, and human floats on currents of air. Is he serious? At least partially. He gives lectures to medical students on the meaning of wings from an engineering perspective, a surgeon's perspective, and a patient's perspective. He has also thought of cochlear implants to enhance normal hearing, fins to make us fish-like, and echolocation devices so that we can better navigate the night. He does not understand the limits we place on hands. He once met a Vietnamese man with two thumbs on one hand. This man was a waiter, and his two thumbs made him highly skilled at his job. "Now," says Rosen, "if that man came to me and said, 'I want you to take off my extra thumb,' I'd be allowed, but I wouldn't be allowed to put an extra thumb on a person, and that's not fair."

We can call Rosen ridiculous, a madman, a monster, a marvel. We could dismiss him as a techno geek or a fool or just plain immature. But then there are the facts. First of all, Rosen is an influential man, an associate professor of surgery at Dartmouth Medical School and the director of the Plastic Surgery Residency Program at the medical center. He was senior fellow at the C. Everett Koop Institute from 1997 to 1998, and he has also served on advisory panels for the navy and for NASA's Medical Care for the Mission to Mars, 2018. Rosen consults for the American Academy of Sciences committee on the role of virtual-reality technology, and he is the former director of the Department of Defense's Emerging Technology Threats workforce. In other words, this is a man taken seriously by some serious higher-ups. "Echolocation devices," Rosen explains, "implanted in a soldier's head, could do a lot to enhance our military capacity." And this isn't just about the army's fantasies of the perfect soldier. Rosen travels worldwide (he gave over a dozen presentations last year) and has had substantial impact not only scalpeling skin but influencing his colleagues' ethics in a myriad of ways. "He has been essential in helping me to conceptualize medicine outside of the box," says Charles Lucey, MD, a former colleague of Rosen's at the Dartmouth Medical School. John Harris, a medical-ethics specialist in Manchester, England, writes in *Wonderwoman and Superman* that "in the absence of an argument or the ability to point to some specific harm that might be involved in crossing species boundaries, we should regard the objections *per se* to such practices . . . as mere and gratuitous prejudice." Rosen himself says, "Believe me. Wings are not way off. It is not a bad idea. Who would have thought we'd ever agree to hold expensive, potentially dangerous radioactive devices up to our ears for hours on end, day after day, just so we could gossip. That's cell phones for you," he says. And smiles.

Rosen has a nice smile. It's, to be sure, a little boyish, but it's charming. Sometimes Rosen is shy. "I mumble a lot," he acknowledges. "I don't really like people. I don't really like the present. I am a man who lives in the past and in the future only."

Now we leave the emergency room. The snow has stopped. The roads are membraned with ice. The sun is setting in the New Hampshire sky, causing the hills to sparkle as though they're full of little lights and other electric things. We drive back to his house, slowly. The emergencies are over, the patients soothed or suffering, he has done what can be done in a day, and still his nose runs. He coughs into his fist. "Truth be told," he says to me, "I didn't start out wanting to be a surgeon, even though I always, ALWAYS, had big ideas. In kindergarten, when the other kids were making these little ditsy arts-and-crafts projects, I was building a room-size Seventh Fleet ship." He goes on. As a child he wanted to be an artist. In high school he became obsessed with Picasso's *Guernica* and spent months trying to replicate it in the style of Van Gogh. As a freshman at Cornell, he made a robotic hand that could crack his lobster for him, and from then on it was hands, fingers, knees, and toes. His interests in the technical aspects of the body drew him away from the arts and eventually into medical school, which was, in his mind, somewhere between selling out and moving on.

We pull into his driveway. Rosen lives in a sprawling ranch-style house. He has a pet hen, who waits for us in the evergreen tree. His second wife, Stina Kohnke, is young and, yes, attractive. I'm afraid to ask how old she is; he looks to be at least fifty-three and she looks twenty-three, though maybe that's beside the point. Nevertheless, it all gets thrown into my mental stew: grandiose man, military man, medicine man, wants to make wings, young thing for a mate. Rooster and hen. Maybe there is no story here. Maybe there's just parody. All breadth, no depth. Except for this. Everyone I tell about Rosen and his wings, his *fin de siècle* mind, widens his or her eyes, leans forward, and says, "You're kidding." People want to hear more. *I* want to hear more. His ideas of altering the human form are repugnant and delicious, and that's a potent combination to unravel. And who among us has not had flying dreams, lifted high, dramatically free, a throat-catching fluidity in our otherwise aching form, above the ocean, all green, like moving marble?

Rosen and his wife have invited me for dinner. I accept. Stina is an artist. Her work is excellent. "Joe is an inspiration for me," she says. "He brings home pictures of his patients, and I sculpt their limbs from bronze." In her studio, she has a riot of red-bronze deformed hands clutching, reaching, in an agony of stiffness. She has fashioned drawer pulls from gold-plated ears. You go to open the breadbox, the medicine cabinet, the desk drawer, and you have to touch these things. It's at once creepy and very beautiful.

We sit at their stone dining-room table. Behind us is a seventy-gallon aquarium full of fish. Cacti, pink and penile, thrust their way into the odd air. Stina, homesick for her native California, has adorned the living room with paper palm trees and tiny live parakeets. We talk. Stina says, "Joe and I got married because we found in each other the same aesthetic and many moral equivalents. We found two people who could see and sculpt the potential in what others found just ugly."

"How did you two meet?" I ask.

"Oh, I knew Stina's sister, who was an art professor . . . That sort of thing," mumbles Rosen.

"I kissed him first," says Stina. She reaches across the table, picks up Rosen's hand, and wreathes her fingers through his. She holds on tightly, as if she's scared. I study Stina. She is conventionally pretty. She has a perfect Protestant nose and a lithe form, and a single black bra strap slips provocatively from beneath her blouse. Rosen, a man who claims to love the unusual, has picked a very usual beauty.

"Look!" Stina suddenly shouts. I jump, startled. "Look at her ears!" she says to Rosen.

Before I know it they are both leaning forward, peering at my ears. "Oh, my God," says Stina, "you have the most unusual ears."

Now, this is not news to me. I have bat ears, plain and simple. They stick out stupidly. In the fifth grade, I used to fasten them to the sides of my skull with pink styling tape in the hope of altering their shape. I have always disliked my ears.

Rosen uncurls his index finger and touches my left ear. He runs his finger along the bumpy, malformed rim. "You're missing the *scapha,*" he says. "It's a birth defect."

"I have a birth defect?" I say. I practically shout this, being someone who desires deeply not to be defective. That's why I take Prozac every day.

"Joe," says Stina, "are those not the most amazing ears. They would be so perfect to sculpt."

"They're just a perfect example," Rosen echoes, "of the incredible, delectable proliferation of life-forms. We claim most life-forms gravitate toward the mean, but that's not true. Lots of valid life exists at the margins of the bell curve. You have beautiful ears," he says to me.

"I have nice ears?" I say. "Really?"

This is just one reason why I won't dismiss Rosen out of hand. Suddenly, I see my ears a little differently. They have a marvelous undulating ridge and an intricately whorled entrance, and they do not stick out so much as jauntily jut; they are ears with an attitude. Rosen has shifted my vision without even touching my eyes. He is, at the very least, a challenger of paradigms; he calls on your conservatism, pushes hard.

That night, I do not dream of wings. I dream of Sweeny and his oozing face. I dream he comes so close to me that I smell him. Then I wake up. Sweeny is very sick. He is going to die soon. Earlier in the day, I asked Rosen when, and Rosen said, "Oh, soon," but he said it as if he didn't really care. Death does not seem to interest Rosen. Beauty, I think, can be cold.

Part II: Monster and Marvels

TODAY, ROSEN AND I are attending a conference together in Montreal. Here, everyone speaks French and eats baguettes. The conference room is old-fashioned, wainscoted with rich mahogany, ornate carvings of creatures and angels studding the ceiling, where a single light hangs in a cream-colored orb. Around the table sit doctors, philosophers, graduate students: this is a medical-ethics meeting, and Rosen is presenting his ideas. On the white board, in bold black lines, he sketches out his wings, and then the discussion turns to a patient whose single deepest desire was to look like a lizard. He wanted a doctor to split his tongue and scale his skin, and then put horns on his head. "You wouldn't do that, would you?" a bespectacled doctor asks. "Once," says Rosen, dodging in a fashion typical of him, "there was a lady in need of breast reconstruction who wanted blue areolas. What's wrong with blue areolas? Furthermore, rhinoplasty has not reached its real potential. Why just change the nose? Why not change the gene for the nose, so that subsequent generations will benefit from the surgery? Plastic surgery, in the future, can be about more than the literal body. It can be about sculpting the genotype as well."

The bespectacled doctor raises his hand. "Would you make that man into a lizard?" the doctor asks again. "What I want to know is, if a patient came to you and said, 'I want you to give me wings,' or 'Split my tongue,' would you actually do it?"

"Look," says Rosen, "we genetically engineer food. That's an issue."

"You're not answering my question," the doctor says, growing angry. Other people are growing angry, too. "Do you see any ethical dilemmas in making people into pigs, or birds?" another attendee yells out. This attendee is eating a Yodel, peeling off the chocolate bark and biting into a swirl of cream.

Rosen darts and dodges. "There is such a thing as liberty," he says.

"Yes," someone says, "but there's such a thing as the Hippocratic oath, too."

This goes on and on. At last a professor of anthropology says, "Just tell us, clearly, please. Would you give a human being wings, if the medical-ethics board allowed it?"

Rosen puts down his black marker. He rubs his eyes. "Yes," he says, "I would. I can certainly see why we don't devote research money to it. I can see why the NIH would fund work on breast cancer over this, but I don't have any problem with altering the human form. We do it all the time. It is only our Judeo-Christian conservatism that makes us think this is wrong. Who here," he says, "doesn't try to send their children to the best schools, in the hopes of altering them? Who here objects to a Palm Pilot, a thing we clasp to our bodies, with which we receive rapid electronic signals? Who here doesn't surround themselves with a metal shell and travel at death-defying speeds? We have always altered ourselves, for beauty or for power, and so long as we are not causing harm what makes us think we should stop?"

For a group of intelligent people everyone looks baffled. What Rosen has said is very right and very wrong, but no one can quite articulate the core conflicts. After all, we seem to think it's okay to use education as a way of neuronally altering the brain, but not surgery. We take Prozac, even Ritalin, to help transform ourselves, but recoil when it comes to wings. Maybe we're not recoiling. Maybe wings are just a dumb idea. No one in his right mind would subject himself to such a superfluous and strenuous operation. Yet socialite Jocelyne Wildenstein has dedicated much of her life to turning herself into a cat, via plastic surgery. She has had her lips enlarged and her face pulled back at the eyes to simulate a feline appearance. An even more well-known case is Michael Jackson, who has whitened himself, slimmed his nose, and undergone multiple other aesthetic procedures. The essential question here is whether these people are, and forever will be, outliers, or whether they represent the cutting edge of an ever more popular trend. Carl Elliott, a bioethicist and associate professor at the University of Minnesota, recently wrote in *The Atlantic* about a strange new "trend" of perfectly healthy folks who desire nothing more than to have a limb amputated, and about the British doctor who has undertaken this surgery, believing that if he doesn't amputate the patients will do it themselves, which could lead to gangrene. Elliott wonders whether amputation obsession will morph into another psychiatric diagnosis, whether, like hysteria, it will "catch on." The metaphor of contagion is an interesting one. Multiple-personality disorder "caught on"; hysteria caught on. Why then might not an unquenchable desire for wings or fins catch on, too? In any case, we use medical/viral metaphors to explain trends, and, in the case of plastic surgery, we then use medical means to achieve the trend's demands.

Rosen himself now repeats to the conferees, "We have always altered ourselves for beauty or for power. The chieftains in a certain African tribe remove their left ears, without Novocain. Other tribes put their bodies through intense scarification processes for the sake of style. In our own culture, we risk our bodies daily to achieve status, whether it's because we're bulimic or because we let some surgeon suck fat from us, with liposuction. Wings will be here," Rosen says. "Mark my words."

He suddenly seems so confident, so clear. We should do this; beauty is marvelous and monstrous. Beauty is difference, and yet, to his patients in the ER just two weeks back, he kept saying, "Let nature take its course." Perhaps he is more ambivalent than he lets on.

LATER THAT EVENING, over dinner, conferees gossip about Rosen. "He's a creep," someone says. "A megalomaniac," someone else adds. For a creep or a megalomaniac, though, he's certainly commanding a lot of attention. Clearly, his notions are provocative. "The problem with wings," says someone, "is that only rich people would have them, would be able to afford them. Our society might begin to see rich people as more godly than ever."

I order a glass of wine. The waitress sets it on the table, where it blazes in its goblet, bright as a tulip. With this wine, I will tweak not only my mind but all its neuronal projections as well. My reflexes will slow down and my inhibitions will lift, making it possible for me to sound either very stupid or very smart. Is this wine an ethical problem? I ask the group that.

"Wine is reversible," someone says. "Wings aren't."

"Well, suppose they *were* reversible," someone says. "Supposing a surgeon could make wings that were removable. Then would we be reacting this way?"

"It's a question of degree," a philosopher pipes up. He is bald and skinny, with bulging eyes. "Rosen is going to the nth degree. It's not fair to lump that in with necessary alterations, or even questionably necessary alterations. Without doubt, it is very clear, diagnostically, that wings are not necessary."

I think about this. I think about what Rosen might say to this. I can imagine that his answer might have something to do with the fluidity of the concept of necessary. Four years ago, cell phones weren't necessary. Now they seem to be. Furthermore, he might say, if a person wants wings, if wings won't hurt a person, if they will help a person enjoy life and feel more beautiful, and if, in turn, the winged woman or man helps us to see beauty in what was before unacceptable, as we adjust and then come to love the sight of her spreading and soaring, then isn't this excellent? Later on, in my hotel room, I stand in front of the mirror, naked. My body contains eons. Once, we were single cells, then fish, then birds, then mammals, and the genes for all these forms lie dormant on their cones of chromosomes. We are pastiches at the cellular, genetic level. This may be why I fear open spaces, blank pages, why I often dream my house opens up into endless rooms I never knew were there, and I float through them with a kind of terror. It is so easy to seep, to be boundless. We clutch our cloaks of skin.

Back in Boston, I try to ascertain clearly, logically, what so bothers people about Rosen's ideas. At first glance, it might seem fairly obvious. I mean, wings. That's playing God. We should not play God. We should not reach for the stars. Myth after myth has shown us the dangers of doing so—Icarus, the Tower of Babel; absolute power corrupts absolutely. Bill Joy, chief scientist at Sun Microsystems, says, as our technological capabilities expand, "a sequence of small, individually sensible advances leads to an accumulation of great power and, concomitantly, great danger." Rosen's response to this: "So are we supposed to stop advancing? And who says it's bad to play God? We already alter the course of God's 'will' in hundreds of ways. When we use antibiotics to combat the flu, when we figure out a way to wipe smallpox off the very face of the earth, surely we're altering the natural course of things. Who says the natural course of things is even right? Maybe God isn't good."

The second objection might have to do with our notions of categorical imperatives. Mary Douglas wrote in her influential anthropological study *Purity and Danger* that human beings have a natural aversion to crossing categories, and that when we do transgress we see it as deeply dirty. In other words, shoes in themselves are not dirty, but when you place them on the dining-room table they are. When you talk about crossing species, either at the genetic or the anatomical level, you are mucking about in long-cherished categories that reflect our fundamental sense of cleanliness and aesthetics.

Rosen's response to this, when I lob it at him in our next meeting: "Who says taboos are anything but prejudice at rock bottom? Just because it feels wrong doesn't mean it is. To a lot of people, racial intermingling and miscegenation feel wrong, but to me they're fine. I'm not a racist, and I'm not a conservative."

The third objection I can come up with has to do with the idea of proteanism. Proteus, a minor mythological figure, could shape-shift at will, being alternately a tiger, a lizard, a fire, a flood. Robert Lifton, one of, I think, the truly deep thinkers of the last century, has explored in his volumes how Proteus has become a symbol for human beings in our time. Lacking traditions, supportive institutions, a set of historically rooted symbols, we have lost any sense of coherence and connection. Today it is not uncommon for a human being to shift belief systems several times in a lifetime, and with relatively little psychological discomfort. We are Catholics, Buddhists, reborn, unborn, artists, and dot-commers until the dot drops out of the com and it all comes crashing down. We move on. We remarry. Our protean abilities clearly have their upsides. We are flexible and creative. But the downside is, there is no psychic stability, no substantive self, nothing really meaty and authentic. We sense this about ourselves. We know we are superficial, all breadth and no depth. Rosen's work embodies this tendency, literally. He desires to make incarnate the identity diffusion so common to our culture. Rosen is in our face making us face up to the fact that the inner and outer connections have crumbled. In our ability to be everything, are we also nothing?

For me, this hits the nail on the head. I do not object to Rosen on the basis of concerns about power, or of Mary Douglas's cross-category pollution theory. After all, who, really, would wings reasonably benefit but the window washers among us? And as for the pollution issue, protean person that I am, I could probably adjust to a little chimerical color. Rosen's ideas and aspirations are frightening to me because they are such vivid, visceral examples of a certain postmodern or perhaps, more precisely put, post-authentic sensibility we embrace and fear as we pop our Prozacs and Ritalins and decide to be Jewish and then Episcopalian and then chant with the monks on some high Himalayan mountain via a cheap plane ticket we purchased in between jobs and just before we sold our condo in a market rising so fast that when it falls it will sound like all of the precious china plates crashing down from the cabinet—a mess. What a mess!

Over and over again, from the Middle Ages on, when the theologian Pico wrote, in a direct and influential challenge to the Platonic idea of essential forms—"We have given you, Adam, no visage proper to yourself, nor endowment properly your own . . . trace for yourself the lineaments of your own nature . . . in order that you, as the free and proud shaper of your own being, fashion yourself in the form you may prefer . . . [W]ho then will not look with awe upon this our chameleon . . . "—over and over, since those words at least, we as human beings have fretted about the question of whether there is anything fixed at our core, any set of unalterable traits that make us who we were and are and always will be. Postmodernism, by which I mean the idea of multiplicity, the celebration of the pastiche, and the rejection of logical positivism and absolutism as viable stances, will never die out, despite its waning popularity in academia. Its roots are too deep and ancient. And there has been, perhaps, no field like modern medicine, with all its possibilities and technological wizardry, to bring questions of authenticity to the burning forefront of our culture. At what point, in altering ourselves, would we lose our essential humanity? Are there any traits that make us essentially human? When might we become monsters or marvels, or are we already there? I vividly remember reading a book by a woman named Martha Beck. She had given birth to a Down's syndrome child and she wrote in a few chilling sentences that because of one tiny chromosome, her child, Adam, is "as dissimilar from me as a mule is from a donkey. He is, in ways both obvious and subtle, a different beast." Is it really that simple, that small? One tiny chromosome severs us from the human species? One little wing and we're gone?

As for me, I am an obsessive. I like my categories. I check to make sure the stove is off three times before I go to bed. I have all sorts of other little rituals. At the same time, I know I am deeply disrooted. I left my family at the age of fourteen, never to return. I do not know my family tree. Like so many of us, I have no real religion, which is of course partly a good thing but partly a bad thing. In any case, last year, in some sort of desperate mood, I decided to convert from Judaism to Episcopalianism, but when it came time to put that blood and body in my mouth I couldn't go through with it. Was this because at bottom I just AM a Jew and this amness has profundity? Or was this because I don't like French bread, which is what they were using at the conversion ceremony? In any case, at the crucial moment of incorporation, I fled the church like the proverbial bride who cannot make the commitment.

I want to believe there is something essential and authentic about me, even if it's just my ears. And although my feelings of diffusion may be extreme, I am certainly not the only one who's felt she's flying too fast. Lifton writes, "Until relatively recently, no more than a single major ideological shift was likely to occur in a lifetime, and that one would be long remembered for its conflict and soul searching. But today it is not unusual for several such shifts to take place within a year or even a month, whether in the realm of politics, religion, aesthetic values, personal relationships. . . . Quite rare is the man or woman who has gone through life holding firmly to a single ideological vision. More usual is a tendency toward ideological fragments, bits and pieces of belief systems that allow for shifts, revisions, and recombinations."

What Lifton has observed in the psyche Rosen wants to make manifest in the body. I ask Rosen, "So, do you believe we are just in essence protean, that there is nothing fundamental, or core, to being human?"

He says, "Lauren, I am a scientist. My original interests were in nerves. I helped develop, in the 1980s, one of the first computer-grown nerve chips. The answer to your question may lie in how our nervous systems operate."

Part III: The Protean Brain

FIRST, A LESSON. In the 1930s, researchers, working on the brains of apes, found that the gray matter contained neural representations of all the afferent body parts. Ape ears, feet, skin, hands, were all richly represented in the ape brain in a series of neural etchings, like a map. Researchers also realized that when a person loses a limb—say, the right arm—this portion of the neural map fades away. Sometimes even stranger things happen. Sometimes amputees claimed they could feel their missing arm when, for instance, someone touched their cheek. This was because the arm map had not faded so much as morphed, joined up its circuitry with the cheek map, so it was all confused.

It was then discovered, not surprisingly, that human beings also have limb maps in their brains. Neurologists conceptualized this limb map as "a homunculus," or little man. Despite my feminist leanings, I am enchanted by the idea of a little man hunched in my head, troll-like, banging a drum, grinning from ear to ear. Of course the homunculus is not actually shaped like a human; it is, rather, a kind of human blueprint, like the drawing of the house in all its minute specificity. Touch the side of your skull. Press in. Buried, somewhere near there, is a beautiful etching of your complex human hand, rich in neural web-work and delicate, axonal tendrils designed to accommodate all the sensory possibilities of this prehensile object. Move your hand upward, press the now sealed soft spot, and you will be touching your toe map. Your eye map is somewhere in your forehead and your navel map is somewhere in your cerebellum, a creased, enfolded series of cells that recall, I imagine, ancient blue connections, a primitive love.

Today, Rosen is giving a lecture. I have come up to New Hampshire to hear him, and, unlike on the last visit, the day is beautiful and bright. Rosen explains how brains are partly plastic, which comes from the Greek root meaning to mold, to shape. When we lose a limb, the brain absorbs its map or rewires it to some other center. Similarly, Rosen explains, when we gain a limb, the brain almost immediately senses it and goes about hooking it up via neural representation. "If I were to attach a sonographically powered arm to your body," Rosen explains, "your brain would map it. If I were to attach a third thumb, your brain would map it, absolutely. Our bodies change our brains, and our brains are infinitely moldable. If I were to give you wings, you would develop, literally, a winged brain. If I were to give you an echolocation device, you would develop in part a bat-brain."

Although the idea of a brain able to incorporate changes so completely may sound strange, many neurological experiments have borne out the fact that our gray matter does reorganize according to the form and function of our appendages. Because no one has yet appended animal forms to the human body, however, no studies have been done that explore what the brain's response to what might be termed an "evolutionary insult" would be. Assuming, probably wrongly but assuming nevertheless, that human beings represent some higher form of species adaptation, at least in terms of frontal-lobe intelligence, the brain might find it odd to be rewiring itself to presumably more primitive structures, structures we shed a long time ago when we waded out of the swamps and shed our scales. Rosen's desire to meld human and animal forms, and the incarnation of this desire in people like the cat-woman and the lizard-man, raise some interesting questions about the intersection of technology and primitivism. Although we usually assume technology is somehow deepening the rift between nature and culture, it also can do the opposite. In other words, technology can be, and often is, extremely primitive, not only because it allows people a sort of id-like, limbic-driven power (i.e., nuclear weaponry) but also because it can provide the means to toggle us down the evolutionary ladder, to alter our brains, stuck in their rigid humanness, so that we are at last no longer landlocked.

All this is fascinating and, of course, unsettling to me. Our brains are essentially indiscriminate, able to morph—like the sea god Proteus himself—into fire, a flood, a dragon, a swan. I touch my brain and feel it flap. Now I understand more deeply what Rosen meant when he said, "Plastic surgery changes the soul." To the extent that we believe our souls are a part of our brains, Rosen is right. And, all social conflict about its place in the medical hierarchy aside, plastic surgery is really neurosurgery, because it clearly happens, at its most essential level, north of the neck. When a surgeon modifies your body, he modifies your oh-so-willing, bendable brain.

I get a little depressed, hearing this lecture. It seems to me proof at the neuronal level that we have the capacity to be, in fact, everything, and thus in some sense nothing. It confirms my fear that I, along with the rest of the human species, could slip-slide through life without any specificity, or "specieficity." Last year, I had my first child. I wonder what I will teach her, what beliefs about the body and the brain and the soul I really hold. I think, "I will show her pictures of her ancestors," but the truth is, I don't have any pictures. I think, "I will teach her my morals," but I don't know exactly what my morals are, or where they came from. I know I am not alone. Like Rosen, perhaps, I am just extreme. Now I feel a kind of kinship with him. We are both self-invented, winging our way through.

Rosen comes up to me. He is finished with his talk. "So do you understand what I mean," he asks, "about the limitlessness of the brain?"

"Does it ever make you sad?" I say. "Does it ever just plain and simple make you scared?"

Rosen and I look at each other for a long time. He does seem sad. I recall him telling me once that when he envisions the future fifty years out, he hopes he is gone, because, he said, "While I like it here, I don't like it that much." I have the sense, now, that he struggles with things he won't tell me. His eyes appear tired, his face drained. I wonder if he wakes in the middle of the night, frightened by his own perceptions. Strange or not, there is something constant in Rosen, and that's his intelligence, his uncanny ability to defend seemingly untenable positions with power and occasional grace. In just three weeks he will travel to a remote part of Asia to participate in a group called Interplast, made up of doctors and nurses who donate their time to help children with cleft lips and palates. I think it's important to mention this—not only Bin Laden, bandwidth, anthrax, and wings but his competing desire to minister. The way, at the dinner table, he tousles his children's hair. His avid dislike of George W. Bush. His love of plants and greenery. Call him multifaceted or simply slippery, I don't know. All I do know is that right now, when I look at his face, I think I can see the boy he once was, the Seventh Fleet ship, the wonder, all that wonder.

"Do you and Stina want to go out for dinner? We could go somewhere really fancy, to thank you," I say, "for all your time."

"Sure," says Rosen. "Give me a minute. I'll meet you in the hospital lobby," and then he zips off to who knows where, and I am alone with my singular stretched self on the third floor of the Dartmouth-Hitchcock Medical Center. I wander down the long hallways. Behind the curtained cubicles there is unspeakable suffering. Surely that cannot be changed, not ever. Behind one of these cubicles sits Sweeny, and even if we learn to see him as beautiful, the bottom-line truth is that he still suffers. Now I want to touch Sweeny's dying face. I want to put my hand right on the center of pain. I want to touch Rosen's difficult face, and my baby daughter's face as well, but she is far from me, in some home we will, migrants that our family is, move on from sometime soon. I once read that a fetus does not scar. Fetal skin repairs itself seamlessly, evidence of damage sinking back into blackness. Plastic surgery, for all its incredible advances, has not yet been able to figure out how to replicate this mysterious fetal ability in the full-born human. Plastic surgery can give us wings and maybe even let us sing like loons, but it cannot stop scarring. This is oddly comforting to me. I pause to sit on a padded bench. A very ill woman pushing an IV pole walks by. I lift up my pant leg and study the scar I got a long time ago, when I fell off a childhood bike. The scar is pink and raised and shaped like an *o,* like a hole maybe, but also like a letter, like a language, like a little piece of land that, for now, we cannot cross over.

ATUL GAWANDE

Crimson Tide

FROM *THE NEW YORKER*

Blushing can be an endearing quirk or an embarrassing moment of uninvited self-exposure, but for most of us it is only a minor irritant. For some people, though, it is a chronic condition that can diminish their quality of life. Scientists are not sure why we blush—or how to treat its most severe form. Is a cutting-edge surgical technique the answer? Atul Gawande, himself a surgeon, follows one woman who found out.

In January of 1997, Christine Drury became the overnight anchorwoman for Channel 13 News, the local NBC affiliate in Indianapolis. In the realm of television news and talk shows, this is how you get your start. (David Letterman began his career by doing weekend weather at the same station.) Drury worked the 9 P.M. to 5 A.M. shift, developing stories and, after midnight, reading a thirty-second and a two-and-a-half-minute bulletin. If she was lucky and there was breaking news in the middle of the night, she could get more airtime, covering the news live, either from the newsroom or in the field. If she was very lucky—like the time a Conrail train derailed in Greencastle—she'd get to stay on for the morning show.

Drury was twenty-six years old when she got the job. From the time she was a girl growing up in Kokomo, Indiana, she had wanted to be on television, and especially to be an anchorwoman. She envied the confidence and poise of the women she saw behind the desk. One day during high school, on a shopping trip to an Indianapolis mall, she spotted Kim Hood, who was then Channel 13's prime-time anchor. "I wanted to be her," Drury says, and the encounter somehow made the goal seem attainable. In college, at Purdue University, she majored in telecommunications, and one summer she did an internship at Channel 13. A year and a half after graduating, she landed a bottom-rung job there as a production assistant. She ran the teleprompter, positioned cameras, and generally did whatever she was told. During the next two years, she worked her way up to writing news and then, finally, to the overnight anchor job. Her bosses saw her as an ideal prospect. She wrote fine news scripts, they told her, had a TV-ready voice, and, not incidentally, had "the look"—which is to say that she was pretty in a wholesome, all-American, Meg Ryan way. She had perfect white teeth, blue eyes, blond hair, and an easy smile.

During her broadcasts, however, she found that she could not stop blushing. The most inconsequential event was enough to set it off. She'd be on the set, reading the news, and then she'd stumble over a word or realize that she was talking too fast. Almost instantly, she'd redden. A sensation of electric heat would start in her chest and then surge upward into her neck, her ears, her scalp. In physiological terms, it was a mere redirection of blood flow. The face and neck have an unusual number of veins near the surface, and they can carry more blood than those of similar size elsewhere. Stimulated by certain neurological signals, they will dilate while other peripheral vessels contract: the hands will turn white and clammy even as the face flushes. For Drury, more troubling than the physical reaction was the distress that accompanied it: her mind would go blank; she'd hear herself stammer. She'd have an overwhelming urge to cover her face with her hands, to turn away from the camera, to hide.

For as long as Drury could remember, she had been a blusher, and, with her pale Irish skin, her blushes stood out. She was the sort of child who almost automatically reddened with embarrassment when called on in class or while searching for a seat in the school lunchroom. As an adult, she could be made to blush by a grocery-store cashier's holding up the line to get a price on her cornflakes, or by getting honked at while driving. It may seem odd that such a person would place herself in front of a camera. But Drury had always fought past her tendency toward embarrassment. In high school, she had been a cheerleader, played on the tennis team, and been selected for the prom-queen court. At Purdue, she had played intramural tennis, rowed crew with friends, and graduated Phi Beta Kappa. She'd worked as a waitress and as an assistant manager at a Wal-Mart, even leading the staff every morning in the Wal-Mart Cheer. Her gregariousness and social grace have always assured her a large circle of friends.

On the air, though, she was not getting past the blushing. When you look at tapes of her early broadcasts—reporting on an increase in speeding-ticket fines, a hotel food poisoning, a twelve-year-old with an I.Q. of 325 who graduated from college—the redness is clearly visible. Later, she began wearing turtlenecks and applying to her face a thick layer of Merle Norman Cover Up Green concealer. Over this she would apply MAC Studiofix foundation. Her face ended up a bit dark, but the redness became virtually unnoticeable.

Still, a viewer could tell that something wasn't right. Now when she blushed—and eventually she would blush nearly every other broadcast—you could see her stiffen, her eyes fixate, her movements become mechanical. Her voice sped up and rose in pitch. "She was a real deer in the headlights," one producer said.

Drury gave up caffeine. She tried breath-control techniques. She bought self-help books for television performers and pretended the camera was her dog, her friend, her mom. For a while, she tried holding her head a certain way, very still, while on camera. Nothing worked.

Given the hours and the extremely limited exposure, being an overnight anchor is a job without great appeal. People generally do it for about a year, perfect their skills, and move on to a better position. But Drury was going nowhere. "She was definitely not ready to be on during daylight hours," a producer at the station said. In October of 1998, almost two years into her job, she wrote in her journal, "My feelings of slipping continue. I spent the entire day crying. I'm on my way to work and I feel like I may never use enough Kleenex. I can't figure out why God would bless me with a job I can't do. I have to figure out how to do it. I'll try everything before I give up."

WHAT IS THIS peculiar phenomenon called blushing? A skin reaction? An emotion? A kind of vascular expression? Scientists have never been sure how to describe it. The blush is at once physiology and psychology. On the one hand, blushing is involuntary, uncontrollable, and external, like a rash. On the other hand, it requires thought and feeling at the highest order of cerebral function. "Man is the only animal that blushes," Mark Twain wrote. "Or needs to."

Observers have often assumed that blushing is simply the outward manifestation of shame. Freudians, for example, viewed blushing this way, arguing that it is a displaced erection, resulting from repressed sexual desire. But, as Darwin noted and puzzled over in an 1872 essay, it is not shame but the prospect of exposure, of humiliation, that makes us blush. "A man may feel thoroughly ashamed at having told a small falsehood, without blushing," he wrote, "but if he even suspects that he is detected he will instantly blush, especially if detected by one whom he reveres."

But if it is humiliation that we are concerned about, why do we blush when we're praised? Or when people sing "Happy Birthday" to us? Or when people just look at us? Michael Lewis, a professor of psychiatry at the University of Medicine and Dentistry of New Jersey, routinely demonstrates the effect in classes. He announces that he will randomly point at a student, that the pointing is meaningless and reflects no judgment whatever about the person. Then he closes his eyes and points. Everyone looks to see who it is. And, invariably, that person is overcome by embarrassment. In an odd experiment conducted a couple of years ago, two social psychologists, Janice Templeton and Mark Leary, wired subjects with facial-temperature sensors and put them on one side of a one-way mirror. The mirror was then removed to reveal an entire audience staring at them from the other side. Half the time the audience members were wearing dark glasses, and half the time they were not. Strangely, subjects blushed only when they could see the audience's eyes.

What is perhaps most disturbing about blushing is that it produces secondary effects of its own. It is itself embarrassing, and can cause intense self-consciousness, confusion, and loss of focus. (Darwin, struggling to explain why this might be, conjectured that the greater blood flow to the face drained blood from the brain.)

Why we have such a reflex is perplexing. One theory is that the blush exists to show embarrassment, just as the smile exists to show happiness. This would explain why the reaction appears only in the visible regions of the body (the face, the neck, and the upper chest). But then why do dark-skinned people blush? Surveys find that nearly everyone blushes, regardless of skin color, despite the fact that in many people it is nearly invisible. And you don't need to turn red in order for people to recognize that you're embarrassed. Studies show that people detect embarrassment *before* you blush. Apparently, blushing takes between fifteen and twenty seconds to reach its peak, yet most people need less than five seconds to recognize that someone is embarrassed—they pick it up from the almost immediate shift in gaze, usually down and to the left, or from the sheepish, self-conscious grin that follows a half second to a second later. So there's reason to doubt that the purpose of blushing is entirely expressive.

There is, however, an alternative view held by a growing number of scientists. The effect of intensifying embarrassment may not be incidental; perhaps that is what blushing is for. The notion isn't as absurd as it sounds. People may hate being embarrassed and strive not to show it when they are, but embarrassment serves an important good. For, unlike sadness or anger or even love, it is fundamentally a moral emotion. Arising from sensitivity to what others think, embarrassment provides painful notice that one has crossed certain bounds while at the same time providing others with a kind of apology. It keeps us in good standing in the world. And if blushing serves to heighten such sensitivity this may be to one's ultimate advantage.

The puzzle, though, is how to shut it off. Embarrassment causes blushing, and blushing causes embarrassment—so what makes the cycle stop? No one knows, but in some people the mechanism clearly goes awry. A surprisingly large number of people experience frequent, severe, uncontrollable blushing. They describe it as "intense," "random," and "mortifying." One man I talked to would blush even when he was at home by himself just watching somebody get embarrassed on TV, and he lost his job as a management consultant because his bosses thought he didn't seem "comfortable" with clients. Another man, a neuroscientist, left a career in clinical medicine for a cloistered life in research almost entirely because of his tendency to blush. And even then he could not get away from it. His work on hereditary brain disease became so successful that he found himself fending off regular invitations to give talks and to appear on TV. He once hid in an office bathroom to avoid a CNN crew. On another occasion, he was invited to present his work to fifty of the world's top scientists, including five Nobel Prize winners. Usually, he could get through a talk by turning off the lights and showing slides. But this time a member of the audience stopped him with a question first, and the neuroscientist went crimson. He stood mumbling for a moment, then retreated behind the podium and surreptitiously activated his pager. He looked down at it and announced that an emergency had come up. He was very sorry, he said, but he had to go. He spent the rest of the day at home. This is someone who makes his living studying disorders of the brain and the nerves, yet he could not make sense of his own condition.

There is no official name for this syndrome, though it is often called "severe" or "pathological" blushing, and no one knows how many people have it.

One very crude estimate suggests that from one to seven per cent of the general population is afflicted. Unlike most people, whose blushing diminishes after their teen-age years, chronic blushers report an increase as they age. At first, it was thought that the problem was the intensity of their blushing. But that proved not to be the case. In one study, for example, scientists used sensors to monitor the facial color and temperature of subjects, then made them stand before an audience and do things like sing "The Star-Spangled Banner" or dance to a song. Chronic blushers became no redder than others, but they proved significantly more prone to blush. Christine Drury described the resulting vicious cycle to me: one fears blushing, blushes, and then blushes at being so embarrassed about blushing. Which came first—the blushing or the embarrassment—she did not know. She just wanted it to stop.

In the fall of 1998, Drury went to see an internist. "You'll grow out of it," he told her. When she pressed, however, he agreed to let her try medication. It couldn't have been obvious what to prescribe. Medical textbooks say nothing about pathological blushing. Some doctors prescribe anxiolytics, like Valium, on the assumption that the real problem is anxiety. Some prescribe beta-blockers, which blunt the body's stress response. Some prescribe Prozac or other antidepressants. The one therapy that has been shown to have modest success is not a drug but a behavioral technique known as paradoxical intention—having patients actively try to blush instead of trying not to. Drury used beta-blockers first, then antidepressants, and finally psychotherapy. There was no improvement.

By December of 1998, her blushing had become intolerable, her on-air performance humiliating, and her career almost unsalvageable. She wrote in her diary that she was ready to resign. Then one day she searched the Internet for information about facial blushing, and read about a hospital in Sweden where doctors were performing a surgical procedure that could stop it. The operation involved severing certain nerves in the chest where they exit the spinal cord to travel up to the head. "I'm reading this page about people who have the exact same problem I had, and I couldn't believe it," she told me. "Tears were streaming down my face." The next day, she told her father that she had decided to have the surgery. Mr. Drury seldom questioned his daughter's choices, but this sounded to him like a bad idea. "It shocked me, really," he recalls. "And when she told her mother it shocked her even worse. There was basically no way her daughter was going to Sweden and having this operation."

Drury agreed to take some time to learn more about the surgery. She read the few articles she could find in medical journals. She spoke to the surgeons and to former patients. After a couple of weeks, she grew only more convinced. She told her parents that she was going to Sweden, and when it became clear that she would not be deterred her father decided to go with her.

THE SURGERY IS known as endoscopic thoracic sympathectomy, or ETS. It involves severing fibres of a person's sympathetic nervous system, part of the involuntary, or "autonomic," nervous system, which controls breathing, heart rate, digestion, sweating, and, among the many other basic functions of life, blushing. Toward the back of your chest, running along either side of the spine like two smooth white strings, are the sympathetic trunks, the access roads that sympathetic nerves travel along before exiting to individual organs. At the beginning of the twentieth century, surgeons tried removing branches of these trunks—a thoracic sympathectomy—for all sorts of conditions: epilepsy, glaucoma, certain cases of blindness. Mostly, the experiments did more harm than good. But surgeons did find two unusual instances in which a sympathectomy helped: it stopped intractable chest pain in patients with advanced, inoperable heart disease, and it put an end to hand and facial sweating in patients with hyperhidrosis—uncontrollable sweating.

Because the operation involved open-chest surgery, it was rarely performed. In recent years, however, a few surgeons, particularly in Europe, have been doing the procedure endoscopically, using scopes inserted through small incisions. Among them was a trio in Göteborg, Sweden, who noticed that many of their hyperhidrosis patients not only stopped sweating after surgery but stopped blushing, too. In 1992, the Göteborg group accepted a handful of patients who complained of disabling blushing. When the results were reported in the press, the doctors found themselves deluged with requests. Since 1998, the surgeons have done the operation for more than three thousand patients with severe blushing.

The operation is now performed around the world, but the Göteborg surgeons are among the few to have published their results: ninety-four per cent of patients experienced a substantial reduction in blushing; in most cases it was eliminated completely. In surveys taken some eight months after the surgery, two per cent regretted the decision, because of side effects, and fifteen per cent were dissatisfied. The side effects are not life-threatening, but they are not trivial. The most serious complication, occurring in one per cent of patients, is Horner's syndrome, in which inadvertent injury of the sympathetic nerves to the eye results in a constricted pupil, a drooping eyelid, and a sunken eyeball. Less seriously, patients no longer sweat from the nipples upward, and most experience a substantial increase in lower-body sweating in compensation. (A decade after undergoing ETS for hand sweating, according to one study, the proportion of patients who were satisfied with the outcome dropped from an initial ninety-six per cent to sixty-seven per cent, mainly because of compensatory sweating.) About a third of patients also notice a curious reaction known as gustatory sweating—sweating prompted by certain tastes or smells. And, because sympathetic branches to the heart are removed, patients experience about a ten-per cent reduction in heart rate; some complain of impaired physical performance. For all these reasons, the operation is at best a last resort, something to be tried, according to the surgeons, only after nonsurgical methods have failed. By the time people call Göteborg, they are often desperate. As one patient who had the operation told me, "I would have gone through with it even if they told me there was a fifty-per cent chance of death."

On January 14, 1999, Christine Drury and her father arrived in Göteborg, a four-hundred-year-old seaport on Sweden's southwest coast. She remembers the day as beautiful, cold, and snowy. The Carlanderska Medical Center was old and small, with ivy-covered walls and big, arched, wooden double doors. Inside, it was dim and silent; Drury was reminded of a dungeon. Only now did she become apprehensive, wondering what she was doing here, four thousand miles away from home, at a hospital that she knew almost nothing about. Still, she checked in, and a nurse drew her blood for routine lab tests, made sure her medical records were in order, and took her payment, which came to six thousand dollars. Drury put it on a credit card.

The hospital room was reassuringly clean and modern, with white linens and blue blankets. Christer Drott, her surgeon, came to see her early the next morning. He spoke with impeccable British-accented English and was, she said, exceedingly comforting: "He holds your hand and is so compassionate. Those doctors have seen thousands of these cases. I just loved him."

At nine-thirty that morning, an orderly came to get her for the operation. "We had just done a story about a kid who died because the anesthesiologist had fallen asleep," Drury says. "So I made sure to ask the anesthesiologist not to fall asleep and let me die. He kind of laughed and said, 'O.K.' "

WHILE DRURY WAS UNCONSCIOUS, Drott, in scrubs and sterile gown, swabbed her chest and axillae (underarms) with antiseptic and laid down sterile drapes so that only her axillae were exposed. After feeling for a space between the ribs in her left axilla, he made a seven-millimeter puncture with the tip of his scalpel, then pushed a large-bore needle through the hole and into her chest. Two litres of carbon dioxide were pumped in through the needle, pushing her left lung downward and out of the way. Then Drott inserted a resectoscope, a long metal tube fitted with an eyepiece, fiber-optic illumination, and a cauterizing tip. It is actually a urological instrument, thin enough to pass through the urethra (though never thin enough, of course, for urology patients). Looking through the lens, he searched for her left sympathetic trunk, taking care to avoid injuring the main blood vessels from her heart, and found the glabrous cordlike structure lying along the heads of her ribs, where they join the spine. He cauterized the trunk at two points, over the second and third ribs, destroying all the facial branches except those that lead to the eye. Then, after making sure there was no bleeding, he pulled the instrument out, inserted a catheter to suction out the carbon dioxide and let her lung re-expand, and sutured the quarter-inch incision. Moving to the other side of the table, he performed the same procedure on the right side of her chest. Everything went without a hitch. The operation took just twenty minutes.

WHAT HAPPENS WHEN you take away a person's ability to blush? Is it merely a surgical version of Merle Norman Cover Up Green—removing the redness but not the self-consciousness? Or can a few snips of peripheral nerve fibers actually affect the individual herself? I remember once, as a teen-ager, buying mirrored sunglasses. I lost them within a few weeks, but when I had them on I found myself staring at people brazenly, acting a little tougher. I felt disguised behind those glasses, less exposed, somehow freer. Would the surgery be something like this?

Almost two years after Drury's operation, I had lunch with her at a sports bar in Indianapolis. I had been wondering what her face would look like without the nerves that are meant to control its coloring—would she look ashen, blotchy, unnatural in some way? In fact, her face is clear and slightly pinkish, no different, she said, from before. Yet, since the surgery, she has not blushed. Occasionally, almost randomly, she has experienced a phantom blush: a distinct feeling that she is blushing even though she is not. I asked if her face reddens when she runs, and she said no, although it will if she stands on her head. The other physical changes seemed minor to her. The most noticeable thing, she said, was that neither her face nor her arms sweat now and her stomach, back, and legs sweat much more than they used to, though not enough to bother her. The scars, tiny to begin with, have completely disappeared.

From the first morning after the operation, Drury says, she felt transformed. An attractive male nurse came to take her blood pressure. Ordinarily, she would have blushed the instant he approached. But nothing of the sort happened. She felt, she says, as if a mask had been removed.

That day, after being discharged, she put herself to the test, asking random people on the street for directions, a situation that had invariably caused her to redden. Now, as her father confirmed, she didn't. What's more, the encounters felt easy and ordinary, without a glimmer of her old self-consciousness. At the airport, she recalls, she and her father were waiting in a long check-in line and she couldn't find her passport. "So I just dumped my purse out onto the floor and started looking for it, and it occurred to me that I was doing this—and I wasn't mortified," she says. "I looked up at my dad and just started crying."

Back home, the world seemed new. Attention now felt uncomplicated, unfrightening. Her usual internal monologue when talking to people ("Please don't blush, please don't blush, oh God, I'm going to blush") vanished, and she found that she could listen to others better. She could look at them longer, too, without the urge to avert her gaze. In fact, she had to teach herself not to stare.

Five days after the surgery, Drury was back at the anchor desk. She put on almost no makeup that night. She wore a navy-blue woollen blazer, the kind of warm clothing she would never have worn before. "My attitude was, This is my début," she told me. "And it went perfectly."

Later, I viewed some tapes of her broadcasts from the first weeks after the surgery. I saw her report on the killing of a local pastor by a drunk driver, and on the shooting of a nineteen-year-old by a sixteen-year-old; she was, in fact, more natural than she'd ever been. One broadcast in particular struck me. It was not her regular nighttime bulletin but a public-service segment called "Read, Indiana, Read!" For six minutes of live airtime on a February morning, she was shown reading a story to a crowd of obstreperous eight-year-olds as messages encouraging parents to read to their children scrolled by. Despite the chaos of kids walking by, throwing things, putting their faces up to the camera, she persevered, remaining composed the entire time.

Drury had told no one about the operation, but people at work immediately noticed a difference in her. I spoke to a producer at her station who said, "She just told me she was going on a trip with her dad, but when she came back and I saw her on TV again, I said, 'Christine! That was unbelievable!' She looked amazingly comfortable in front of the camera. You could see the confidence coming through the TV, which was completely different from before." Within months, Drury got a job as a prime-time on-air reporter at another station.

A FEW SNIPS of fibers to her face and she was changed. It's an odd notion, because we think of our essential self as being distinct from such corporeal details. Who hasn't seen a photo of himself, or heard his voice on tape, and thought, That isn't me? Burn patients who see themselves in a mirror for the first time—to take an extreme example—typically feel alien from their appearance. And yet they do not merely "get used" to it; their new skin changes them. It alters how they relate to people, what they expect of others, how they see themselves in others' eyes. A burn-ward nurse once told me that the secure may become fearful and bitter, the weak jut-jawed "survivors." Similarly, Drury had experienced her trip-wire blushing as something entirely external, not unlike a burn—"the red mask," she called it. Yet it reached so deep inside her that she believed it prevented her from being the person she was meant to be. Once the mask was removed, she seemed new, bold, "completely different from before." But what of the person who all her life had blushed and feared blushing and had been made embarrassed and self-conscious at the slightest scrutiny? That person, Drury gradually discovered, was still there.

One night, she went out to dinner with a friend and decided to tell him about the operation. He was the first person outside her family whom she had told, and he was horrified. She'd had an operation to *eliminate her ability to blush?* It seemed warped, he said, and, worse, vain. "You TV people will do anything to improve your career prospects," she recalls him saying.

She went home in tears, angry but also mortified, wondering whether it *was* a freakish and weak thing to have done. In later weeks and months, she became more and more convinced that her surgical solution made her a sort of imposter. "The operation had cleared my path to be the journalist I was trained to be," she says, "but I felt incredibly ashamed over needing to remove my difficulties by such artificial means."

She became increasingly fearful that others would find out about the operation. Once, a co-worker, trying to figure out what exactly seemed different about her, asked her if she had lost weight. Smiling weakly, she told him no, and said nothing more. "I remember going to a station picnic the Saturday before the Indy 500, and thinking to myself the whole time, Please, please let me get out of here without anyone saying, 'Hey, what happened to your blushing?' " It was, she found, precisely the same embarrassment as before, only now it stemmed not from blushing but from its absence.

On television, self-consciousness began to distract her again. In June of 1999, she took up her new job, but she was not scheduled to go on the air for two months. During the hiatus, she grew uncertain about going back on TV. One day that summer, she went out with a crew that was covering storm damage in a neighboring town where trees had been uprooted. They let her practice her standup before the camera. She is sure she looked fine, but that wasn't how she felt. "I felt like I didn't belong there, didn't deserve to be there," she says. A few days later, she resigned.

More than a year has passed since then, and Drury has had to spend this time getting her life back on track. Unemployed and ashamed, she withdrew, saw no one, and spent her days watching TV from her couch, in a state of growing depression. Matters changed for her only gradually. She began, against all her instincts, admitting to friends and then former co-workers what had happened. To her surprise and relief, nearly everyone was supportive. In September 1999, she even started an organization, the Red Mask Foundation, to spread information about chronic blushing and to provide a community for its sufferers. Revealing her secret seemed to allow her finally to move on.

That winter, she found a new job—in radio, this time, which made perfect sense. She became the assistant bureau chief for Metro Networks radio in Indianapolis. She could be heard anchoring the news every weekday morning on two radio stations, and then doing the afternoon traffic report for these and several other stations. Last spring, having regained her confidence, she began contacting television stations. The local Fox station agreed to let her be a substitute broadcaster. In early July, she was called in at the last minute to cover traffic on its three-hour morning show.

It was one of those breakfast "news" programs with two chirpy co-anchors—a man and a woman—in overstuffed chairs, cradling giant coffee mugs. Every half hour or so, they'd turn to Drury for a two-minute traffic report. She'd stand before a series of projected city maps, clicking through them and describing the various car accidents and construction roadblocks to look out for. Now and then, the co-anchors would strike up some hey-you're-not-our-usual-traffic-gal banter, which she managed comfortably, laughing and joking. It was exciting, she says, but not easy. She could not help feeling a little self-conscious, wondering what people might think about her coming back after her long absence. But the feelings did not overwhelm her. She is, she says, beginning to feel comfortable in her own skin.

One wants to know whether, in the end, her troubles were physical or psychological. But it is a question as impossible to answer as whether a blush is physical or mental—or, for that matter, whether a person is. Everyone is both, inseparable even by a surgeon's blade. I have asked Drury if she has any regrets about the operation. "Not at all," she says. She even calls the surgery "my cure." At the same time, she adds, "People need to know—surgery isn't the end of it." She has now reached what she describes as a happy medium. She is free from much of the intense self-consciousness that her blushing provoked, but she accepts the fact that she will never be entirely rid of it. In October, she became a freelance part-time on-air reporter for Channel 6, the ABC affiliate in Indianapolis. She hopes the job will become full time. "You know, I don't have a face for radio," she says.

Lisa Belkin

The Made-to-Order Savior

FROM *THE NEW YORK TIMES MAGAZINE*

Medical science often outpaces ethics. Technology now exists that allows parents to choose embryos with certain genetic traits. As the journalist Lisa Belkin discovers, the right manipulation of in vitro techniques can produce children who would make perfect donors for older siblings suffering from the grave disease known as Fanconi anemia. Is it morally acceptable to do so? Faced with the anguish of possibly losing a child, what parent wouldn't?

Henry Strongin Goldberg was the first to arrive in Minneapolis. His parents decorated his room on the fourth floor of the Fairview-University Medical Center with his inflatable Batman chair, two Michael Jordan posters, a Fisher-Price basketball hoop and a punching bag hanging from the curtain rod over the bed. They took turns sleeping (or not) in his room for more than a month. It was too risky for his little brother to visit, but there was a playground across the courtyard, and if Henry, who was 4, stood at the window and Jack, who was 3, climbed to the top of the slide, the boys could wave to each other.

Henry had lost his hair by the time 6-year-old Molly Nash moved in down the hall on the bone-marrow transplant unit. Soon she, too, was bald. The two children had always looked alike, just as all children with this type of Fanconi anemia look alike, with their small faces and small eyes and bodies that are tiny for their age. The "Fanconi face" is one more reminder of the claim of the disease. Over time, Fanconi children also come to sound alike, with a deep, mechanical note in their voices, the result of the androgens they take to keep the illness at bay. Once their scalps were bare, Henry and Molly looked nearly identical. But there was one invisible difference between them—a difference that could mean everything.

These two families, the Strongin-Goldbergs and the Nashes, had raced time, death, threats of government intervention and (although they cringe to admit it) each other, to make medical history. The best chance to save a Fanconi child is a bone-marrow transplant from a perfectly matched sibling donor. Many Fanconi parents have conceived second children to save their first, hoping that luck would bring them a match. These two couples became the first in the world not to count on luck. Using in-vitro fertilization, then using even newer technology to pick and choose from the resulting embryos, they each spent years trying to have a baby whose marrow was guaranteed to be an ideal genetic fit.

One family would succeed and one would fail. One child would receive a transplant from a perfectly matched newborn brother and the other from a less well-matched stranger. One would have an excellent chance of survival; the fate of the other was not as clear. Their parents, now friends, would find themselves together in the tiny lounge at the end of the transplant hall, waiting for the new cells to take root, sharing pizza and a pain that only they could understand.

When the rest of the world learned about the baby born to be a donor, there were questions. Is it wrong to breed a child for "spare parts"? ethicists asked. If we can screen an embryo for tissue type, won't we one day screen for eye color or intelligence? There was talk in the news media of "Frankenstein medicine" and threats by Congress to ban embryo research, which had made this technique possible.

It is the kind of talk heard with every scientific breakthrough, from the first heart transplant to the first cloned sheep. We talk like this because we are both exhilarated and terrified by what we can do, and we wonder, with each step, whether we have gone too far. But though society may ask, "How could you?" the only question patients and families ask is, "How could we not?"

Which is why there is virtually no medical technology yet invented that has not been used. It is human nature to do everything to save a life and just as human to agonize over everything we do. The story of Molly and Henry is the story of groundbreaking science. It is also the story of last-ditch gambles on unproven theories, of laboratory technique cobbled from instinct and desperation, of a determined researcher who sacrificed his job and more trying to help and of a frantic drive through a hurricane to deliver cells on time. In other words, it is simply the story of what it now takes, in the 21st century, to save one child.

Back at the beginning, it was Molly who arrived first. She was born on July 4, 1994, at Rose Medical Center in Denver, and from the start it was clear that something was terribly wrong. She was missing both thumbs, and her right arm was 30 percent shorter than her left. Her parents, Lisa and Jack, saw her, but could not hold her, before she was whisked off to the ICU, where doctors would eventually find two separate malformations of her heart. (She was also deaf in one ear, but that would not be known until later.) Lisa, wide awake and distraught at 4 A.M. in the maternity ward, made a phone call to the nearby university hospital where she worked as a neonatal ICU nurse caring for babies just like this one, and asked a friend to bring her the book of malformations. Flipping from page to page, she landed on a photo of a Fanconi face and saw in it the face of her newborn daughter.

Named for the Swiss physician who first identified it in 1927, Fanconi anemia causes bone marrow failure, eventually resulting in leukemia and other forms of cancer. Until very recently, children with Molly's form of FA rarely lived past the age of 6, the age Molly is right now. Fanconi is a recessive disorder, which means both parents must pass along one copy of the mutated gene in order for a child to develop the disease. Among the general population, one of every 200 people has a Fanconi mutation. Every ethnic group carries its own genetic baggage, however, and among Ashkenazi Jews like the Nashes and Strongin-Goldbergs, the incidence is 1 in 89, meaning that if both parents are Ashkenazi Jews the chance of having an affected baby is 1 in 32,000. But Lisa, with all her medical training, had never heard of the disease, and Jack, a Denver hotel manager, certainly had not, either.

The holes in Molly's heart closed by themselves, but her other problems remained. She failed to eat, she failed to grow and she was always sick. She had already been through three major surgeries by Oct. 25, 1995, when Henry Strongin Goldberg entered the world at the George Washington University Hospital in Washington. Doctors had warned his parents that he would be quite small, but Laurie Strongin and Allen Goldberg were not worriers, because life had never given them anything to worry about. "Our family history," Laurie says wistfully, "was blue, sunny skies."

Henry was born with an extra thumb on his right hand and a serious heart defect that would require surgery to fix. His parents were devastated, but within days the prognosis worsened. "Fanconi anemia," Laurie wrote in her journal. "If only it was just the heart and thumb. Please take me back a minute ago and make me feel lucky that it is only the heart and the thumb. Fanconi anemia. Rare. Fatal. Henry."

Laurie had spent her career working for nonprofit organizations; Allen had spent his in the computer industry. Both in their early 30's, they were new to parenting and to Fanconi anemia, but they both knew how to navigate a medical database, and within days they found Arleen Auerbach, a researcher at Rockefeller University in New York and the keeper of the Fanconi patient registry in the United States and Canada, a list that contains about 800 names. Although Molly's parents and Henry's parents still knew nothing of each other, the Nashes had found Auerbach, too, because all Fanconi children eventually find their way to her cluttered Manhattan office.

The rarer the disease, the more it needs a single champion, someone to keep the lists, track the trends, follow the research of others while relentlessly pursuing his or her own. Arleen Auerbach is that person for Fanconi anemia—a sweet, grandmotherly type at the core but with sharp outer edges, armor born of years spent delivering bad news.

She had little but bad news for the Nashes and the Strongin-Goldbergs when they first called. Of the eight separate genes that can mutate and cause Fanconi anemia, Molly and Henry both had Type C, which bares its teeth early and kills often. Had these children been born as recently as 1982, Auerbach explained, there would have been no possible treatment. Bone-marrow transplants—obliterating the faulty immune system and then replacing it with a donated one—used to be fatal for Fanconi patients, because their cells were fragile and crumbled during the chemotherapy and radiation that cleared the way for the actual transplant.

Then, in 1982, doctors in France found that if Fanconi patients were given a significantly lower dose of the chemotherapy drug Cytoxan they could survive. The chances of their survival were increased even further if the donor was a sibling who was a perfect match. The reason for this is found in a web of six proteins that together are known as human leukocyte antigen, or HLA, which is the radar by which bodies recognize what is "self" and what is "intruder." HLA is key to the immune system, and since a bone-marrow transplant is a replacement of the immune system, the HLA of the donor must be as close as possible to that of the recipient, or the new immune system can reject its new container, a life-threatening condition known as graft-versus-host disease.

Over the years it was discovered that the rate of success for sibling transplants was even higher if the sibling was a newborn, because then the transplanted cells could come from "cord blood" taken from the umbilical cord and placenta at birth. These are purer, concentrated, undifferentiated cells, meaning that they are less likely to reject their new body. Back in 1995, when Auerbach first spoke to the Nashes and the Strongin-Goldbergs, the survival odds of a sibling cord-blood transplant were 85 percent, while the odds of a nonrelated bone-marrow transplant were 30 percent and the odds of a nonrelated transplant for patients with Henry and Molly's particular mutation were close to zero.

If there was one thing working in their favor, Auerbach told them, it was that their children's disease was diagnosed so early in life. Fanconi anemia is rare, and few doctors have ever seen a case, which means the condition is often missed or mistaken for something else. Auerbach has seen too many children with this same Fanconi mutation whose blood fails, with little prior warning, at age 5. Those parents don't have time to do the only thing there is to do, the one thing the Nashes and Strongin-Goldbergs could do—have a baby.

Ten weeks into a pregnancy, Auerbach explained, a chorionic villus sampling test can determine whether the fetus is healthy and if it is a compatible donor. Couples regularly abort when they learn that the unborn child has Fanconi, Auerbach says; having seen the devastation wrought by the disease on one of their children, they refuse to allow it to claim another. Few couples abort, however, when they learn that the baby is healthy but not a donor. "Only three that I know of terminated for that reason," she says. "They were getting older, their child was getting sicker and they were running out of time." Far more common, she says, is for couples to keep having children, as many as time will allow, praying that one will be a match.

Timing a child's transplant means playing a stomach-churning game of chicken with leukemia. The younger a patient is when undergoing a transplant, the better the outcome, because the body is stronger and has suffered fewer infections. On the other hand, the longer the transplant can be delayed, the greater the odds of conceiving a sibling donor, and the better the chance that transplant technology will have improved. The risk of waiting is that every Fanconi patient will develop leukemia, and once that happens a transplant is all but impossible. "You want to wait as long as you can," Auerbach says, "but not so long that it's too late."

GOOD DOCTORS LEARN from their patients, and so it was when Dr. John Wagner answered his telephone one afternoon seven years ago. A lanky, easy-going man, Wagner is scientific director of clinical research in the Marrow Transplant Program at the University of Minnesota, and he says he believes he has performed more bone-marrow transplants on Fanconi children than any other doctor in the country. The caller who set him thinking, however, was not the parent of a Fanconi patient, but rather the father of a toddler with thalassemia, another rare blood disease. The man was calling to inquire about a sibling cord-blood transplant. "You have another child who is a match?" Wagner asked. "No," came the reply. "But we will."

The father went on to explain that he and his wife were using a relatively new technique known as pre-implantation genetic diagnosis, or PGD, to guarantee that their next child would be free of thalassemia. PGD is an outgrowth of in-vitro fertilization; sperm and egg are united in a petri dish, and when the blastocyst (it is still technically too small to be called an embryo) reaches the eight-cell stage, it is biopsied (meaning one of those cells is removed and screened). Only blastocysts found to be healthy are returned to the womb. Then the waiting game begins—more than two months until it is possible to know if the fetus is a transplant match, then an agonizing choice if it is not. Why, the caller wondered, can't the donor-compatibility tests be done before the embryos are implanted?

Wagner was intrigued by the possibility. Why use PGD just as prevention, he wondered, when it could be used as treatment? Why not, in effect, write a prescription that says "one healthy baby who is going to be a perfect donor"?

Wagner called Mark Hughes, who pioneered the technique and who was working with this family. Hughes is known as a brilliant researcher, simultaneously passionate and wary, a scientist and physician who chose the field of genetics because it combined the intellectual rigor of the lab with the emotional connection to flesh-and-blood patients. In 1994, at about the time he first spoke to Wagner, Hughes was recruited to work at the National Institutes of Health and also as director of Georgetown University's Institute for Molecular and Human Genetics, where his salary was paid in part by the government. At that time he was also a member of a federal advisory committee that developed guidelines for the type of single-cell embryo analysis that was central to PGD. But no sooner had those guidelines been developed than Congress banned all federal financing of embryo research, and Hughes was forced to continue his research with private funds only.

Under the current Bush administration there is talk of banning all embryo research, even work supported by private funds. For that reason—and for reasons that will become clearer as this tale unfolds—Hughes has developed a healthy distrust of the limelight and refused to be interviewed for this story. As Wagner and Auerbach tell it, Hughes had certainly thought of the possibility of using PGD to determine HLA type long before Wagner called, but he had several concerns.

The ones that weighed heaviest were ethical. It could be argued that using PGD to eliminate embryos with disease helps the patient—in this case, the embryo, the biopsied organism—by insuring that it is not born into a life of thalassemia or cystic fibrosis or Duchenne muscular dystrophy or any of the other agonizing illnesses for which Hughes was screening. Using the same technique to select for a compatible donor, however, does not help the "patient" whose cells are being tested. "It helps the family," says Arleen Auerbach, "and it helps the sibling with Fanconi, but it does not help the embryo."

What Wagner proposed, therefore, would be stepping into new territory. If society gives its blessing to the use of one child to save another, then what would prevent couples from someday going through with the process but aborting when the pregnancy was far enough along that the cord blood could be retrieved? Or what would prevent couples whose child needed a new kidney from waiting until the fetal kidney was large enough, then terminating the pregnancy and salvaging the organs? What would stop those same couples from waiting until the child was born and subjecting it to surgery to remove one kidney? Once the technology exists, who decides how to use it?

Ethicists think in terms of a slippery slope. But is the potential for abuse in *some* circumstances reason not to pursue research that can be lifesaving under the *right* circumstances? Unlike donating a kidney, or even donating bone marrow, donating cord blood involves negligible harm to the newborn donor. The stem cells are collected at birth, directly from the placenta, not from the baby. That is one reason why Wagner argued that HLA testing is ethically defensible. A second reason, he said, was that it is indefensible not to try.

"I'm here as the patient's advocate," he says, meaning Molly and Henry and all the other children in need of transplants. "It's my obligation to push the envelope because I see how bad the other side can be. I see the results of a sibling transplant; they're the easiest transplant to do. And then I walk into the room of the patient who had an unrelated donor, I see that their skin is sloughing off, the mucous membranes are peeling off and they have blood pouring out of their mouths. You cannot imagine anything so horrible in your entire life, and you're thinking, I did this—because there was nothing else available for me to do."

That was apparently what Hughes's gut told him, too, and he agreed to try to develop a lab procedure to screen HLA at the single-cell level. His participation came with certain conditions. First, that the mother must be younger than 35, because younger women produce more eggs, increasing the odds of a healthy match. Second, that he would work only with families who carried a specific subset of the Type C mutation, known as IVS4, because it is the most common. And, last of all, the child being created must be wanted. Only families who had expressed a wish for more children would be approached for this procedure. Hughes did not want to create a baby who was nothing but a donor.

Arleen Auerbach immediately thought of two couples who were the right age, fit the specific genetic profile and who had always planned to have a houseful of children. Her first phone call was to Lisa and Jack Nash in Denver. Without a moment's hesitation, they said yes. Her second call was to Laurie Strongin and Allen Goldberg in Washington.

"If I told you that you could potentially go into a pregnancy knowing that your baby was healthy and a genetic match for Henry, would you be interested?" she asked.

Two hours earlier, Laurie had taken a home pregnancy test. It was positive. If early test results were negative for Fanconi she would carry to term, she answered, even if the baby were not the right HLA type to save Henry's life.

Henry was only 5 months old. His heart surgery had gone smoothly, he was happy and looked deceptively healthy. Fate seemed to be on his side. "If this baby's not a match, we'll try it your way in nine months," Laurie remembers telling Auerbach. "We still thought," she says, "that we had a lot of time."

HENRY BECAME A BIG BROTHER in December 1995. Jack Strongin Goldberg was free of Fanconi and was not even a carrier of the disease, so there was no chance that he might pass it on to his own children. His HLA, however, was as unlike Henry's as a biological brother's could possibly be. Laurie and Allen admit that they were briefly disappointed when they heard this last piece of news, three months into the pregnancy. Then they brushed off their psyches and called Mark Hughes, telling him they would be ready to try PGD at the start of the following year.

As baby Jack was being born, Lisa Nash was undergoing the shots and monitoring that are part of in-vitro fertilization. Theirs would be a very difficult case, Hughes had told them. Of the cluster of genes that together determine HLA type, science, at the time, could look at only three. As it happened, Lisa and Jack's patterns were almost identical on those three genes, making it nearly impossible to sort hers from his. That genetic quirk, he warned, could lead to the wrong results. The science to fix this didn't exist yet, he said, and he was figuring it all out as they spoke.

Hughes was also struggling with other problems, ones that had nothing to do with the Nashes' DNA. On the day that Lisa's eggs were retrieved by laparoscopy and fertilized in a dish, the headline in *The Washington Post* read: "NIH Severs Ties With Researcher Who Experimented on Embryos." Hughes had been accused of using federal funds for embryo research, in violation of the Congressional ban. Hughes denied that government money was used for that portion of his work and argued that in any case his research was not even on embryos since all that ever arrived in his lab was DNA extracted from a biopsied cell.

Lisa Nash did not become pregnant.

Mark Hughes resigned from his positions with NIH and Georgetown University rather than agree to stop his research.

The turn of events was devastating for Hughes. He was out of a job and forced to uproot his two young sons and his wife, who was fighting a battle of her own, against breast cancer. Those close to him say he talked of quitting medicine entirely, so frustrated and angry was he that the rug had been pulled out from under him.

The turn of events was also devastating for the Nashes. "We called him two, three times a week," Jack Nash remembers, and as he speaks a frantic note creeps into his voice. "But he wouldn't return our calls. Months went by, then a year." Over those months they learned that Hughes was moving halfway across the country to a new, privately financed lab where he could continue his work. Then they learned that Hughes's wife was critically ill, that her cancer had spread, that the prognosis was grim. The one thing they did not learn was when and if their quest to save Molly might begin again.

They now understand that science solves the simplest equation first, then moves on to the more difficult ones; their complicated genetic makeup meant their case had to wait. Added to that was the fact that the initial decoding of their DNA had been done at Hughes's former lab in Washington, and he no longer had access to the data. They now also understand that Hughes was in this to save lives, and that having to come to the phone and say that he couldn't, that he didn't know how to match an HLA type for Molly, was more than he could bear. But at the time they didn't understand. At the time they were angry.

"When we manage to speak to him he says we have to give him a few more months to get the lab set up," Jack says. "Meanwhile Molly's counts are dropping and he's the only one who can do this, and he won't help."

LIFE FOR A CHRONICALLY ILL CHILD is a jumble of numbers. The average platelet count in a healthy child: 150,000 to 450,000. The lowest that platelets are allowed to drop before Dr. Wagner urges a transplant: 40,000. Where Henry's platelets hovered when Jack was born: 100,000. The cost of each in-vitro cycle: $11,000. The amount paid by insurance: officially, $0, because the in-vitro fertilization was not being done to treat infertility, nor was it being done to directly treat Henry. The amount the Strongin-Goldbergs raised for Fanconi anemia research at the fund-raiser they held on Henry's first birthday: $67,500. The odds of a blastocyst being healthy: 3 in 4. The odds of a blastocyst being a match: 1 in 4. The odds of a blastocyst being a match and also being healthy, and of Laurie becoming pregnant and delivering before Henry had to have a transplant: God only knows.

Since the day Henry's FA was diagnosed, life for Laurie and Allen was filtered through these numbers, through the lens of Fanconi anemia. "Every ensuing pregnancy," she wrote in her journal after baby Jack was born, "will be marred by the fact that the little baby in my belly could have a fatal disease. Every job that Allen and I consider has to offer medical insurance without excluding pre-existing conditions and with compassion and flexibility. Every relationship has to offer quiet understanding of our travails accompanied by the capacity to give without expecting too much in return."

While Mark Hughes worked to set up his new lab at Wayne State University School of Medicine, near Detroit, the Nashes and the Strongin-Goldbergs were at home, waiting in two very different ways. A crisis can strip a family down to its skeleton of strengths and faults, peeling the niceties away and revealing the bare core of who they are. Henry's parents, for instance, effervescent, embracing and fiercely optimistic from the start, became more so as the clock ran out. They took on Hughes's problems as their own, bonding with him deeply, knowing that they needed him to bond back if they were to save Henry. Molly's parents, in turn, are determined and intense, and they did not waste emotional energy that might be spent protecting their daughter. They were demanding of Hughes, but no more demanding than they were of themselves or of anyone else who could help Molly.

Until the spring of 1997, the two families had still not met. In May of that year, when Hughes was promising both of them that he would be able to resume work soon, a retreat for Fanconi families was held near Portland, Me. The Strongin-Goldbergs went there determined to meet the anonymous couple Arleen Auerbach had mentioned—the couple who had already tried HLA screening with Hughes. Armed with two facts—that the couple had a daughter, and that they lived in Colorado—the Strongin-Goldbergs skimmed the directory and found a family who fit that description.

When Laurie Strongin shook Lisa Nash's hand for the first time she felt an instant bond with the only other mother in the world whose life paralleled her own. Lisa was more reserved. Up to that moment she hadn't realized that the elusive Hughes was working with a second family. Six months later, however, by the time of Laurie's initial in-vitro attempt, the women had paddled past their opening awkwardness and were close telephone friends. When Henry, now two, talked about his future, he spoke in gradations: first he would be "better," then "super better," then "super-duper better." When all this was over, Lisa and Laurie promised each other, when their children were both "super-duper better," the two families would travel to Disney World to celebrate.

In January 1998, when Hughes was finally ready for them, Laurie took the train up to New York City for her appointment with Dr. Zev Rosenwaks, the baby-making guru at the in-vitro fertilization clinic at New York Weill Cornell Medical Center. Henry's platelet count was 71,000 that morning. Eighteen days later, after 18 shots of Lupron, a brutal migraine, hot sweats and cold chills, Laurie's body refused to cooperate, and the in-vitro fertilization process for that cycle had to be abandoned. That week Henry's platelet count dropped to 31,000, its lowest level up to that point.

Doctors often suggest that in-vitro fertilization patients wait a month or more between attempts, but Laurie didn't have a month, and in early February she was in New York again. This time the numbers were on her side. She produced 24 eggs, and 21 of them were mature enough to be fertilized. Statistically that meant six should be perfect matches for Henry, and three or four of those six should also be disease free.

Sixteen blastocysts survived the biopsy. Allen refused to entrust the cells to anyone, so he flew them to Detroit himself. At the airport he handed his Styrofoam hope chest to a waiting Mark Hughes, then got on the next plane back to New York. The following evening, Laurie was at the Rosenwaks clinic ready for the re-implantation when word came from Hughes. Of the 16 blastocysts tested, two were absolutely perfect matches to Henry. Both those matches had Fanconi anemia.

"I'm struggling to come to terms with how much pain I can withstand," Laurie wrote in her journal. She and Allen shared that pain long-distance with the Nashes, who still had not heard when and if Hughes would begin to work with them again. Jack and Lisa were supportive, but also envious and confused. "Were they the family of choice because he liked them better?" Jack remembers wondering. "Is this personal? Does he have something against us, and he's taking that out on Molly? Things like that definitely go through your mind."

The Nashes sent frantic e-mail messages to Hughes, telling him what he already knew—that Molly's counts were dropping and that they were running out of time. In August 1998, when Molly's platelet count had fallen to 30,000, they received his answer. He couldn't help them, he wrote in an e-mail message. Their case was too complicated, both genetically and politically. The genetic analysis he'd so painstakingly done on them belonged to the NIH. "We tried to get the lab at Georgetown to help us, since they were key in our being able to do this for you the first time around," Hughes wrote. The lab has been ordered by the "Catholic administration" of the university "not to get involved 'in any way.' "

Hughes continued: "Go ahead without us. You are anxious, and we understand that very well. But I cannot make this work today and I don't know when I will be able to do so. I am sorry. Science sucks sometimes."

Reeling, Lisa and Jack called Laurie and Allen, who were about to begin their third in-vitro cycle—one that would produce 26 eggs, 24 of which were mature and 21 of which would fertilize. Of those, three would be perfect, healthy matches for Henry. The Strongin-Goldbergs would not share these details with the Nashes because they had come to understand that other people's good news is sometimes too difficult to hear.

TAKING MARK HUGHES'S ADVICE, the Nashes did go on without him. They'd decided to jump into a cross-your-fingers pregnancy when they learned, almost by accident, of a private clinic in Chicago that had been quietly doing PGD for nearly 10 years, though never for Fanconi anemia. This news was "like opening a door," say the Nashes, who had not realized that other labs in the country besides Hughes's were providing PGD. If this Chicago lab could test for cystic fibrosis and Tay-Sachs, they wondered, why not Fanconi? And if it had the equipment to screen DNA for disease, why not also screen for HLA?

Lisa and Jack brought Molly along on their trip to the Reproductive Genetics Institute, on the theory that doctors couldn't say no with their adorable but ashen-cheeked child in the room. Her platelets were half what they had been a month earlier. She was weak and tired. They could not have walked into a more receptive office. A year earlier, Charles Strom, then the head of the institute's genetics lab, had heard Mark Hughes speak at a genetics meeting about his attempts to screen DNA for an anonymous couple who were trying to have a child who would be a cord-blood match. "It was like a revelation to me," says Strom, a broad, genial bear of a man now at Quest Diagnostics in California, who could, at that time, perform PGD for 35 diseases but had never thought of HLA screening. "This is what pre-implantation genetics should be about."

A few in the audience expressed their disapproval, he remembers, fearing that this was a step on the road to eugenics. Strom, on the other hand, was enthusiastic. "I stood up and said I thought this was great," he says. "I'm trained not just as a geneticist, but as a pediatrician, and I was tired of watching kids die. I thought this would be the future, and from then on, I was basically waiting for someone to ask me to do it." So when the very same "anonymous" couple arrived and asked, Strom said yes.

He immediately discovered what Hughes had struggled with for years—the "nightmare" caused by the near-identical patterns in the HLA portion of Jack and Lisa's DNA. But he and his team tried something new—they looked farther down the strand, beyond the three known genes, to a spot where it was easier to differentiate one parent from the other. This increased the risk of being wrong, but Molly's blood counts were dropping, and they did not have time to waste. "This isn't what we want to do, but it will probably work," Strom told the Nashes two months after they first met.

It is one thing to screen embryos; it is another to become pregnant, and adding HLA screening to Fanconi anemia screening lowers the odds even more. Only one in six blastocysts is likely to be both healthy and a matched donor, and that one might not be the quality that the reproductive endocrinologist would have chosen under ideal circumstances. Lisa spent all of 1999 trying to defy those odds. In January she produced 12 eggs, two of which were healthy matches; she became pregnant, then miscarried. In June she produced only four eggs, one of which was a match but did not result in a pregnancy. In September she produced eight eggs, six of which had Fanconi anemia; the single healthy match was implanted, but again, her pregnancy test was negative.

In October the Nash family traveled to Minneapolis for Molly's twice-yearly checkup with Dr. Wagner. Her platelets were down to 10,000. In every measurable way she was failing, and she needed a bone-marrow transplant. "You have to stop," Wagner told her parents. It was time to proceed with a transplant from a nonrelated donor. "There comes a point where I have to say: 'It's over. You've done it. You've done the best you could.' "

He began to search for a donor. Lisa and Jack went ahead with the in vitro that had been scheduled for December. "I couldn't hear the word no," Lisa says. " 'No' meant Molly could die."

Because they knew it was the last try, and because they needed to feel certain that they had done the best they could, the Nashes insisted on one change of procedure for this final try. It troubled them that Lisa was producing so few eggs per cycle, and they wondered if a different in-vitro fertilization clinic might do better. They approached Dr. William Schoolcraft, an infertility doctor in Colorado known for pushing the envelope. He changed Lisa's hormone regimen and in December 1999 retrieved 24 eggs from her ovaries. For two days the Nashes fantasized about twins and even triplets. Then Strom called to say that there was only one match.

It all came down to one embryo that, statistically, had less than a 30 percent chance of taking hold and staying put. "All it takes is one, all it takes is one," Lisa reminded herself as she drove to Dr. Schoolcraft's office nine days later for a pregnancy test. Minutes after she left, her cell phone rang.

"You're pregnant," said the nurse on the other end.

It was too soon, however, for a happy ending. And indeed, seven weeks into the pregnancy Lisa had just gotten out of the shower when deep red blood began flowing down her legs. The drive to Schoolcraft's office was a blur, but the memory of the picture on the ultrasound screen is vividly clear: a large gap where the placenta had separated from the uterine wall, and the flub-dub pulses of a tiny, living, beating heart.

Lisa went home and went to bed. She was permitted to get up three or four times a day to use the bathroom and once a week for an appointment with Schoolcraft, nothing more. Every time she stood up she began to bleed. Molly, too weak to really play, was on her own manner of bed rest, and mother and daughter spent entire days lying upstairs together.

In March, Molly's blood tests showed signs of pre-leukemia. Wagner sent more data to the national bone-marrow bank, escalating his search for an unrelated donor. In April, Molly's platelets fell to 3,000. She began to need blood transfusions but fought whoever tried to insert the needle; one particularly rocky weekend Strom flew to Denver from a business meeting in Los Angeles, because he was the only one Molly would permit to start the IV. April became May; May turned to June. Along the way, Lisa asked her doctors what could be done should she spontaneously lose the baby. They began to discuss whether stem cells could be harvested from a fetal liver. And all the while, Lisa was still bleeding—clawing her way through the pregnancy, trying to hold onto her baby while holding off her daughter's transplant.

BACK WHEN THE Nashes were deciding whether to go ahead with Molly's transplant or try somehow to wait until summer, the Strongin-Goldbergs were making their own impossible choice: whether or not to give up. Their optimism back in August 1998, when they had three healthy embryos, had long since faded. That in-vitro attempt did not result in a pregnancy. Neither did attempt No. 4, in November, when 30 eggs failed to provide a single healthy match.

Attempt No. 5, in February 1999, was almost more than they could bear. Laurie produced 17 eggs and was waiting to be summoned to the clinic when she received another call instead. Allen had taken to scanning the Detroit newspapers online, knowing that Mark Hughes's wife was dying, but not wanting to pester his friend. The morning of Laurie's retrieval, Allen found the news he'd been dreading in the obituary section. Laurie's ovaries were past the point of no return, so Rosenwaks went ahead with the retrieval and fertilization without any idea who would screen the blastocysts. "I couldn't imagine doing this without our friend on the other end and didn't even know if it was possible," Laurie wrote.

But saving Henry had come to mean as much to Hughes as to Laurie and Allen. The researcher had watched his own life nearly destroyed in defense of this work, and he promised he would be there. Fourteen blastocysts survived the biopsy, and on the morning of Feb. 11, 1999, just a few days after Hughes's wife's death, Allen loaded his Styrofoam box with vials and dry ice and boarded the 11:10 A.M. flight to Detroit. He took the container to the lab, where he was moved to tears to find a large picture of his own son hanging on the wall. Underneath it was the question, "Can we help save Henry's life?"

Allen was certain that this attempt would work. There was bittersweet poetry in the timing: death preceding life and preventing death. But when Hughes called the Strongin-Goldbergs in New York, his news was not the stuff of poetry. There was only one match. It did not result in a pregnancy.

Attempt No. 6 took place in June 1999. Twenty-eight eggs, two healthy matches. No pregnancy. Attempt No. 7 came in the middle of Hurricane Floyd. Allen drove his Styrofoam box through the eye of the storm—1,200 miles in 26 hours—and delivered the cells, alive, at 2 A.M. Laurie became pregnant, then miscarried.

Their eighth try took place in February 2000. Laurie was in New York, at the clinic, the morning that Allen raced Henry to the hospital with pneumonia so serious doctors warned it could kill him. Laurie agonized over whether to come home (canceling the in-vitro cycle) or stay where she was. If she left, she was certain Henry would die, because he would have lost this chance for a sibling donor. If she stayed, then she was equally certain that Henry would die—of pneumonia, in a Georgetown hospital, without his mother.

She stayed. Henry received two blood transfusions and was pumped full of three intravenous antibiotics. Laurie produced 21 eggs and only one implantable match. "I did not get pregnant," she says, "and I still haven't recovered from the experience."

As the Strongin-Goldbergs dragged themselves from one attempt to the next, the technology of bone-marrow transplants was changing. Specifically, Wagner was testing a new method of removing T-cells from donor blood. T-cells are the ones that recognize the host as foreign, leading to graft-versus-host disease. Simultaneously, Wagner was using fludarabine, an immunosuppressant that appears to encourage the new cells to engraft, or take root. Based on a tiny sample of patients, Wagner's best guess was that these adjustments to the protocol showed promise, apparently increasing the odds of surviving an unrelated bone-marrow transplant from 30 percent to 50 percent in a Fanconi anemia patient. This was still far lower than the 85 percent odds of a sibling cord-blood transplant, but better than it had been before.

Laurie went through one last, disappointing in-vitro cycle, then she and Allen grabbed those new 50–50 odds. Wagner warned that it was time to stop, and they knew, from looking at Henry, that he was probably right. Henry had had two platelet and two red-cell transfusions in the past two months, and he had been on Anadrol, a steriod to boost his blood counts, for two and a half years. There comes a point at which a child is too sick for a transplant, and Henry, like Molly, was all but there. In two and a half years of desperate trying, Laurie had 353 injections, produced 198 eggs and had no successful pregnancy. During the same time period, Henry's platelets fell from a high of 103,000 to a low of 10,000.

"We gave it all we had," Laurie wrote when her last pregnancy test was negative and the family was leaving for Minneapolis, for Henry's transplant. "We worked with the world's best doctors. We hoped. We believed. We were brave. We persevered. And despite all that it didn't work. I am left with my belief system intact. I believe in love and science. Nothing more, nothing less."

A BONE-MARROW TRANSPLANT is a medical resurrection. First doctors all but kill a patient; then they bring him back to life. Treacherous and risky, in the end it all comes down to one squishy plastic bag of pale brown liquid which could easily be mistaken for rusty water from a tap. Henry's bag of marrow was collected from an anonymous donor somewhere in the United States on the morning of July 6, 2000, and was flown to the Fairview-University Medical Center, arriving in Room 5 of the bone-marrow transplant floor around dinnertime. A nurse came in with a Polaroid, snapped a few pictures, then added the bag to Henry's leafy IV tree. There was no blaring of trumpets, no rolling of drums. From 8:15 to 8:30 Central Daylight Time, the fluid dripped soundlessly.

Molly Nash's bag was collected with more drama. Lisa's pregnancy had managed to hold. For months Molly's baby brother had been trying to arrive prematurely, and now that he was due, he didn't seem eager to arrive at all. By the evening of August 29, Lisa had been in labor for 52 hours, insisting she be allowed to continue because she knew that more cord blood could be collected during a vaginal birth. Finally, when it looked as if the baby was in distress, he was delivered by C-section. Dr. Strom—his godfather—collected the cord blood. Lisa cradled both the newborn Adam and the warm intravenous bag in her arms.

"God created Adam in his image," Lisa says, explaining how she chose her son's name. "Adam was the first. And from Adam—from his rib, which is full of marrow—God created woman, which is fitting because God used our Adam to give Molly a second chance at life."

When he was nine days old, Adam flew with his parents and his sister to Minneapolis. Molly settled into the room down from Henry's for the standard four-month stay—a surreal time when it seems as if every child in the world is having a bone-marrow transplant, because every child that you see is. Molly went through all that Henry had gone through a month before her, and yet everything was different. She had a higher chance of engraftment and a far lower chance of rejection. Her parents were rubbed emotionally raw watching her suffer in order to live. But then they looked at Henry, whose parents feared he was showing early signs of graft-versus-host disease—something Molly would almost certainly never get. They looked beyond Henry, too, at the eight patients who died in the bone-marrow transplant unit during Molly's endless summer.

In the end, Molly's life was saved. That is the Nashes' answer to people who question their right to manipulate nature. Their right springs from the difference between 30 percent and 85 percent; the difference between Molly and Henry. That is also their answer to those who would urge the government to ban all embryo research because it harms unborn children. The research, they say, saves children like Molly.

"We did what we needed to do to keep our daughter from dying," Lisa Nash says. "That is what any parent would do. Isn't this what parents are supposed to do? How can anything be wrong with that?"

Yes, ethicists say, it is exactly what any parent would do, and that is why it is troubling. Parents are being asked to make a choice not only on behalf of their living child, but also on behalf of their unborn child, and that can be an impossible position when the choices get hard. If Molly were closer to death, for instance, would her parents have terminated the pregnancy and used stem cells from Adam's fetal liver to save her?

"We know people will do anything to save their child," says Jeffrey Kahn, an ethicist at the University of Minnesota, where there was much debate about the decisions of the transplant team at the hospital next door. "Now we are learning what 'anything' really means."

Susan M. Wolf, a professor of law and medicine at the University of Minnesota, says she believes that this case is emblematic of the whole of reproductive technology, which she describes as "a multibillion dollar industry based solely on consumer demand." While it might seem logical in each isolated case to let the parents decide, all those single choices add up to a hodgepodge of technology scattered throughout private clinics and laboratories, with no one authorized to say no.

Wagner and Strom agree. They say they do not believe that they, or any other individual doctor, should have the responsibility of sorting through this thicket alone. "As the technology progresses," Strom says, "I see the possibility that someone will come to us and say: 'While you're screening for Tay-Sachs, how about making sure he's not going to have heart disease, too? And while you're at it, why not check for the gene that predisposes him to lupus or makes him immune to HIV?' "

"It has the potential to be abused," agrees Wagner. But the response to that potential, he warns, should not be to ban the research or suspend federal financing of the procedure. "It's not going to go away," he says. "We can't put our heads in the sand and say it doesn't exist. I have a stack of requests this high from all over the world, couples asking if they can come use this technology."

Compounding the problems caused by the current ban on federal financing, he says, is the accompanying lack of federal rules. "It's all been forced into the private sector," he says, "where there are no controls. There should be controls. There should be limits. It is up to us, as a society, to decide what they are."

Since her transplant, Molly Nash has gone back to school. More accurately; school has started to come to her, but her visiting teacher has to wear a mask during lessons. Her ballet teacher comes for in-home classes, too, and Molly twirls and pliés and giggles. Her hair is beginning to grow back. Instead of taking 44 pills every day she only takes 10. She is still fed through a stomach tube that her mother hooks up four times a day, and she doesn't have much of an appetite, which is characteristic of Fanconi anemia. The transplant did not cure her of that disease; it merely erased her risk of developing imminent leukemia. She is still likely to suffer Fanconi's other complications, particularly cancers of the mouth and neck. But those will not show themselves for many years, and, her mother says, "maybe they will have a cure by then."

Henry Strongin Goldberg has been ill almost since the day he left the hospital in Minnesota. While Molly's platelet count is 381,000, Henry's is 15,000. He spent months looking yellow and feeling miserable, moaning instead of talking, the result of a near fatal liver infection that is common in transplant patients because of the drugs they are given to suppress their immune system.

In January, for the first time in his tortured life, his parents were struck full force by the thought that he was dying. "All I can think about," Allen said then, "is how much I'll miss him."

Since then, things have gotten even worse. Allen lost his job at an Internet start-up in January, and although he is now working again, the family has burned through its savings. Laurie, who takes home $600 every other week, has spent months sleepwalking through work, hanging on partly out of a need to have one foot tenuously in the real world but also because Henry needed health insurance. Henry's liver slowly improved, but he then began to lose weight at an alarming rate—20 percent of his body weight within weeks—and his skin began to disintegrate, turning red, scaly and raw. Several painful skin biopsies were inconclusive, suggesting that this was either an allergy to a medication or a sign of graft-versus-host disease.

While Henry was at the clinic having his skin examined, one doctor noticed that he was dragging his left leg when he walked. Two weeks later his left side became so weak that he could not lift himself to a sitting position in bed. He was rushed back to Minneapolis, where a scan showed a mass of unknown origin in his brain. Doctors operated but were unable to determine the cause. Whatever it was, it may have spread to his chest. Just last week, Henry was rushed to the hospital again—his sixth hospitalization in the past 12 weeks—where doctors found lesions in his lungs.

Of the 21 Fanconi patients who have received transplants within the past two years at Fairview under the new drug protocol that gave the Strongin-Goldbergs so much hope, 13 have survived so far. Of those, Henry is in the greatest danger. The first anniversary of his transplant is this coming Friday, a milestone that no longer seems like a victory.

What might have been another red-letter day will come in October, when the Nashes and the Strongin-Goldbergs had planned to meet in Disney World. The Strongin-Goldbergs will not be there. After years of technology and intervention, Laurie became pregnant the old-fashioned way, and her baby is due this fall. Tests show him to be a healthy boy who is not an HLA match for Henry.

Margaret Talbot

A Desire to Duplicate

FROM *THE NEW YORK TIMES MAGAZINE*

How far away is human cloning? Late in 2001, Advanced Cell Technology, a biotechnology firm, announced it had taken mature human cells and transferred them into human egg cells, creating an embryo, the important first step in cloning. While scientists agree that major hurdles remain, researchers around the world, clandestinely or otherwise, and despite widespread disapproval, continue to work on cloning human beings. Margaret Talbot, one of the most perceptive observers of American culture writing today, investigates a mystical sect who may be the unlikely producers of the first cloned human being.

In 2000, a 10-month-old baby boy died in the hospital after a minor operation went wrong. The baby's parents, an American couple, had two other children and probably could have had another if they wished; neither parent was infertile, and both were healthy and in their 30's. But they did not want another child. They wanted this child. And before long, they began to believe that the longing they felt was telling them something quite specific— that their dead baby's genes were crying out, as a ghost might, to express themselves again in this world. The idea preoccupied them that their little son's genotype deserved another chance, that it had disappeared by mistake and could be brought back by intention.

Now if all this had happened, say, five years ago, their conviction might have soon faded away. The couple might have told their friends or family about this secret dream of resurrecting their baby's genes, and been talked out of it, or comforted in some other way. But it happened last year—four years into the cloning revolution sparked by Dolly the sheep, at a moment when optimism about the miracles of biotech was running high and when it was not at all hard to find other people who shared a kind of metaphysical faith in the power of genes. One such group, a science-loving, alien-fixated sect called the Raëlians, for whom cloning is a central tenet, was eager to put its faith into action. Last June, the grieving couple and the Raëlians found one another with results that could—and should—reopen the whole debate over whether human beings ought ever to be cloned, and for what purpose.

For it turned out that the couple, who had been well off to begin with, now had an infusion of cash: a promised malpractice settlement from the hospital where their baby died. They were willing to finance the Raëlians in an all-out effort to clone the boy from cells they had frozen. And while they are not likely to succeed, the fact is that with at least 50 young female followers eagerly volunteering as egg donors and surrogate mothers, the Raëlians can't be ruled out, either. Cloning mammals is a wildly inefficient process that can require hundreds of attempts both to create an embryo and to implant it successfully. Only two or three out of every hundred attempts to clone an animal typically result in a live offspring. But for that very reason, cloning is partly a numbers game, in which luck and the ready availability of many donor eggs and borrowed wombs can play as significant a role as technical expertise. "When you look at what would be critically required to clone a human being, surrogates and a large number of eggs are key ingredients, and the Raëlians have those," said Gregory Stock, the director of UCLA's Program on Medicine, Technology and Society. "They certainly have what's necessary to make a solid attempt." Besides, said Stock, "what they're doing is of symbolic significance. If they don't succeed, someone else will in the next five years." Indeed, on January 26, 2001, an American fertility specialist announced what may be an even more credible effort. Panos Zavos, a professor of reproductive physiology at the University of Kentucky, said he and the Italian infertility doctor Severino Antinori would be leading a project to clone a human baby in the next 12 to 24 months.

What the Raëlians have already tapped into, though, is a well-spring of desire for human cloning—a desire whose fulfillment would surely take many of us off guard. In the years since Dolly, public discussions of cloning have shifted away from the specter of multiple human replicants to less disturbing possibilities, like the creation of genetically identical tissue grown for people with Parkinson's and other diseases. The initial revulsion at the very notion of cloning—what bioethicists call the "yuck factor"—has dwindled as more mammals have been cloned and as the prospect of someday replicating household pets seems to render the whole concept somehow cuter and more benign. Legislative efforts to ban cloning for reproductive purposes have stalled—only four states (California, Rhode Island, Louisiana and Michigan) have passed laws against it—and the federal moratorium merely precludes government money from going to it. Meanwhile, bioethicists, the professionals who promise to guide us through these troubled waters, have by and large embraced cloning, convinced that access to it constitutes a "reproductive right," a natural extension of technologies intended to help the infertile. Indeed, the people who openly express a desire to clone these days tend not to be megalomaniacs vowing to manufacture their own Mini-Me's (although Arnold Schwarzenegger did recently tell a reporter that he would "have nothing against cloning myself"). They're infertile couples who want a biologically related child and have exhausted other means, or bereaved parents yearning to "replace" a child they've lost.

Moreover, it is not unprecedented for fringe groups to serve as incubators for concepts that would not be acceptable in mainstream science: think of the Aum Shinrikyo sect and its ventures in biological warfare. The Raëlians are not a tiny group—they claim 55,000 members worldwide, though the number is probably closer to 25,000, according to Susan Palmer, a sociologist who has studied them. And they are not without resources. Since 1974, they have raised $7 million toward the construction of an "embassy" where alien visitors could be welcomed to our planet in style. Their followers, who hold fast to the ideal of everlasting life created through technology, are a devoted lot. Their leader has, in the words of Charles Cameron, a researcher with the Center for Millennial Studies at Boston University, "done an extremely good job of placing himself astride a powerful tide of hope and fear—the longings of people who want to find emotional and religious meaning in science and biotechnology."

For all of these reasons, I decided to take the Raëlians and their cloning project seriously. It wasn't always easy. It was hardest when I waded into the teachings of Raël, a French-born former race-car driver who is the movement's leader. In 1973, Raël says, he had an encounter with a four-foot-tall alien ("his skin was white with a slightly greenish tinge, a bit like someone with liver trouble") whose flying saucer had landed atop a volcano in southern France. From this creature, he heard the message that humans had been created in a laboratory by advanced beings from another planet who had mastered genetics and cell biology. Subsequent visits to the spacecraft, during which Raël enjoyed the sensual attentions of six "voluptuous and bewitching" female robots, convinced the fun-loving prophet that the aliens did indeed have a superior civilization.

It was easier to take them seriously when I met with Brigitte Boisselier, a French chemist who is the "scientific director" of Clonaid, the Raëlians' cloning venture. It took a while to get to Boisselier, and I still don't know exactly where she lives. I was told by the group's unfailingly polite PR people that she "travels a lot." But eventually, a meeting was set, for a hotel lobby in Syracuse. The Raëlians have a knack for drawing in pleasant, attractive, professionally successful people in scientific or technical fields—computer analysts, robotics engineers, lab technicians. Boisselier, who worked for many years as a research chemist at the French company Air Liquide and who now teaches chemistry at Hamilton College in upstate New York, was of that type. She has two Ph.D.'s, one from the University of Dijon and one from the University of Houston. With her handsome leather satchel and discreet gold jewelry, her dark hair pulled back from her pretty face in a neat bun, she projected, as she walked up to greet me, an air of cool, academic professionalism.

We sat down in the hotel restaurant, and Boisselier ordered a chicken Caesar salad and a decaf coffee. I don't know what I expected—somebody spacier or in silver spandex, maybe. Boisselier was neither, and yet the conversation we had that day was deeply strange nonetheless. She said she had a lab up and running, "not offshore, not in the Bahamas; somewhere in the United States." She wouldn't say where, except that it wasn't in one of the states that had outlawed cloning: "I'm no fool." She had assembled a team of three—a geneticist, a biochemist and an ob-gyn currently affiliated with an in-vitro fertilization clinic—and she said that the first two were working on the project full time, experimenting with cattle cloning initially and then moving on to human cells sometime this winter.

Was Boisselier worried about miscarriages or fetal abnormalities? "We will monitor the developing embryo and the pregnancy very closely," she said calmly. "We want a healthy baby." And she said that the surrogates who had volunteered to carry the cloned embryo—one of whom is Boisselier's own 22-year-old daughter, Marina Cocolios—were prepared to undergo abortions if defects were revealed by ultrasound or amniocentesis. If one pregnancy failed, another surrogate would automatically step into line; there would be no need to wait another month, as you would have to if you were dependent on the cycles of just one woman.

In Montreal a few weeks later, I spoke to Cocolios and another surrogate, Sylvie Tremblay, a 38-year-old computer troubleshooter at Laval University. For shiny-eyed devotion to the cause, you could hardly beat these two. "There is nothing else that is a higher priority in my life than this," said Tremblay. "If my boyfriend didn't agree with my taking part, I would drop him like that." Cocolios, an arts student, smiled shyly and said that she had "always been obsessed by having a baby—to carry life, to give birth. But my life is so sped up I don't even have time for my cat." She sighed girlishly. "I think it is so beautiful how this couple loved that child and wanted to bring back his genetics. I would offer this pregnancy as a gift to the whole of humanity."

Maybe Boisselier was thinking of her passionate volunteers when she said that she "didn't think there are major obstacles, actually, since cloning has been demonstrated in other mammals, and the technique has been refined. The main problem is the transfer of the DNA and making sure there are no defects when that transfer occurs." Still, the Raëlians had decided to use surrogates rather than encouraging the dead baby's mother to carry the child, because they didn't want her to have to go through a miscarriage "and lose that baby all over again."

For the time being, Boisselier told me, the couple wanted to remain anonymous: "Even their neighbors, friends and family don't know they're doing this." The family trusted her to represent their story—all the details about them here were provided by her—without revealing their identity. The couple weren't Raëlians themselves, she said; far from it. They "went to church every Sunday" and were well known in their community. "You have to understand, they are still grieving. It's hard for them to talk about the baby."

But what, or who, I wondered, did they think they would be getting, if by chance the Raëlians succeeded in growing another embryo from one of the dead boy's cells? "Well, you see, if they just have another child, it will be a different one," Boisselier said matter-of-factly. "And they say this child was unique. He was taken from us because of some malpractice at the hospital. He should be around us laughing and so on. He deserves to live again. And through cloning, there is a way for this genetic code to express itself so he can laugh and play and become whoever he was meant to become."

Did the parents realize that even a baby who shared the original's genotype would not be the same person? He'd be gestated in a different womb. He'd be subject to different environmental influences. And he'd be reared by parents who had been irrevocably altered by the loss of a baby.

Yes, said Boisselier, in her unflappable way, the parents realized that "this baby will have no memory of the 10 months they had been living together. They know the baby will not be exactly the same as the first one. But they are still working to get him back. They think this baby should be alive."

It was Halloween the day we met, and all around us, costumed people were eating their lunches. A big man in a toga was talking to a vampire. A girl-ghost stood by the bar, looking disconsolate. Boisselier paid no attention; she was going on in her low, sibilant voice about parents and children and bringing back the dead. It was one of those speeches that are meant to be self-evidently persuasive, in which each separate assertion sounds reasonable enough—but the links between them are all wrong. "You mentioned that you have a son," she said evenly. "Imagine he died. Would you consider cloning? I think it might cross your mind. You have seen that child going through many stages, you have spent hours just to love that child and then suddenly it's not there anymore. You give a lot as parents. And I think that is probably part of the gift that you can give to a child. If a child is dead before its time, then it's another gift to give that life back to a child."

Clonaid has a list of a hundred people who have expressed interest in its services, most of them would-be parents with severe infertility problems, a handful of them homosexual couples. "In some cases," Boisselier conceded, "they could go to a sperm bank and so on, but they felt better about having a child with their own genes. I think it's probably written in us to have a succession of our genes." But lately, she was getting calls mainly from the parents of children who have died—infants, teenagers, young adults—though few of the callers had the foresight, or whatever one would call it, to freeze the tissue that would make cloning the dead even theoretically possible.

Boisselier finished her salad and started in on a story about a woman she knows who was badly injured in a car accident, and who, though she eventually recovered physically, lost all memory of her previous life. At first, I couldn't see where this was going. "Her family could hardly recognize her," she explained. "Because she has no memory of the past, no emotions connected to it." That must have been hard for the family, I said. "Yes, they were surprised," said Boisselier, who then added—and this turned out to be the point of the story—"This lady was a good example of what a clone would be." That is, she looked the same, but lacked the experiences and associations that had made her dear to the people who wanted to bring her back in the first place. It was an indictment of the whole concept, and yet Boisselier didn't seem remotely aware of this, didn't seem to realize that if a person was unique and therefore by definition irreplaceable, then the quest to replace him was misguided and the encouragement of it cruel. Either our genes determine us or they don't, but if they don't, then cloning for the purpose of replicating a person is pointless. It holds forth the promise of unprecedented control—manufacturing a human being who will share specific traits with a preapproved model—but it cannot deliver.

YET THE ODD THING, as I found out over the next few weeks, is how many people know this and don't know it at the same time. They will tell you that they realize cloning does not produce a copy of the original person, but something more like a later-born identical twin, and yet say that they would want to do it anyway. They'd want to do it so that they could know as much as possible in advance about their unborn children, so they wouldn't have to take their chances on sexual reproduction, so they could perpetuate their own genes or so they could hope against hope to get back somebody very, very much like somebody they had lost.

In 1998, a 37-year-old man named Matthew Vuchetich fell out of a tree he was trimming and died. Matthew was the youngest of four children and the only son in a close-knit, Catholic family where he was generally regarded as a dazzler. (Nobody in the family has any connection to the Raëlians.) "He was brilliant; I mean that," said his sister Margo, a nurse-practitioner who lives, as Matthew did, in Atlanta. "He had a near-photographic memory. He talked over my head many a time." Matthew had a double degree in biology and chemistry and had worked in a lab at Emory. "But he couldn't stand to be inside four walls," his sister said, so he started a tree-trimming business that allowed him to be outdoors all day. Matthew was, she laughed, "a beautiful physical specimen, too. Pumped up. We joke sometimes that he's up in heaven saying, 'Hey, I'll never get old and y'all will.' "

Within two days of Matthew's death, it occurred to his mother, Marion, that he ought to be cloned. Marion Vuchetich is a retired middle-school biology teacher, 77 years old, still sharp, energetic and, as she said to me, "open to new things." Matthew had been an organ donor, so Marion called the organ bank and asked that a few square inches of her son's skin be preserved. Marion says she believes that human cloning will happen in the not-so-distant future, and so she holds onto the hope that Matthew will come back in some form to this world, even if it's after she herself has died. "My son was very interested in things scientific," said Marion. "He once said, 'If I should die, use everything you can from me and throw the rest away.' He would certainly have said, 'Go ahead, Mom.' "

Matthew's sister Margo said the cloning idea "is pretty much my mother's own thing. But my sisters and I felt, let her do it. It's probably something she needs to do emotionally, intellectually, to get through this. He was the sparkle of her eye and she cries over him every day. I think she's trying to hold onto any thread of him."

In the last year or so, some, but by no means all, of Marion Vuchetich's determination to clone Matthew has subsided. Sometimes she thinks saving Matt's cells was just a sensible thing to do for other people in the family—maybe someday a donor organ could be cloned from them. But she has kept up with cloning research by staying in touch with experts like Gregory Pence, a professor of ethics at the University of Alabama at Birmingham and an advocate of cloning. "Marion knows this wouldn't be Matthew exactly," said Pence, who gets about "a call a month" from people hoping to replicate a dead loved one. "She just wants something back. And who am I to tell her that's the wrong thing for her? A lot of people believe in the theory of the proper way to grieve—Kübler-Ross's stages, that sort of thing. But in practice, heterosexual couples have replacement children all the time. They have a stillborn baby; the woman gets pregnant again right away. But when somebody wants to use cloning for the same purpose, they're criticized."

Of course, if Matthew could be cloned, the result would be a baby—and then a child and then an adult—created to fulfill inescapably precise and poignant expectations. It's hard to imagine a human being who would be less of an end in himself. Marion Vuchetich said she understood that a clone "wouldn't be a duplicate of Matthew." Still, she referred to this theoretical clone as Matthew: "If I were a younger woman, I wouldn't hesitate for a minute. But I don't know if I want someone else to raise Matthew." And if Matthew could be cloned, I asked, what did she think she would get back? "His mind," she said unhesitatingly. "My son had an IQ of 165. I feel like the world lost something."

Over the last few years, as human cloning has receded from legislative agendas and public discussion, it has become a subterranean fantasy for all kinds of people. But it seems to have most powerfully caught the imagination of certain people in mourning, people who find in it an outlet for the lacerating need just to have their beloved here again. Internet chat groups devoted to cloning help stoke their hopes, as do companies like Southern Cross Genetics, an Australian start-up that offers to store DNA for future cloning.

The Web site of a group called the Human Cloning Foundation is a reservoir of such longings. There's the Mexican doctor who keeps posting urgent messages, pleading for information about how to clone his dead son: "I had a child 3 years old but he died last week, I need that somebody help me to find someone that make human clonings at this moment any place around the world." There's Eileen, who posts: "Person is dead for a month, can his hair help? . . . Pls help a 17 year old boy." Or David, whose 8-year-old son has just died: "My wife and myself miss him so much. . . . How would it be if there was a way to start all over again?" On other Web sites, I found messages from Dianne, a woman who wants to clone her dead father ("My father was a remarkable man, and I intend to see that he goes on in this world"); a Lebanese family eager to clone a nephew killed in a car accident (his mother "is desperately in need of getting him back at any cost"); a woman named Fiona who says she was "shattered by the death" of her twin sister. The day Fiona heard that a sheep had been cloned, she writes, "My first thought was, 'I can get my twinnie back.' " Only with time did it occur to her that "it would be impossible to re-create my sister in her totality as a person. . . . The best that could be done is the production of a blank page with my sister's genetic format." Most of us know this: that without our particular accretion of memories we wouldn't be who we are; that a baby born in 2001 couldn't possibly be the same person as someone born 30 years ago; that if you could have a baby who would truly grow into an eerie simulacrum of a dead loved one, it would be painful to look at him. But grief can derail what we know—as can the feeling, fed by genetic breakthrough and the glorification of them, that maybe genes really *are* us.

UNTIL A FEW YEARS AGO, would-be cloners had little to encourage them—a sci-fi novel here, a *Star Trek* episode there. But in February 1997, Ian Wilmut and his colleagues at the Roslin Institute in Scotland announced that they had successfully used a technique called somatic cell nuclear transfer to create a cloned sheep they named Dolly, after Dolly Parton. (It was a geek's joke: Dolly was cloned from a mammary cell.) Wilmut and his team had transferred the DNA-containing nucleus from the cell of an adult ewe to a donor egg whose own nucleus had been removed, leaving only the outer membranes and the cytoplasm. To fuse the adult nucleus and the hollowed egg together and to activate development of the embryo, a process usually set in motion by the helpful sperm, the researchers applied an electrical pulse, essentially shocking it to life. The hard part—what Wilmut managed that had been thought impossible before Dolly—was to show that the DNA in the adult nucleus, which was already serving its mature, specialized purpose, could essentially be tricked into dividing and otherwise behaving like a brand-new fertilized egg. The next, not particularly complicated step, was to implant the embryo in the uterus of yet another sheep, which served as a surrogate mother. (It could have been implanted in the adult ewe who contributed the nucleus, but in this case, that ewe was by then dead.) Since the nucleus of each cell in the body contains the genetic instructions for the whole, the resulting offspring, Dolly, was one that shared an identical genotype with the original—a clone. Human cloning by somatic cell nuclear transfer would work essentially the same way.

Public reaction was immediate, and initially at least, laced with bafflement and even horror. Nearly everyone who commented on Dolly made the imaginative leap to human cloning. And while there were those, like the sociobiologist Richard Dawkins, who wrote that "it would be mind-bogglingly fascinating" to watch "a younger edition of myself growing up in the 21st century," or the freewheeling physicist Richard Seed, who claimed he couldn't "wait to clone himself three or four times," most people were disturbed by the idea of making genetic copies. If a woman cloned herself and reared the child, she would be her own daughter's identical twin. If she had a husband, he would eventually find himself with a daughter who uncannily resembled his wife. Would this lead to confusion, even incest? And how could a cloned child live out his life freely, knowing he was the recipient of a preworn, consciously selected genotype? Wouldn't it be horrifying to know so much from such an early age about your own fate—what diseases you'd be likely to get, what personality flaws? What sort of narcissism would cloning unleash in us? What new enticement would it offer to tinker with our genes and produce "superior" babies by design? Would cloning, with its seeming guarantees, gain an edge on sexual reproduction, with all of its unknowns? Would babies no longer be conceived but manufactured? What would it say about us if we wanted that?

In June 1997, President Clinton, following the recommendations of his National Bioethics Advisory Commission, which had concluded that human cloning would be unsafe and therefore unethical "at this time," signed a five-year moratorium on the use of federal funds for human cloning research.

Animal cloning, however, proceeded apace. In the past four years, scientists have cloned more sheep, as well as goats, cows, pigs and in the most efficient cloning experiment yet, dozens of mice over six generations from a single mouse. The goal of most cloning projects is to create herds of genetically identical animals that could produce drugs in their milk or replacement organs for humans. But researchers working on the Missyplicity Project at Texas A&M are using the $2.3 million offered them by a bereft pet owner to try to copy his dead dog, Missy. Each of these breakthroughs makes human cloning that much more technically feasible. And the prospect of cloning a pet—what might be called the first instance of sentimental cloning—makes it, for some people, all that much more emotionally feasible.

Despite concerted effort, no one has yet succeeded in cloning a monkey, and this failure could mean it will be harder to achieve in humans than it has been in other mammals. But Don Wolf of the Oregon Regional Primate Research Center isn't so sure. "From a logistical point of view, it's actually *more* difficult to clone monkeys than humans," said Wolf. "We don't have 750 labs across the country doing assisted reproduction in monkeys. Some of the clinical work we've been struggling with—how best to grow monkey embryos, how to transfer them directly to the uterus—are old hat in the IVF world."

Nearly all of the animal cloning efforts, however, have led to high rates of fetal and neonatal mortality in the resulting offspring. Those who compare cloning to current IVF techniques—arguing that lots of those fail, too—neglect to mention that IVF failures consist mostly of unsuccessful implantations, not the sudden deaths of young babies.

"All sorts of things go wrong," said George Seidel, a cloning researcher at Colorado State University. Cloned cattle and sheep are often born dangerously large. "Normally you might expect a 100-pound birthweight in a calf, but with a clone, you might get 160 pounds," said Seidel. Because such outsize calves don't have room to wriggle around in the uterus, they can be born lame or with limb deformities. "Sometimes the kidneys aren't right, they're just plain put together wrong—or the heart is, or the lungs, or the immune system," he added. "It can be a unique abnormality in each case. They can die within a few days after birth, or sometimes they just can't make it after you cut the umbilical cord." Nobody really knows why.

Only if such problems are surmounted, said Seidel, would experimenting with human cloning be ethical: "We shouldn't be deliberately producing babies with abnormalities. We're talking about an abnormality rate of maybe 30 percent in cloned animals. In human babies, the normal rate of congenital defects is about 2 percent, and we wouldn't tolerate a jump to 3 percent." Indeed, virtually all of the scientists who have tried to clone other mammals say that we don't know enough at this point to try it in humans, and that to do so would amount to hugely risky experimentation on prospective people. Citing such safety concerns (as well as the possible psychological impact on children), the ethics committee of the American Society for Reproductive Medicine issued a report in November saying that cloning as a treatment for infertility did not currently meet "standards of ethical acceptability."

Besides, though cloned animals can be normal and healthy-appearing—some cloned mice and cattle even seem "improved," in the sense that they appear to age more slowly—what's normal in a barnyard animal isn't all that high a standard. "The fact that you can get a sheep or a mouse that looks normal," said Stuart Newman, a developmental biologist at New York Medical College, "doesn't mean that some subtle things haven't gone wrong in brain development that you wouldn't necessarily notice in a sheep, but you would in a human. Yes, you can clone a mouse—but can you take him to the opera?" Cloned humans might show higher rates of cancer or other diseases, but we'd only find out by cloning them and waiting to see if disaster strikes.

None of this means, however, that cloning services won't someday be marketed to desperate people—or even that human cloning isn't going on somewhere right now. "It's relatively easy to set up a lab and find someone competent to carry out the procedures," warned Roger Gosden, an infertility researcher at McGill University. "Regrettably, we will probably wake up one day to the news that someone, somewhere, has used somatic-nuclear-transfer technology to produce a human clone."

Severino Antinori, the Rome-based doctor known for pushing the limits of assisted reproduction—he helped a 62-year-old become pregnant in 1994—has been saying for a year that he wants to clone babies. Others in the field think he might not have a hard time of it. "From a technical point of view, cloning a human would be a very simple thing," said Jacques Cohen, the director of reproductive medicine at St. Barnabas Medical Center in Livingston, N.J. "I'm not advocating its use, but on the scale of things we've attempted, I just wouldn't expect it to be a big challenge."

Moreover, while cloning for human reproduction is now banned in Japan and in most of Europe, there are countries where assisted reproductive technology has become increasingly sophisticated and where no legislation forbids cloning. South Korean researchers, for example, have been active in animal cloning, and a few years ago, a team at Kyunghee University Hospital in Seoul claimed to have created an embryo from the nucleus of a 30-year-old woman, then destroyed it after developing it to the four-cell stage in vitro. (The researchers never published their results, however, casting doubt on the project.) "I wouldn't be surprised if you heard of it happening in some province of India or Pakistan that wanted to show it had a place in the world," said Gregory Pence. China is another possibility, especially since its IVF industry has lately taken off and its one-child policy has encouraged eugenic thinking.

Already, bioethicists who favor cloning have begun outlining the categories of people who might consider it. Indeed, for the last several years, those in the profession who have taken up the subject of human cloning seem to have been more concerned with identifying its worthwhile applications than with raising serious alarms about it. "Bioethicists are the most enabling community of all," said George Annas, a professor of health law at Boston University and one of the few bioethicists who has called for a ban on human reproductive cloning. "There's a libertarian strain among bioethicists—autonomy and individual rights are so important to them that it's virtually impossible for them to look beyond that." Indeed, the pro-cloning bioethicists I talked to often resorted to a sort of consumer logic; there's a market out there that wants this, and who am I to say they can't have it?

"I could imagine three main groups who'd be interested in cloning," said Ronald Green, a professor of ethics at Dartmouth (and an adviser to the biotech company Advanced Cell Technologies). The first would be couples in which the woman lacks viable eggs and the man lacks viable sperm, "and cloning is the only way they can have a child who is biologically related to them." The second "would be lesbian parents—and, to a lesser extent, gay men, since they'd still need a female surrogate." Unlike sperm donation, cloning would allow such a couple to sidestep "a genetic third party who, years down the line, might want to gain access to the child." The third group, said Green, would be made up of "people with serious genetic disorders that are not amenable to other modes of prevention like genetic screening—because maybe the specific mutation isn't known or many different genes are involved—and who still want to have their own biological child." Gregory Pence envisions an even larger market that would include people disappointed with the current array of assisted-reproduction technologies (which, after all, only succeed 25 percent of the time); heterosexuals wary of using the eggs or sperm of a stranger; and people "too sophisticated" to take a chance on the "lottery" of sexual reproduction with their husbands or wives. A button given to me by one of the Raëlians puts it succinctly: "Cloning—Reproduction Without Compromise."

For some people, cloning just seems like a chance to have a baby with some kind of genetic connection—even if it's only to one parent, even if the connection is uncomfortably close, even if they're a little vague on what a clone is. Desirée Boen, one of the dozens of infertile people who have posted messages to the Human Cloning Foundation Web site seeking help, said she was "really, really interested" in trying to clone a baby. Boen, a 25-year-old teacher's aide and former nurse from Orlando, Fla., has two children, 9 and 6, but because of a hysterectomy she underwent a year ago, she can't have any more with her new husband. Unlike many infertile people intrigued by cloning, Boen tried to adopt—but it was a private adoption, and "the mom seemed so flaky and unsure that I backed out of it." She said she would consider having a baby with the help of a surrogate mother, "except that you hear such bad stories about them changing their minds and keeping their babies." And because she lacks ovaries, Boen would need donor eggs as well. "I have a niece who's 19, and she offered one of her eggs," she said in a phone conversation. "But you know it gets pretty complicated when it's inside the family." When I asked if she would prefer to clone one of her cells or her husband's, Boen was puzzled, then admitted she thought "you could use both." But after a moment, she said, "Oh, I'd take my husband's, I guess." She said her husband was a great guy and she'd like him to have a boy just like him—one whom he could enjoy "in the cutesy phase."

Cloning for reproduction could be done, at least for a while, without most of us knowing it, said Mark Eibert, a lawyer who has submitted testimony to Congress in favor of cloning. "People who wish to have children may not want those children to become the focus of a media circus," he said. As for the health risks involved, Eibert pointed to the precedent of an infertility treatment called ICSI (Intracytoplasmic Sperm Injection), in which doctors inject slow-moving or otherwise inefficient sperm directly into the egg's cytoplasm. ICSI was developed in the early 90's and rushed into clinical use in fertility clinics across the country. "Even though it had virtually no animal experimentation and virtually nothing was known about its safety, it took off," Eibert explained. "Once you get normal healthy children on the ground, cloning will take off, too." And if the Raëlians are the first to pull it off? So be it. "From a PR point of view, I'd be happy if people like that weren't the first," he said. "But then again, if a Hare Krishna scientist was the first to invent a cancer cure, I think other scientists would be interested enough to pay attention. After a while, the history books wouldn't even mention the religion."

THE FACT IS, for all of the Raëlians' eccentricities, there is something about them that is perfectly attuned to their times. The Raëlians are enthusiastic about e-mail and sex and down on smoking and homophobia. And most of all, they love, love, love genetic engineering. They find it, as they like to say, in their French pop singer way, "very beautee-ful." In this sense, you could see them not as bizarros inflamed by a singular vision but simply as the most fervent proponents of a genetic essentialism that is fairly widely shared these days. To put it another way, the Raëlians are just a bunch of people who took literally the cliché that science is replacing religion.

The group's spiritual headquarters, where I visited Raël on a rainy November day, is a monument to its delirious scientism. Centre UFOland sits plop in the middle of rural Quebec, about an hour outside Montreal. Its neighbors are wheat and cattle farms; back down the sodden dirt road, on the main highway, are a smattering of trailer parks, a tractor factory and a couple of fast-food places specializing in gravy-soaked french fries. The building itself is shaped like a giant swoosh, and constructed, ingeniously, out of bales of hay slathered over with concrete. Inside is a museum devoted to extraterrestrial phenomena and to genetics—a curious amalgam of sci-fi and science. The high-ceilinged main exhibit room contains a life-size replica of the flying saucer Raël boarded back in 1973 for his visit with the aliens. It's made of plywood spray-painted silver and is pretty much empty except for a couple of inflatable vinyl chairs. (When visitors are around, a propane torch, suggesting the sound of a landing spaceship, gives it a Vegasy sort of grandeur.) Nearby, a 26-foot-high model of a double-helix spins slowly, illuminated by a spotlight so that its Lego-colored molecules gleam. There are plenty of pictures of little green men at UFOland, but there are also accurate models of cells and straightforward explanations of the Human Genome Project, the cloning of Dolly and so on.

I got a tour of the place from an uncharacteristically saturnine Raëlian named Michel. We walked from one dark and chilly exhibit room to the next, throwing on the lights as we went. Michel explained that "primitive man," which is to say everyone who lived before the age of genomics and of Raël, "did not understand the chemical nature of life. They didn't understand that the DNA is the soul." Michel permitted himself a little cackle at primitive man's expense. "They didn't understand that reincarnation can only happen through science—through cloning—so they imagined some, some. . . ." He paused, searching for the right term of opprobrium. "Theology."

Raël's office is next door to the museum, and I was escorted there by Sylvie

Chabot, the angular, henna-haired business consultant who handles most of the Raëlians' publicity. We had to walk through a set of electronic doors that hissed shut behind us, then remove our shoes and wait for the second set of automatic doors to open.

Raël admitted us, smiling a crinkly Clinton-like smile and wearing his characteristic samurai-style top-knot, white pants, wide-shouldered white tunic and gold medallion. His sealed lair is dominated by a large white bed with a tiger-print throw on it. The walls are covered with photographs of his companion, Sophie, a stunning young redhead, in which she is usually bare-breasted and nibbling on a rose or some such. I kept wondering why it is that futuristic prophets so often have to wear jumpsuits and medallions and whether we'll all have to wear them in the future. (My hairdresser had specifically instructed me to ask what was up with the top-knot, but I subsequently read in Raël's book that hair and beards are antennae helping the brain transmit messages, so I figure he just wanted taller antennae.) I also thought about how, if you are a futuristic prophet, your life is suffused with pop culture and you can't help looking and acting like guys from *Deep Space Nine*, just as real-life Mafiosi can't help looking and acting like guys from *The Sopranos.*

None of this, however, was exactly the point. The point was cloning, a subject upon which Raël was more than happy to discourse at length. Of course, he said, the cloning project they were undertaking now through Clonaid—the making or remaking of a baby for his devastated parents—was really a piffle, a tiny step toward the ultimate goal of eternal life through cloning. The couple's hopes, and of course their money, were a necessary ingredient, but the goal was much larger than them; nothing short, in fact, of the defeat of aging and of death. "The next step will be to make it possible to clone directly an adult person and not a baby," Raël explained, and then to figure out how to "upload" his memory into the new body.

In the meantime, baby cloning was a good place to start. "And then," he said, "comes a lot of new technology that we support—genetic engineering, genetic modification of human beings, improvement of human beings. You can call it eugenics, but not in a bad way, like the Nazi way of thinking before, which results in a superior race. No, cloning would be available to all human beings, to improve their characteristics and possibilities."

When I asked about resistance to cloning and qualms about the genetic engineering of humans, Raël heaved a gentle sigh of pity. "It was the same when Louise Brown, the first test-tube baby, was born. It was all Frankenstein and monsters. And now you have hundreds of test-tube babies made every day, and nobody asks anything about it because they know it's not bad. And that's why I am hoping that Clonaid will be the first company to make a cloned baby. And then everyone will see on CNN, maybe *Larry King Live,* a beautiful family, a smiling baby, and we know it will be smiling because it will be a copy of the one we know, and people will say, 'Ah, that's beautiful!' and public opinion will change. It was the same at the beginning of fire, and with the steam engine and electricity. All human progress."

In some ways, Raël is merely the surreal version of other more respectable biotech utopians—academics like Gregory Stock of UCLA, who told me that new reproductive technologies are the beginning of the end of sex as the way we reproduce. "We will still have sex for pleasure, but we will almost certainly see our children as too important to leave to a random meeting of sperm and egg." Or Lee Silver, a molecular biologist at Princeton, who sanguinely predicted that parents will one day be able to choose for their children genes that increase athletic ability, genes that increase musical talents and ultimately, genes that affect cognitive abilities. "Why shouldn't parents be able to give their child something that other children already have?" (Like Raël, few of the mainstream biotech utopians seem overly concerned about the willful creation of genetic haves and have-nots.) Or brainy business guys like the former Microsoft executive Nathan Myhrvold, who has said that resistance to cloning is "just another form of racism," a kind of "discrimination against people based on a genetic trait—the fact that somebody has an identical DNA sequence."

But more than anyone else, perhaps, Raël has hit upon a certain psychological truth: namely, that a common response to the disquieting feeling that science is accelerating beyond our capacity to comprehend it—let alone control it—is to declare oneself fervently, if confusedly, on its side. And that can also mean believing that somewhere, some wiser and higher force is guiding the latest discoveries and their uses, absolving us of the responsibility to judge them.

"Most traditional people are lost, spiritually lost, when it comes to space exploration, genetic engineering, genetically modified food, computers," said Raël. By contrast, the Raëlian movement was "the most fanatically pro-science of all the religions" and, therefore, "the best adapted to the new century." He continued: "Science and technology are beautiful, but if you don't link it to spirituality, you can easily become unbalanced or depressed and go to drugs and suicide. When you realize, on the other hand, that technology is not only technology but an extension of our spiritual life, it changes everything."

In the midst of such futurological abstractions, it can be easy to forget that if the Raëlians or some group like them succeed in their cloning project, they will be introducing an actual new person into this world. Raël himself didn't seem to have given a great deal of thought to the question of how a cloned child, being forced to play out some complicated re-enactment of a parent's or dead relative's life, might actually feel—how that sense of uncharted destiny that we think of as a kind of birthright might be foreclosed for a purposefully replicated child.

Sitting in Raël's gleaming white office, it occurred to me that it doesn't really matter in the end that a perfect "replacement" human can never be created. What matters is that some people think it can. What matters is the contempt shown for death and limits and our own peace with ceding the world to those who come after us. What matters is the insidious idea that if someone can be replaced, then he is no longer singular, which is to say priceless.

And what about the resulting child? As the philosopher Hans Jonas wrote in the 1970's, with cloning, we endanger the "right of each human life to find its own way and be a surprise to itself."

But never mind. According to Raël, what we want, we ought to have. The intensity of the desire is the proof of its virtue. That, and the ability to finance it. "These people we are helping, they want this child," he said, smiling. "They are willing to pay millions of dollars to have one. You can't be more welcome than that."

SALLY SATEL

Medicine's Race Problem

FROM *POLICY REVIEW*

Genetically, we now know that race is skin-deep: The average genomes of two traditional "races" do not differ by as much as the genomes of two individuals chosen at random. However, it is also the case that certain populations are at heightened risk for certain genetic diseases—such as the prevalence of sickle-cell anemia among those of African ancestry. Sally Satel, a psychiatrist and commentator on political trends in medicine, ponders whether the genuinely humane goal of doing away with racial categories conflicts with some avenues of potentially life-saving research.

On June 26, 2000, the White House announced to the world that the human genome had been sequenced (with the final, polished version due perhaps by 2002). Though the precise functions of all our genes have yet to be deciphered—Nobel laureate David Baltimore foresees a "century of work ahead of us"—it is clear that we are on the threshold of knowledge that could revolutionize the way we predict, diagnose, and treat disease.

The momentous discovery was lauded for something else as well: It supposedly laid to rest the idea that race is a biological category. "Researchers have unanimously declared there is only one race—the human race," said *The New York Times* in an article headlined "Do Races Differ? Not Really, DNA Shows." Much heralded was the finding that 99.9 percent of the human genome is the same in everyone regardless of race. "The standard labels used to distinguish people by 'race' have little or no biological meaning," claimed the *Times.* Said Stephen Jay Gould, evolutionary biologist at Harvard: "The social meaning [of race] may finally liberate us from [that] simplistic and harmful idea."

That point has found its way into the rhetoric of politicians. As former President Clinton has said, "in genetic terms all human beings, regardless of race, are more than 99.9 percent the same. . . . The most important fact of life on this earth is our common human ancestry." We were reminded by former GOP vice presidential candidate Jack Kemp that "the human genome project shows there is no genetic way to tell races apart. For scientific purposes, race doesn't exist," he averred. In Hollywood, too, notice was taken. When a television talk show host asked actor Rob Reiner about the sparring his character did with Archie Bunker on the long-running television program *All in the Family,* Reiner (aka "Meathead") explained that Archie's signature bigotry stemmed from ignorance. "We are all the same, though," Reiner said. "The human genome project taught us that." The human genome project, in the view of many, helped pound the final coffin nail in place: Race was at long last dead.

It is noble, of course, to celebrate spiritual kinship within the family of man, and now even to suggest that race has biological meaning, it seems, can be something close to professional suicide. The mere mention of race and biology together sends many physicians and scientists scrambling to protest (too much) against a possible connection. The facts, however, paint a more complex picture, one with clinical implications: Race does have biological dimensions, and if we regard it solely as a social construct, we may forfeit opportunities to enlarge our medical treatment repertoire.

Wanting It Both Ways

THE SENTIMENTS FUELING the impulse to regard race as an arbitrary biological fiction should be taken seriously, especially given the shameful history of race and biology. People properly shudder at the memory of the Tuskegee Syphilis Experiment, in which hundreds of black sharecroppers were never told they had the disease nor offered penicillin for its treatment. Many now worry that genetic determinism might be used as the sole explanation for social differences between races or, worse, as justification for new eugenics movements and programs of ethnic cleansing.

Nevertheless, the corrective is not obfuscation or outright censorship of inquiry. It is a clear-eyed understanding of the intertwining of race and biology. Denying the relationship flies in the face of clinical reality, and pretending that we are all at equal risk for health problems carries its own dangers.

We were reminded in May 2001 of the controversy about the role of race in biomedical research when *The New England Journal of Medicine (NEJM)* published two papers describing the responses of black and white patients to medications used to treat heart disease. The researchers sought to compare racial groups in light of the well-documented observation that, on average, African Americans with high blood pressure and other cardiovascular conditions do not fare as well as whites when given the same medications. Reaction from the race-is-a-social-artifact school was swift.

"There is a lot of scientific racism that's accepted as normal, [but] it's not valid science," Richard S. Cooper of Loyola University Medical School told the *Chronicle of Higher Education.* Even J. Craig Venter, the geneticist and entrepreneur whose company played a key role in mapping the human genome, expressed his dismay to *The New York Times:* "It is disturbing to see reputable scientists and physicians even categorizing things in terms of race . . . there is no basis in the genetic code for race."

Discrediting the idea of racial distinctiveness is a common strategy. "It has become clear that human populations are not unambiguous, clearly demarcated, biologically defined groups," states the American Anthropological Association, for example. Well enough, but no serious scholar believes that races or ethnic groups (blacks, Eskimos, Asians, whites) represent genetically discrete, static classifications. To claim as much is to set up a straw man. As early as 1775, Johann Friedrich Blumenbach, who originated the concept of "race," remarked that "innumerable varieties of mankind run into each other by insensible degrees." Considered the father of physical anthropology, Blumenbach wrote the landmark treatise *On the Natural Variety of Mankind* (1795) in which he introduced five races (Caucasian, Asiatic, American, Mongolian, and Malay) and saw them as part of a single species, *Homo sapiens.* Indeed, two centuries later the human genome project confirmed what Blumenbach had first articulated. Robert S. Schwartz, a member of *NEJM*'s own editorial staff, attacked the straw man in an editorial response entitled "Racial Profiling in Medical Research." He insisted that, genetically speaking, "humans—all humans—are so similar that to try to divide them up into neat little categories and label them yellow race, white race, black race, and so on, is insupportable." Instead of investigating biological variables associated with race, Schwartz wrote, "physicians everywhere must teach the immorality of racial discrimination in clinical practice."

Schwartz wants it both ways. On one hand he warns against the "dangers inherent in practicing race-based medicine," yet on the other he eagerly encourages the analysis of race as a variable in "research to root out social injustice in medical practice." Schwartz need not worry. The Department of Health and Human Services as well as numerous non-profits, such as the MacArthur and Robert Wood Johnson Foundations, are dedicating vast resources to investigating social causes of illness in minorities. Many of these causes are valid, among them culture-based habits of diet and lifestyle; attitudes about health and health care; and access to care. But other presumed causes, such as bias within the health care system or the impact of social hierarchy and income inequality on health, are susceptible to politicization. Not surprisingly, the quality of research targeting those factors is often dubious.

Despite the rhetoric about the bankruptcy of the race concept, it is notable that the federal government is pouring millions of dollars into studying racial differences in course-of-disease and in treatment response. In 1993 Congress passed and President Clinton signed a law requiring researchers whose projects are funded by the National Institutes of Health to include subjects of all major races. There are reasonable criticisms of the law as policy—for example, the expense entailed in amassing large enough study samples for valid statistical comparison—but there can be little question that one of its purposes was to illuminate potential differences in treatment response that are mediated by physiology, i.e., racial differences.

The practical clinical implications are real. For example, when doctors transplant kidneys or bone marrow, they have more difficulty finding a tissue match for African Americans because there are more possible antigen (protein) combinations on their cell surfaces than on the cells of white patients; some of those antigens are very rare in the population at large. In treating pain, doctors often give low doses of narcotics to Asian patients, given their tendency toward an acute sensitivity to the effects of those drugs. Or consider the higher mortality rate from breast cancer among African-American women. While obstacles to timely diagnosis and treatment may partly account for this difference, it cannot tell the entire story, as African-American women have a 50 percent higher incidence of breast cancer before the age of 35, a greater likelihood of developing more aggressive tumors as well as the highest incidence of pre-menopausal cancer.

Medicine, like society as a whole, seems torn about how to regard race. But a sober look at the relationship between race and genes can clarify the purpose and promise of research based on the notion that vulnerabilities to certain diseases and response to treatment can sometimes vary by race. "Race" is an approximate category, to be sure, but one with undeniable biological significance.

Small Differences Loom Large

ONE OF THE MOST heralded (and misunderstood) findings to emerge from the human genome project is that fact about the 99.9 percent genetic similarity between races. This is technically true. Then again, we share a very high percentage of our genetic makeup, about 98.4 percent, with chimpanzees, our closest primate relative, and we are clearly distinct from mice despite the fact that only 300 of the 60,000–80,000 genes in man appear to have no counterpart in the mouse. The resemblances extend into the plant kingdom, where mustard weed has 26,000 genes, about two thirds of the gene count of human beings. Even a one-celled organism—the yeast—has about 6,200 genes. No surprise is it, then, that there is quite enough room within that 0.1 percent of variation among humans for the existence of gene patterns, or frequencies, that cluster by group, or race.

The best way to appreciate how even tiny DNA changes can translate into big differences is to consider the single-gene diseases. Within a single human gene there are hundreds to thousands of irreducible building blocks called nucleotides. The order of a gene's nucleotides forms the recipe for a protein, each gene coding for a unique protein—for example, an enzyme that breaks down a target substance or forms a structural element of a cell or tissue.

These single-gene (monogenic) diseases—among them Tay Sachs, muscular dystrophy, and cystic fibrosis—show that alteration of a handful of nucleotides, or sometimes only a single one, can have profound physiological effects. In cystic fibrosis, the critical abnormality is in the protein that helps move ions and fluid along their normal travel routes in and out of cells. As a result, that movement is disrupted and the patient's organs become packed with thick mucus. Death before adulthood is common. The most common specific defect that causes cystic fibrosis? Three out of the 250,000 nucleotides of the gene that codes for the protein are missing: three out of thousands of nucleotides in that gene and out of three billion in the entire genome.

Single-gene diseases are relatively rare, accounting for about 2 percent of all diseases that afflict humans (most human conditions, such as heart disease, diabetes, or high blood pressure, are likely caused by several variant genes acting together) but they clearly illustrate how very small genetic variations can translate into palpable differences, sometimes between individuals and sometimes between racial or ethnic groups. This is not to say that every nucleotide variation spells disaster. Often it means that the enzyme produced by the gene will be less efficient at its job, for example, in breaking down other chemicals that are made by the body or are taken into the body, like medications. The clinical manifestations of this can vary, ranging from nonexistent to subtle to striking.

Consider the Canadian tuberculosis epidemic of the 1950s. All TB patients received the standard treatment: many months of triple-drug therapy, including a medication called isoniazid. It turned out that a sizeable fraction of Canadian Eskimos had a variant form of a liver enzyme that metabolized isoniazid so quickly that the drug was effectively used up before it could attack the tuberculosis bacteria. Many of the Eskimos metabolized isoniazid much faster than the general population and thus fared poorly under what was an inadvertent two-drug regimen. Many succumbed to TB and the partly treated, still-living tuberculosis bacteria themselves mutated into drug-resistant forms that went on to infect others in the general Canadian population. The significance of these results is that to ignore race under such circumstances is practically akin to withholding treatment.

The investigation of medication effects in black and white patients was published in that *NEJM* issue. The patients under study all had chronic heart failure, a condition in which the heart muscle is weak and cannot pump blood efficiently. The researchers were interested in improving treatment for heart failure and noted that while a class of medication called ACE-inhibitors was generally considered a standard treatment for heart failure, accumulating evidence suggested that black patients, on average, did not derive as much benefit as whites. The study, in which black and white patients were randomized to treatment with an ACE-inhibitor drug or placebo, found that black and white patients taking placebo fared no worse than black patients taking an ACE-inhibitor with regard to blood pressure control and hospitalization for worsening heart failure.

How to explain the results? The working theory revolves around a molecule called nitric oxide (NO), a gas that is normally produced by the cells that line our blood vessels. The NO gas rapidly spreads through the lining cells to the underlying muscle cells that surround blood vessels and regulate their constriction, an important dynamic in the control of blood pressure.

Specifically, NO dampens contraction of the muscle cells, thus relaxing the vessels and lowering blood pressure. For unexplained reasons, black individuals are more likely to have nitric oxide insufficiency arising from either reduced production by the lining cells, enhanced inactivation, or both. Since ACE-inhibitor drugs appear to exert their blood pressure–lowering effect by interacting with NO, patients with deficits (innate or acquired) will not have as vigorous a response to the medication as those with higher NO levels.

To be sure, not all black patients have low nitric oxide activity. But it is helpful for physicians to be aware of the likelihood, so that they can make better judgments about risk-benefit ratios of particular treatments. One group of scientists has taken the relationship between race and nitric oxide to its logical end: producing a medication specifically for African-American patients. Jay N. Cohn, a professor of medicine at the University of Minnesota School of Medicine, holds the patent on BiDil, a combination diuretic and vessel dilator that is believed to replenish stores of NO. Last March, the Food and Drug Administration authorized the testing of BiDil and the clinical trials will enroll only black patients with heart failure. The Association of Black Cardiologists is helping recruit patients for the trials in the hope that doctors will be able to offer another treatment option for black patients. Their rationale is obvious: Black patients suffer heart failure at twice the rate of whites, and those afflicted are twice as likely to die. "It is in the name of science that we participate," says B. Waine Kong, of the association. The Congressional Black Caucus voiced support as well.

That BiDil's therapeutic strategy relies on a crude predictor of drug response (race) is something its developers readily acknowledge. Some critics, however, are offended by the very idea. "I challenge any member of our species to show where this kind of analysis has come up with something useful," Richard Cooper says. It has "no proven value" according to Schwartz, who hopes the elucidation of the human genome "will force an end to medical research that is arbitrarily based on race." Arbitrary? No proven value? These protestations crumble in the face of one simple fact: In large-scale clinical trials BiDil provided a selective benefit to black patients with heart failure; white patients who took the drug had the same results as those on placebo.

Much as it might offend—"Skin Deep: Shouldn't a Pill Be Colorblind?" asks the headline in *The New York Times* coverage of BiDil—skin color can sometimes be a surrogate for genetic differences that influence disease and the response to treatment. "Right now we have only skin color to identify populations," Cohn explained. "You'd have to blindfold yourself to say we are not going to pay attention to obvious differences."

Pharmacogenomics

THE ULTIMATE PURPOSE of work like Cohn's and other biological realists is to identify factors that may be genetic in origin. First, researchers hope that identifying particular genetic markers within certain ethnic groups will yield insight into the genetic basis of disease and reveal why certain conditions are more prevalent in some groups. Second, the ultimate goal is to understand differences between *individuals,* not between races or ethnic groups. Pharmacologists talk enthusiastically of one day being able to customize medications to individuals based on their unique genetic profiles. While there is debate about the technical and commercial feasibility of this practice (dubbed pharmacogenomics), the ability to map an individual's genome efficiently is at least several decades away.

Until then, researchers can reasonably use race or ethnicity to direct them in identifying genetic markers that may predispose some patients to greater vulnerability to illness or to less robust treatment response. Dr. Cohn's work makes a persuasive case that there is a clinical imperative to do so. Knowing that a particular illness is more likely to afflict a certain group allows scientists to focus their attention on specific defects or the constellation of abnormalities that produce it. As far as treatment is concerned, "racial differences in response to drugs should alert physicians to the important underlying genetic determinants of drug response," explains Alastair J.J. Wood, a pharmacologist at the Vanderbilt University School of Medicine. And why not? The entities we call "racial groups" essentially represent individuals united by a common descent—a huge extended family, as evolutionary biologists like to say. Blacks, for example, are a racial group defined by their possessing some degree of recent African ancestry (recent because, after all, everyone of us is out of Africa, the origin of *Homo sapiens).*

In a sense, then, the purpose of examining race as a variable in biological research is to be able to transcend the use of race, to perfect a bottom-up kind of genetic analysis that is routinely performed in other arenas, such as forensic medicine. Its specialists can use tissue, hair, blood, or other fluids remaining at a crime scene to look for "population-specific genetic markers" in the DNA to help determine the race or ethnicity of a victim or suspect from his traces. Anthropologists can calculate, with startling precision, a person's ancestry from different parts of the world by analyzing the DNA from a few cells. Renowned population geneticist Luigi Luca Cavalli-Sforza has reconstructed human migration out of Africa within and between the continents in his landmark book, *The History and Geography of Human Genes* (Princeton, 1994).

Forensic and anthropological uses of genetic analysis show that race is not built on any particular trait. It is built on ancestry. Every present-day population harbors clues to its ancient roots. Members of a group would have more genes in common than members of the population at large. This is called, not so mysteriously, population genetics—the field of inquiry that examines changes in frequencies of particular genes within a population over time. And its legitimacy actually seems to be a point of agreement for both social constructionists and those who comfortably examine race in biomedical research.

Indeed, members of the "social construct" camp do not deny the basic observation that certain illnesses cluster by race. They recognize that among Americans, sickle-cell anemia, for example, is most common among African Americans. They are quick to point out, however, that other people whose ancestors come from the Mediterranean region are at risk as well, albeit less so than those of African heritage. This doesn't change the fact that African-American descendants are vulnerable, but it seems to reduce the anxiety that blacks alone will be saddled in the public imagination as having a "defective" trait. (Ironically, depending upon the environment, the sickle-cell condition—another condition in which a single nucleotide has undergone mutation—is less a defect than it is a genetic protection against malaria. When the malaria protozoan infects the red blood cell, the sickle shape makes the cell collapse and die, preventing the multiplication of the protozoan.)

The rancor surrounding the use of race as a variable in biomedical research was captured by the *Chronicle of Higher Education* in its exploration of the question: "How much emphasis should doctors and biomedical researchers place on the role of race as they evaluate health problems and potential treatments?" In "Shades of Doubt and Fears of Bias in the Doctor's Office," Jay S. Kaufman, an epidemiologist at the University of North Carolina at Chapel Hill, registers dismay over the study of race in biomedical research. "We live in a society in which race is the primary axis of social discrimination," he told the reporter. Referring to *The New England Journal of Medicine* articles, he opined, "An article that shows that there is some scientific foundation for the distinction is very comforting to people." As far as doctors' behavior is concerned, Kaufman says that these types of studies engender needless focus on race. "A patient walks in the room, the first thing the physician notices is, 'Oh, it's a black person' . . . a [journal] article says, 'oh, yeah, that's good, you should notice that, it makes a big difference.' "

Indeed, government-funded research in health is becoming extremely race conscious, but not in the way Kaufman criticizes. Efforts to uncover the social explanations for differences in health status between minorities and the general population are blossoming, including millions spent on—in the words of *NEJM*'s Schwartz—"research to root out social injustice in medicine." This is a popular theme. The U.S. Commission on Civil Rights, in its 1999 annual report to Congress and the White House, concluded that "racism continues to infect our health care system." The commission's assessment derived mainly from the fact that health outcomes were often poorer in minorities, especially African Americans. As is increasingly common, the commission inferred that bias in the system must be the underlying reason accounting for disparate health.

In academia, combating inequities has become a mission. According to Harvey V. Fineberg, former dean of the Harvard School of Public Health, "a school of public health is like a school of justice." In 1996 the theme of the American Public Health Association's annual meeting was "Empowering the Disadvantaged: Social Justice in Public Health." One of the association's proposed solutions is affirmative action in medical schools, believing as it does that "institutional racism has been an important contributor to racial disparities in health and economic status as well as a barrier to their elimination." The Centers for Disease Control and the National Institutes of Health give grants and sponsor workshops to assess the impact of "racism," "classism," "powerlessness," and income inequality on health. Unfortunately, politics—not rigorous methodology—often fuels these investigative forays.

Not all efforts, thankfully, are grounded in victim politics. Many simply want to close the health disparity gap in the service of improving the nation's health. Every 10 years, the Department of Health and Human Services announces its health agenda; this year the blueprint is called Healthy People 2010 and is devoted to the "elimination of racial and ethnic health disparities." Narrowing the gap between health outcomes in minorities and whites is a worthwhile effort. HHS is targeting six important areas for improvement: infant mortality, cancer screening and management, cardiovascular disease, diabetes, HIV infection/AIDS, and child and adult immunizations. In 2000, Congress allocated over $100 million to fund a new agency within NIH called the Center for Minority Health and Health Disparities.

It should come as little surprise that the anxieties plaguing critics tend to follow the ideological fault lines of the culture war. Take the notion of inborn differences. If we acknowledge that risk for particular illnesses or physiological features clusters by race or ethnicity, the critics worry, we are essentially erecting the scaffolding for social hierarchies. As University of Pennsylvania bioethicist Arthur Caplan put it, there is worry that the human genome would be used "to bolster racial and ethnic prejudice and other exclusivity groupings they believe in."

Also distressing to critics is the notion that intrinsic traits of individuals or groups will lead to unequal outcomes, in this case, in health status. The irony here, of course, is that the very purpose of researching race-related differences in physiology is to develop treatments that can help level the clinical playing field by improving health in people who are at risk. Critics also fret that by paying attention to patterns of biological differences, practitioners and researchers will underestimate or even ignore outright the considerable contributions that environment makes to health. This is an unlikely development given the vast interest, generous research funding, and public prevention campaigns devoted to the role of nonbiological factors such as diet, stress, environmental toxins, attitudes about health care, and access to it.

To acknowledge a biological dimension to race is to risk inflammatory and often groundless accusations of racism. The public's health is best served by a balanced portfolio that employs race as a variable in both biomedical and social health research. In the debate over the proper place of race in scientific inquiry, the meaning of the human genome project has become a powerful Rorschach test. Deliberately ignoring race in biomedical research can lead to inferior or improper treatment. The fact of a tainted history of race and medicine need not prefigure irresponsible activity now or in the future.

JEROME GROOPMAN

The Thirty Years' War

FROM *THE NEW YORKER*

In the past two years breakthrough drugs, such as Gleevec and Herceptin, and innovative strategies, such as Judah Folkmann's method of choking off cancer cells' blood supplies, have brought new promise of curing cancer. As the physician and writer Jerome Groopman points out, however, this sense of imminent victory is nothing new in the frustrating saga of the war on cancer.

To judge from recent headlines, we are on the verge of conquering cancer. The feature story in a May 2001 issue of *Barron's*, entitled "Investing in Health: Curing Cancer," ended by saying that "we are finally winning the war," and predicted that for our children cancer will be just another chronic illness, for which they will simply "pop a few pills every day." The cover of the May 28th issue of *Time* read, "There Is New Ammunition in the War Against Cancer. These Are the Bullets," and Dr. Michael Gordon, an oncologist at the University of Arizona, told *Time* reporters that in 20 years or so he "might just be out of a job." The annual meeting of the American Society of Clinical Oncology in June 2001 was buoyed by a spirit of optimism, and in the days that followed there was a sharp rise in the share prices of biotechnology and pharmaceutical companies that are developing cancer drugs. Meanwhile, Senator Dianne Feinstein, of California, has constituted a committee under the auspices of the American Cancer Society to consider how the government should respond to the challenges of cancer in the new millennium.

Important advances have been made in oncology in recent years, and the current atmosphere of hope is not without foundation. But it is not without precedent, either: ever since 1971, when President Nixon declared war on cancer, oncologists and cancer patients have been caught in a cycle of euphoria and despair as the prospect of new treatments has given way to their sober realities. The war on cancer turned out to be profoundly misconceived—both in its rhetoric and in its execution.

THE MOST AMBITIOUS health initiative ever undertaken by a country on behalf of its citizens began not with scientists, physicians, politicians, or patients but with a middle-aged New Yorker named Mary Lasker. Born in Wisconsin and educated at Radcliffe, Lasker had achieved great success in the fashion industry; her husband, Albert, had made a fortune in advertising. After the Laskers retired, they devoted themselves to Democratic Party politics and health-care issues. Mary Lasker was the quintessential American idealist; she believed that with enough money, influence, energy, and conviction you could accomplish anything. Then, in 1950, Albert Lasker developed intestinal cancer.

For the first half of the twentieth century, cancer was mainly the province of surgeons. Small tumors that had not spread were cut out, along with large amounts of normal tissue, in an attempt to catch any stray malignant cells. Still, microscopic deposits of cancer often remained, and patients were given radiation treatments, intended to destroy the residual cells. A few chemotherapy drugs were used as well, some of them derived from mustard gas, which had been used in the First World War. These treatments were highly toxic but seemed to shrink the cancer, at least temporarily, in patients with diseases such as Hodgkin's lymphoma. But for the vast majority of patients whose cancers had metastasized, or spread beyond the initial site, there was little that could be done.

Lasker had surgery, which did not completely remove the tumor. Intestinal cancer spreads within the abdomen, destroying the liver and often causing great pain. After two years of unsuccessful treatment, Lasker died, and his wife—using her network of political, medical, and business contacts, the advertising savvy that he had embodied, and the considerable resources from his estate—set out to transform the nation's response to the disease that had killed him.

The Laskers had been major contributors to the American Cancer Society, but after her husband's death Mary Lasker came to believe that only the government had the financial and organizational resources to launch a full-fledged crusade against cancer. In order to gain greater credibility in Washington, she cultivated relationships with high-profile academic physicians, most notably Sidney Farber, the scientific director of the Children's Cancer Research Foundation, in Boston. In the late forties, Farber had discovered that chemotherapeutic drugs that blocked folic acid, an essential vitamin, brought about remissions in some children with acute leukemia. His success in fighting this devastating pediatric cancer made him a frequent "citizen witness," invited by Congress to testify on behalf of medical legislation. Farber believed that, if the right drugs were developed, the gains he had seen in children with leukemia could be reproduced and improved upon. Lasker was also impressed by *Cure for Cancer: A National Goal* (1968), a book by a Denver physician named Solomon Garb, who asserted that cancer cures could emerge quickly if scientists stopped searching for new answers and devoted themselves instead to aggressively exploiting existing knowledge.

By the late sixties, however, the governmental largesse that had characterized Lyndon Johnson's Great Society programs had run its course; Nixon was determined to fight inflation, and Congress was under pressure to hold down domestic spending. To overcome this resistance, Mary Lasker organized the first major grass-roots cancer-advocacy group, the Citizens Committee for the Conquest of Cancer. On December 9, 1969, it began a campaign to make eradication of the disease a federal responsibility, running a full-page advertisement in the *Times* that declared, "Mr. Nixon: you can cure cancer." If American determination and ingenuity had put a man on the moon just months before, why shouldn't the nation attempt to conquer cancer by America's two-hundredth birthday? This political gambit quickly gained momentum. By the end of the following summer, both the Senate and the House of Representatives had unanimously passed a resolution to cure cancer by the Bicentennial.

In the ensuing debates over how this was to be accomplished, Farber argued before the House health subcommittee that researchers did not need to fully understand the workings of cancer in order to proceed: "The 325,000 patients with cancer who are going to die this year cannot wait; nor is it necessary, in order to make great progress in the cure of cancer, for us to have the full solution of all the problems of basic research." He pointed out that vaccination, digitalis, and aspirin were unquestionably beneficial, even if doctors didn't know exactly how they functioned: "The history of medicine is replete with examples of cures obtained years, decades, and even centuries before the mechanism of action was understood for these cures." What was needed, he maintained, was a generously funded cancer institute with strong leadership and a clearly articulated battle plan.

Richard A. Rettig points out, in his book *Cancer Crusade* (1977), that Farber's view was not universally accepted. Some scientists argued that a cure for cancer could not come about by directive. One such dissenter was a colleague of Farber's at Harvard, Dr. Francis Moore, the surgeon-in-chief at the Peter Bent Brigham Hospital. Moore, a medical-history buff, invoked what might be called the law of unintended consequences in scientific discovery. If there had been a diabetes institute in the late nineteenth century, for example, it would not have funded Langerhans's research on the pancreas, which led to the discovery of insulin, because the link between diabetes and insulin was not recognized at that time. Similarly, a government institute on polio probably wouldn't have supported the work of Dr. John Enders in the late nineteen-forties, when he was attempting to grow the mumps virus and found the method that ultimately proved essential to producing the polio vaccine. Advances had occurred in medical research, Moore argued, because support had gone to creative researchers in universities, "often young people, often unheard-of people." In fact, he could not recall a single example of a scientific breakthrough of clinical importance which had come from the sort of directed funding that was now being proposed.

Farber dismissed such critics, saying that they were not "cancer people," and were therefore ignorant of the possibilities at hand. Congressmen who expressed doubts about the wisdom of rapidly spending vast sums of money in response to what was essentially anecdotal testimony became targets of the Citizens Committee. They received hundreds of thousands of pleading letters, and committee members threatened to work against their reelection if they didn't reconsider their position.

Lasker also persuaded Congress to convene a panel of experts, with Sidney Farber as the co-chairman, which laid out the battle plan for the war against cancer. The President would henceforth appoint the director of the National Cancer Institute, and the institute's budget would be submitted directly to the White House, bypassing the regular channels of the National Institutes of Health. In December, 1971, Congress passed the National Cancer Act, and

Nixon signed it into law less than two weeks afterward. At a ceremony that made front-page headlines in newspapers across the country, Nixon declared, "This legislation—perhaps more than any legislation I have signed as President of the United States—can mean new hope and comfort in the years ahead for millions of people in this country and around the world."

Three decades later, the high expectations of the early seventies seem almost willfully naïve. This year alone, more than a million new diagnoses of major cancers will be made and about five hundred and fifty thousand Americans will die of cancer, an average of fifteen hundred a day. In the course of a lifetime, one of every three American women will develop a potentially fatal malignancy. For men, the odds are one in two. All the same, the triumphalist rhetoric that animated the war on cancer still shapes public opinion: many people believe that cancer is, in essence, a single foe, that a single cure can destroy it, and that the government is both responsible for and capable of spearheading the campaign. The military metaphors have retained their potency—even though they have proved to be inappropriate and misleading.

IN THE EARLY NINETEEN-SEVENTIES, many researchers believed that a cancer was generally caused by a virus that triggered important changes in a cell's metabolism, and that these changes accounted for a tumor's uncontrolled growth. Abnormalities in the genes of the cancer cell were thought to be incidental, rather than fundamental, to the disease.

The virus hypothesis was plausible because there were some hundred viruses that were known to cause cancer in amphibians, birds, and mammals. These were so-called retroviruses—RNA viruses that made their way into normal animal cells, copied their genes into a DNA form, and then subverted the ordinary functions of the cells for their own reproduction. The origin of most human cancers, the experts contended, would prove to be retroviruses as well, and hundreds of millions of dollars were poured into research to prove this assumption.

Sidney Farber's panel didn't just set the government's bureaucratic approach to cancer; it also dictated the National Cancer Institute's scientific agenda in research and clinical testing. The NCI awarded contracts to refine systems for growing cancer cells in bulk, and for producing enzymes that cut and copied DNA and RNA so that the nucleus of the cancer cell could be dissected and the hidden human cancer viruses exposed. Hundreds of thousands of botanical extracts and chemical poisons were systematically screened against different cancer cells to find the next generation of curative drugs. To test these new drugs, the NCI utilized a vast clinical-trials network of "coöperative groups," which were organized with the help of fifteen new cancer centers across the nation. The NCI also designated funding to train young physicians to become oncologists, the specialists who would prescribe the new drugs.

The NCI clinical-trials network employed three phases of testing new drugs. Phase I sought to determine the toxicity of the treatment and the maximum dose that patients could tolerate. Phase II assessed whether the therapy was of any benefit, and what doses of the drug and schedule of treatment seemed to work; it also established objective standards to measure success rates. Phase III studies compared the safety and benefits of the treatment under review with standard therapies. If the results were conclusively favorable, then the therapy would be submitted to the FDA for approval.

Within two years, the war on cancer was well under way, but the miraculous cures failed to appear. Many of the new chemotherapeutic drugs proved to be so toxic that they were quickly abandoned. In the absence of effective single agents, doctors began using combinations of less effective drugs, given at the highest dose a patient could tolerate. At the same time, the clinical-trials network had to justify its existence. Dr. Vincent DeVita, a prominent cancer specialist and a former director of the NCI, who is now the head of the Yale Cancer Center, recalls his frustration with the NCI when he was working there in the seventies, before he became the director. His research group developed a treatment regimen for an aggressive form of cancer called large-cell lymphoma, using a combination of chemotherapy drugs. In 1975, about forty-one per cent of the patients with this lymphoma were cured using DeVita's regimen. The NCI proceeded to compare the therapy with four similar treatments. "I screamed my head off, saying, 'You are all crazy! None of these regimens is good enough to merit being tested against another,' " DeVita recalled. " 'You will wrap up all the lymphoma research in this country. It will cost five million dollars, and in the end it will show that the treatment we started with is as good as but no better than any of the others.' " And this, indeed, was the result. The need to justify the bureaucracy meant that scores of clinical trials of relatively ineffective but toxic drugs were conducted with little benefit to the patient or to science.

THE SAME YEAR that Nixon signed the National Cancer Act, my grandmother Rose, an energetic sixty-seven-year-old Bronx homemaker, fell ill. She experienced frequent bouts of exhaustion, her lower back ached, and mysterious bruises began to appear on her arms and legs. A blood test showed that her white-blood-cell count was extremely high, and that many of the cells were abnormally large and immature. My grandmother had chronic myelogenous leukemia, or CML, a cancer of the bone-marrow stem cells. Malignant white cells grow uncontrollably, filling the marrow and flooding the bloodstream. The ache in my grandmother's back came from this expanding mass of leukemia pressing within her bones. Her exhaustion was caused by severe anemia, and the depletion of platelets in her blood was preventing it from clotting normally.

The prognosis for leukemia patients was grim, but my family was extremely hopeful; my grandmother's doctors told us that scientists were close to identifying human cancer viruses, and that new treatments would soon be available. Building on the logic of cancer as an infectious disease, researchers thought that, if the body could somehow be made to recognize the cancer cell as aberrant, its immune system would attack the tumor; by injecting the cancer patient with extracts of microbes, they hoped to jump-start that immune response.

After two years of standard chemotherapy, when my grandmother's condition began to deteriorate once more, a specialist recommended moving her to the front lines of immune therapy. She was enrolled in an experimental trial that used an immune booster called MER (methanol-extracted residue), a preparation from tuberculosis-like bacilli. This extract was injected under the skin of her back. Each injection was meant to cause severe inflammation, thereby stimulating her immune system to attack the cancer. I was a medical student at the time, and I remember examining her after the treatments. Her back was studded with raised welts, the size of silver dollars, that ran parallel to her spine. When I touched them, they felt hot and she winced in pain. Nevertheless, Rose resolutely kept every weekly appointment and received every injection. "I'm going to lick this," she said to my mother after each treatment. "They didn't give me a placebo. I'm lucky. I'm getting the cure."

After nearly a year, it was clear that the immune therapy wasn't working. We were not told at the time, but MER wasn't working for anyone else in the study, either. My grandmother's leukemia soon accelerated, then exploded into "blast crisis," which means that hordes of primitive cells invade vital organs like the lungs, the liver, and the kidneys, and the patient becomes susceptible to infections. In 1976, Grandma Rose contracted a bacterial pneumonia and died.

In many respects, my grandmother's experience in Phase I and Phase II trials was typical; most cancer treatments are unpleasant at best, and there is no way to judge the efficacy of a new approach without testing it on human beings. What was unusual was how little scientific basis there was for these particular experiments, and how much sensationalism surrounded them. Everyone involved in the war on cancer—from Mary Lasker and President Nixon to my grandmother's oncologist—had raised the hopes of Americans suffering from the disease to extraordinary heights. The Bicentennial celebrations came and went, and more people were dying of cancer than ever before.

IN 1977, when Dr. Arthur Upton, a radiation expert, was appointed director of the National Cancer Institute, he was immediately attacked for the NCI's failures. "I spent much of my time disabusing the public of the notion that the war on cancer was like the Manhattan Project or the Apollo space program," he told me recently. "It wasn't merely engineering. We didn't know enough about biology to understand the problem and point to solutions." Clinicians argued that not enough money had been earmarked for trials of different therapies, while scientists doing basic research pointed to millions of dollars that had been spent chasing nonexistent viruses. Upton began to question how the NCI's budget was being handled. "Bureaucrats were spending vast sums of money at the NCI without rigorous peer review," he said. "Contracts were awarded without any real scrutiny. I was besieged by scientists who felt money was being wasted." Instead of letting the senior staff at the NCI continue to dictate research objectives and then contract outside laboratories to perform much of the work, Upton gave priority to non-government scientists who applied for grants. These applications were assessed by an independent committee of scientific peers. "I took a lot of heat for it," Upton said. "Bureaucrats didn't like their turf being invaded."

During this period, Dr. Harold Varmus and Dr. Michael Bishop, at the University of California, San Francisco, who were using federal funds to study viruses as the cause of cancer, found evidence to suggest the opposite: that the seeds of our destruction are present within our DNA. These seeds are oncogenes, genes that can cause cancer when they mutate. Although the significance of oncogenes was not immediately understood, by the early nineteen-eighties the notion that most cancers were caused by human retroviruses had been largely discarded. The bulk of cancer research had been built on a false premise.

Yet the idea that the immune system could be stimulated to recognize and attack cancer cells—that the power to heal ourselves lies within our own bodies—remained tantalizing, even after the failure of treatments like MER. Some conjectured that the crude immune boosters like the ones my grandmother received had failed only because the triggers were not powerful enough, and that what was needed was a pure and potent stimulus. Interferon, a natural protein that functions as part of the body's immune system, was believed to be such a trigger.

Three major types of interferon had been identified—alpha, beta, and gamma—and laboratory experiments suggested that all of them might be effective in fighting aggressive, chemotherapy-resistant cancers, such as melanoma, metastatic breast cancer, and kidney cancer. Animal tests were encouraging; in mice, interferon caused tumors to melt away without harming normal tissue. Soon oncologists and journalists were speaking of interferon as the long-awaited panacea, and on March 31, 1980, *Time* ran a cover story on the drug. The American Cancer Society spent two million dollars on interferon that had been purified in Finland from the blood of volunteer donors. Pharmaceutical and biotechnology companies spent hundreds of millions of dollars to genetically engineer alpha, beta, and gamma interferon, then produced the proteins in large quantities. With great fanfare, clinical trials began.

As a cancer researcher in Boston, I participated in Phase II studies of alpha and gamma interferon. As soon as the trials were announced, we were deluged with requests from cancer patients who were desperate to participate; we could admit only a few, and then had to explain to hundreds of others that the rosters were filled.

A woman I will call Nora Dusquette was accepted for treatment in 1983. Nora, a middle-aged schoolteacher, was in the late stages of malignant melanoma. The cancer not only had formed large black deposits in the skin on her arms and back but had spread to her lymph nodes, her lungs, and her liver. We treated her with high doses of interferon by injection three times a week. Nora lived in New Hampshire, and was still teaching full time when she first came to see me, and yet she was more than willing to drive two and a half hours to Boston for her therapy.

Nora experienced intense side effects: fevers, chills, loss of appetite, and extreme fatigue. After her injections, she was rarely able to sleep through the night. She also started losing weight rapidly. Soon she was no longer able to teach or to maintain her household, and she had to rely on family members to take care of her. After four months, it was clear that the treatment wasn't having any impact. The cancer spread to her brain, and a few weeks later she died.

For the vast majority of cancers, it turned out, interferon just didn't work. Alpha interferon did prove successful in the treatment of some rare cancers—especially hairy-cell leukemia—but Nora's experience was typical: like the majority of cancer patients who participate in these kinds of studies, she suffered considerable toxicity with no apparent benefit.

By 1985, hopes had shifted to a protein called interleukin-2, which had been discovered at an NCI laboratory. Interleukin-2 stimulates immune cells called T lymphocytes. Dr. Steven Rosenberg, at the NCI, experimented with removing lymphocytes from cancer patients, stimulating them with interleukin-2, and returning them to the patients. In a few instances, there seemed to be significant shrinkage of metastatic melanoma and kidney cancer. Again, the news media were filled with speculations that a cure had been found. The NCI spent millions of dollars supporting interleukin-2 trials, which were administered to cancer patients both at the NCI and in cancer centers across the nation. The new treatment was also extremely toxic. A few patients experienced severe cardiac and pulmonary complications and died.

In 1987, a fifty-seven-year-old friend and colleague I will call Samuel Driscoll received a diagnosis of kidney cancer. He underwent extensive surgery to excise the primary cancer and the metastatic deposits, which were in his abdomen and lungs, but within a year the cancer had returned. Sam participated in a Phase II study using interleukin-2. Lymphocytes were removed from his blood, treated in the laboratory with the immune-stimulating protein, and then reinfused. As with virtually all other patients in these clinical trials, he suffered severe side effects and had to be hospitalized. He had high fevers, a widespread rash, and difficulty breathing; his body became painfully bloated. Sam did enjoy a six-month remission, during which he continued to teach and do research. Then the cancer recurred—this time in his lungs—and he died of pulmonary failure within a month.

WHY DID DOCTORS welcome therapies with known toxicity and uncertain gain, and why did patients like Sam Driscoll subject themselves to them? Because the conventional therapies were no better. The best way to treat tumors is by detecting them early enough to prevent their growth and spread, but many oncologists don't meet their patients until long after that point. As one doctor said bitterly to me, "What do you say to these people—'Too bad, you flunked prevention'?" Ironically, the nature of the NCI studies meant that in Phase III trials treatments with slight benefits were used on more patients for a longer time; interferon, for example, affords at best a marginal improvement, so it is possible to discern its benefits only in large studies conducted over long periods.

Most of the new cancer drugs were extremely toxic, and the real advances were in finding drugs that would temper their side effects. Platinol, which can cure testicular cancer, caused intense nausea and projectile vomiting that could last for days; in some patients, the retching was so severe that it tore the esophagus. It led researchers to develop potent anti-emetics. Other chemotherapy drugs, like Adriamycin, destroyed so many white blood cells that the patient was susceptible to fatal infections. Proteins like G-CSF were found that could stimulate the bone marrow to produce white cells, thereby greatly reducing the likelihood of such complicating infections. Even some of the toxicity of interleukin-2 was ameliorated. Thus, the horrors of chemotherapy were sometimes made less severe, or, at least, less prolonged. Chemo also became easier to administer, and oncologists could offer some reassurance to their patients that refinements in supportive therapy would reduce the suffering.

By the nineteen-eighties, a huge superstructure had resulted from the government's war on cancer. Some eight billion dollars had been spent. About thirty government-funded comprehensive cancer centers and major regional coöperative treatment groups linked virtually all university hospitals and community-based specialists. The American Society of Clinical Oncology had grown from several hundred members to nearly ten thousand. Cancer treatment had become one of the cash cows of academic and community hospitals, which competed fiercely for patient referrals. Treatment had also become the focus of a wide range of cancer-advocacy groups, whose constituents forcefully lobbied Congress for more funds to address their needs.

There were some success stories; by laboriously cobbling together combinations of chemotherapy agents, researchers had discovered that the majority of patients with Hodgkin's lymphoma and nearly all patients with testicular cancer could be saved. Great strides were also made in treating several pediatric cancers. Unfortunately, all of these types of tumors are relatively unusual. Hundreds of thousands of cancer patients underwent experimental treatments; in most cases, the pain and discomfort caused by the side effects were unaccompanied by genuine benefit, and in some cases the treatments were fatal.

IN 1984, Vincent DeVita, who had become the director of the National Cancer Institute four years earlier, provided Congress with a new goal in the war on cancer: a fifty-per-cent reduction in cancer-related mortality by the year 2000. According to an article by John C. Bailar III and Elaine Smith, which appeared in *The New England Journal of Medicine*, in 1986, this prediction was not justified by clinical data. Bailar, who had worked at the National Cancer Institute as a statistician studying trends in cancer incidence and outcomes, had grown wary of the predictions surrounding the war on cancer. Bailar and Smith's paper, dispassionately analyzing the claims of recent "advances" in treating cancer, demonstrated that there had been a slow and steady *increase* in cancer deaths over several decades. It concluded that "we are losing the war against cancer."

The paper was extremely controversial. Some felt that the authors hadn't given certain treatments adequate time to be properly measured; others felt that their statistics were not meaningful without more specifics. In response, in 1997 Bailar and Heather Gornik published a more sophisticated analysis in the same journal, entitled "Cancer Undefeated." Here the authors examined all American cancer deaths between 1970 and 1994 according to age, sex, and type of disease. They showed that there had been a six-per-cent increase in age-adjusted mortality due to cancer since Congress first acted, at the behest of Mary Lasker. There had been a recent dip, about a quarter of a per cent per year, which they attributed to reduced cigarette smoking and improved screening (thanks to mammograms, colonoscopies, and Pap smears). This said little for the enormous efforts that had been made over the previous decades on the therapeutic front. Bailar believes that, instead of focussing exclusively on fighting this generation of cancers, our work should be directed toward thwarting future generations of tumors through prevention and early detection.

In fact, the principal benefits from the war on cancer have been in other realms. The technologies developed to seek out cancer viruses in the seventies and eighties coalesced in the new field of molecular biology, which opened up the cell and its genetic blueprint to examination for the first time. This revolutionary DNA work also spawned a highly lucrative industry. Using the tools developed through the National Cancer Institute's contracts, biotechnology companies have created lifesaving treatments for heart disease, sepsis, colitis, and countless other serious maladies. Equally dramatic gains were made in AIDS research: the molecular techniques and reagents used to search for human cancer viruses proved essential in identifying HIV and mapping its genes. In addition, the inventory of failed cancer drugs includes agents like AZT, which proved beneficial in treating AIDS. These unintended consequences of the war on cancer make it more difficult to gauge its success or failure.

"The idea of a war sets up a false metric," says Dr. David Golde, who, as the physician-in-chief at Memorial Sloan-Kettering Cancer Center, oversees all the institution's clinical programs. "If a complete victory is not achieved, then it is deemed a failure." Do we examine the impact on patients and their families? Do we ask whether patients' quality of life is improved—do they get more time without cancer, even if the tumor ultimately returns and kills them? Do we measure success and failure in terms of cost and benefit, calculating how much money is spent in treatment and how much economic productivity is gained for the nation? Or do we measure it by knowledge gained, progress in scientific understanding, even if that knowledge is not readily translated into improvements for patients? There is no consensus among cancer specialists on these questions.

IN THE PAST DECADE, cancer research has progressed in a number of different directions. In the area of immune therapy, there have been some promising results from so-called monoclonal antibodies, like Rituxan, which train the immune system to recognize tumor cells. Three years ago, a flurry of excitement greeted some early results in animal studies of angiogenesis, conducted by the researcher Judah Folkman. Folkman's research identified compounds that might prevent a tumor from generating its own blood supply, and so choke its growth. Unfortunately, the first clinical studies have not shown significant shrinkage of tumors.

By far the greatest source of excitement in cancer research, however, has been targeted therapies, an approach to treatment that is tailored to specific kinds of cancers. Unlike most experimental treatments of the past three decades, targeted therapies are based on a growing understanding of the molecular machinery of the diseased cell. The oncogene is the cornerstone of the new approach—which is based on the work of Robert Weinberg, a scientist at MIT.

In the late seventies, Weinberg began conducting just the sort of research that Farber had insisted was no longer necessary: an exploratory investigation of the oncogene's possible relationship to the origin of human cancer. Within a few years, he had found a conclusive link between the mutations of an oncogene and the development of a bladder tumor. Mutations in all our genes occur every minute, because there is an intrinsic error rate when DNA is copied. If the errors are extreme, the cell will self-destruct, but otherwise the aberrant cell survives. Normally, oncogenes provide the blueprint for proteins that signal when a cell should divide, mature, and die; they are often described as the accelerators of the cell's growth. A mutated oncogene may direct a cell to reproduce wildly, and this means, in turn, that more mutations are likely to occur.

In 1986, Weinberg isolated another type of genes, called tumor suppressors. These act as brakes on growth, but, when they mutate, the brakes can fail. Yet a third genetic control, the so-called telomerase gene, helps determine how long a cell can perpetuate itself. Normal cells can divide only a set number of times, but an alteration in the telomerase gene can make a cell immortal. A cell becomes cancerous when several preconditions are met: mutations in oncogenes, or simply an excess of oncogenes, either of which promotes growth; changes in tumor-suppressor genes, which then fail to restrain growth; and changes in telomerase genes, which sustain the mutating cell. In 1999, Weinberg demonstrated this when he produced a cancer cell from a normal cell in the test tube by introducing oncogenes, tumor-suppressor genes, and telomerase into a healthy cell. It was the first time that human cancer had been artificially created.

As the significance of Weinberg's work became clear, researchers in both the private and the public sectors began searching for ways to target these malfunctioning genes. Chemotherapy and radiation have traditionally been crude tools against cancer, indiscriminately smashing not only the diseased cells but the healthy tissue around them. A sophisticated understanding of the cell's workings greatly increases the likelihood that the mutant genes can be shut down without affecting the healthy cells. In 1993, the Swiss drug company Ciba-Geigy synthesized hundreds of thousands of possible targeted therapies, which it then tested against dozens of oncogene proteins. One of the targeted drugs was STI-571.

DOUG JENSON is a sixty-seven-year-old retired systems engineer who lives in the Pacific Northwest. Four years ago, attending a Promise Keepers rally in Washington, D.C., he was overcome by exhaustion. A month later, after lunch at his church, he noted that his urine was "the color of cranberry juice." This was followed by wheezing and shortness of breath. He went to his local doctor, who gave him a blood test. The doctor called back that evening. "I hate to tell you this over the phone," he said, "but your white count is more than three hundred thousand." Jenson's diagnosis was chronic myelogenous leukemia, the same cancer that killed my grandmother. "They told me I had three to five years at the most," Jenson said. He was a robust man, five feet eleven and two hundred and twenty pounds, but his condition deteriorated rapidly. Initially, he was treated with chemotherapy, and that brought his white count down to about fifty thousand. This was a temporizing measure. Later, he began interferon treatments. Interferon has a modest benefit for CML patients, and significant side effects. Jenson became severely anemic, and then suffered a seizure. "The stuff is killing you," his hematologist said. But there seemed to be no alternative. Then, in September of 1998, Doug Jenson was referred to Dr. Brian Druker, a leukemia expert at Oregon's Health Sciences University, who enrolled him in a Phase I study of an experimental drug, STI-571.

Five years earlier, Druker had received a series of compounds from Ciba-Geigy to test on malignant cells, and he worked with the company to choose what appeared to be the best of the series of candidate drugs. Among them was the compound STI-571, which appeared to be an ideal tool: it deftly dismantles three oncogene proteins, including one called Abl, which is the accelerator of chronic myelogenous leukemia. When it was tested in small animals, however, it was found to cause liver damage in dogs, and the pharmaceutical company, with meagre experience in cancer drugs, was leery about beginning human trials. But Druker pushed hard, because of the striking effects in the test tube against CML. In June of 1998, three institutions—Oregon, UCLA, and M. D. Anderson, in Houston—began a Phase I safety study in patients with CML. To date, they have not reached a maximally tolerated dose, meaning a dose of STI-571 that causes significant toxicity in patients. The alarming dog studies proved not to be relevant for human beings.

Jenson had lost seventy pounds during his illness and prior treatment, and he gained it all back when he started taking STI-571. Within a few weeks, his white count fell to fifteen thousand. Not long after that, his anemia was ameliorated, and his platelets returned to normal. Few white cells that show the mutated Abl oncogene remain. "I go up to the health club every day, sometimes twice a day," he said. "I take spinning classes three times a week." The only side effect has been "a little puffiness around the eyes, which comes and goes." How long the benefits will continue is unknown, but, like the vast majority of the more than five hundred patients with CML who have been treated with STI-571 in the Phase II study, Jenson has enjoyed a profound and sustained remission. "It is a journey I don't wish on anyone," he said. "But, all things considered, I've been very, very fortunate."

Druker is cautious about the drug's dramatic results; negative side effects could still manifest themselves in the future, and the leukemia could also become resistant to the drug, precipitating relapse. Still, STI-571 is the most exciting new cancer drug in years. It turns out that the drug blocks oncogenes that may be critical to other kinds of cancer, such as glioblastoma, a type of brain tumor, and gastrointestinal stromal tumor, a rare sarcoma of the intestine. On May 10th, at a Washington press conference, the Secretary of Health and Human Services, Tommy Thompson, announced the FDA's approval of STI-571, after only two and a half months of review. It was the fastest agency clearance ever for a cancer drug.

DeVita believes that the recent advances in cancer genetics will allow us to make enormous strides in treatment almost immediately. For him, STI-571—now known as Gleevec—is the proof of the principle. "I think we have the targets," he said. "It's not difficult to synthesize chemicals that block those targets. And when they come into clinical trials they work—surprisingly well." In the *Barron's* article, he said, "Within 15–20 years, I think cancer will become just another chronic, survivable disease, much like hypertension or diabetes." He predicts that the difficulties will lie not with the science but with the lack of resources for clinical trials to test all the drugs that will soon be discovered.

BUT IS MAKING Gleevec the poster child of imminent targeted cures premature—a replay of Sidney Farber's response to his success with childhood leukemia? After all, curing cancer entails understanding a hundred-odd diseases, which behave in different ways in different individuals. In the lab, the chemicals that are being screened are interacting only with the proteins of oncogenes or tumor-suppressor genes. In the patient, however, they are interacting with a complex living organism; it is impossible to accurately predict the success of chemicals that look promising in the lab.

I spoke with Dr. Glenn Bubley, a cancer researcher at Boston's Beth Israel Deaconess Medical Center, who, in 1997, conducted a clinical trial of a drug called SU-101. "SU-101 was touted as the Second Coming a few years ago," Bubley told me. It is a small molecule that, like STI-571, blocks PDGF-r, an oncogene that is believed to be important in the proliferation of a number of intractable cancers. When human tumors that had high levels of PDGF-r were implanted in mice, treatment with SU-101 blocked their growth. Safety tests in animals, unlike those involving STI-571, were promising, with no major liver toxicity or other red flags, and researchers anticipated dramatic success in the clinical trials.

A patient I will call George Mitsopoulos was a restaurant owner who had prostate cancer. He was still working when he entered the trial, despite the fact that his cancer had spread to his bones. The tumors could no longer be controlled by hormonal therapy, and so he began treatment with great hopes that SU-101 would ameliorate his condition. He quickly discovered, however, that the drug had severe side effects. "I am exhausted," he told Dr. Bubley. "I feel like I can hardly move out of bed." He also had immense difficulty sleeping, because he felt terrible even when he was lying down. Other patients in the trial begged to interrupt the therapy because of the exhaustion it caused, but Mitsopoulos was determined to persevere. The course of treatment had no lasting impact, and he died shortly afterward.

Mitsopoulos's experience with SU-101 proved to be typical. What went wrong? Many cancer cells may have redundant machinery, with several different oncogenes driving growth; if you block one, the others may take up the slack. The pharmaceutical company that had supported the research was not enthusiastic about publishing negative data, Bubley said—even high-profile journals prefer articles with positive results—but he believes that it is equally important to publish accounts of the failures, in order to inject a note of realism into the scientific debate. (An article about the failure of SU-101 in treating prostate cancer was published in May 2001, in *Clinical Cancer Research.)*

Other targeted treatments, like Herceptin, an antibody developed by Genentech, have performed better—but not nearly as well as some clinicians initially expected them to. Herceptin targets Her-2, a protein produced by an oncogene that is found in between twenty and thirty per cent of breast-cancer cases. The early news was exciting, and magazines and morning talk shows reported that Herceptin would make chemotherapy treatment for breast cancer a thing of the past. When Herceptin is used in conjunction with chemotherapy, it nearly doubles the likelihood of significant shrinkage of breast cancers with Her-2. But as a solo treatment for breast cancer its impact has been modest. In 2000, Genentech alerted physicians to a potentially lethal respiratory problem among women who had breast cancer that had spread to their lungs or who had prior lung disease. It can also cause significant heart damage in some women—particularly those receiving Adriamycin, a mainstay chemotherapy drug in the treatment of the disease. This is because in this form of breast cancer, as in some other cancers, the problem lies not with a mutated oncogene but with an excess of normal oncogenes, and targeting them can damage healthy heart tissue as well. *The New England Journal of Medicine* recently described the Herceptin study as "a landmark trial," even though it extended life for an average of only five months, and only in the subset of patients who qualified for the treatment. The description is less an example of hyperbole than a sobering reminder of the fact that no prior therapies had been shown to significantly extend the lives of women with metastatic breast cancer.

These limitations reveal how complex the biology of cancer is, and how little can be predicted about the efficacy of any particular treatment. The statistician John Bailar, for one, remains skeptical of the new therapies. "In the nineteen-fifties, there was huge excitement about laboratory programs to screen for chemotherapy drugs," he says. "We found a few drugs, but not many. Then, in the nineteen-seventies, there were cancer viruses. In the eighties, it was immunotherapy, with biologics like interferon and interleukin-2 as the model magic bullets. Now it's cancer genetics. The rhetoric today sounds just the way it did forty years ago. I have no doubt that there has been a huge increase in knowledge about cancer. The problem is to translate it into public benefits we can measure. I want to see an impact on population mortality rates. If the treatments are really that good, then we'll see it."

After decades of listening to unrealistic predictions, cancer-patient advocates have a jaundiced view of researchers who inflate preliminary anecdotes of success. Fran Visco, the president of the National Breast Cancer Coalition, told me recently that she was dismayed, at a meeting of cancer clinicians, at the way researchers interacted with members of the press. "These clinical scientists receive media training and are scripted by their hospitals," she said. "There are so many agendas here: fame, patient referrals, fund-raising, pharmaceutical grants, academic advancement." Ellen Stovall, the president of the National Coalition for Cancer Survivorship, agreed: "The headlines are dreadful." She referred to the sensationalism surrounding the disease as "the pornography of cancer," adding, "I am excited by the new science, but show me hard data. We need to raise the skepticism barometer."

Many former members of the cancer establishment express similar misgivings. Samuel Broder, who succeeded DeVita in 1988 as the director of the National Cancer Institute, and who is currently the chief medical officer at Celera Genomics, believes that we require new breakthroughs in the lab—particularly in understanding the process of how cancer spreads—before we can be confident of great gains in treatment.

"I call it the iron-lung syndrome," he told me. "If you had demanded that the NIH solve the problem of polio not through independent, investigator-driven discovery research but by means of a centrally directed program, the odds are very strong that you would get the very best iron lungs in the world—portable iron lungs, transistorized iron lungs—but you wouldn't get the vaccine that eradicated polio." He thinks that, given the performance of the targeted therapies available so far, it would be premature to invest more in the federal bureaucracy that oversees clinical trials.

Broder argues that the creation of new therapies is no longer the sole or even the primary provenance of the government. "Initiative and creativity have moved to the private sector," he said. "There is just no way of getting around it, and anyone who tells you otherwise is on a different planet. What was done in the early seventies was necessary, even in retrospect, but that doesn't mean we should do it that way now." Furthermore, pharmaceutical companies prefer to run their own clinical trials: both Gleevec and Herceptin were submitted to the FDA for approval without having entered NCI-sponsored studies. This frees more money for the sort of basic research supported by the National Institutes of Health—the grant system and research laboratories that Broder refers to as the jewels in the crown of the NIH. Any scientist or clinician in the United States can propose a new idea and seek support for testing it. "When the NIH sticks to that," Broder said, "it does an astonishing job, and it is the envy of the world."

Harold Varmus is a former director of the National Institutes of Health, and, like Broder, his experience as the head of a large government institute has made him wary of bureaucratic efforts to direct scientific research. Now, as the president of Memorial Sloan-Kettering, he finds himself in a curious position. "My view here is not very popular—especially among cancer researchers and cancer-focussed senators—but I believe cancer doesn't deserve unique distinction for funding," he said recently. Giving one advocacy group special treatment simply doesn't help the balance of research. Varmus also believes that the greatest advances in new knowledge will come not from cancer genetics alone but from a variety of disciplines working together to understand the complex mechanisms of the cancer cell. After all, genes are merely the blueprints for proteins, and it is the proteins that do the cell's work. An ability to decipher protein shapes—how they change in health and disease—will be important in combatting cancer, and this will require advances in chemistry, in computer science, and in physics.

IN RECENT YEARS, the mission to reëducate Congress and the public about the realities of cancer and to reverse the unrealistic attitudes and expectations that we have inherited from Nixon's war has been taken up by an unlikely advocate—the current head of the National Cancer Institute, Dr. Richard Klausner. "I'm pretty well plugged in to what's going on in research," he remarked. "I hear on the news 'Major breakthrough in cancer!' And I think, Gee, I haven't heard anything major recently. Then I listen to the broadcast and realize that I've never heard of this breakthrough. And then I never hear of it again." Klausner himself has been under considerable pressure to predict the eradication of cancer, because powerful members of Congress have promised that such a prediction could mean millions of additional government dollars for the NCI. But he refuses—not only because to do so is impossible but because it would propagate the scientific fallacy that cancer is a single disease.

The most productive way to move forward in cancer research, Klausner believes, is to call off the war. He prefers to think of cancer as an intricate puzzle—one that we currently lack both the knowledge and the tools to solve. Clues could come from any field, and the reforms that he has undertaken at the NCI reflect the need for such disciplinary openness. He has tackled the vast clinical-trials bureaucracy of the cooperative groups so that they no longer function as a closed shop controlled by inbred committees but are, instead, responsive to any researcher with good ideas. He also recognizes that the NCI should complement the drug companies' efforts rather than duplicate them; to this end, the NCI provides assistance to university-based laboratories that are pursuing molecular targets and candidate drugs but lack sufficient resources to develop and market them. So far, more than fifty compounds and molecular targets have been developed in this manner, and two have entered the first phase of clinical testing.

Klausner refers to these reforms as "an experiment, to see if science can take over the National Cancer Institute, instead of politics and hype." He continued, "Human beings seem to have this endless ability to think they are at the end of history. The only people who now are saying we know enough are people who don't know enough."

FRANCIS MOORE'S CONGRESSIONAL TESTIMONY about science's law of unintended consequences has been amply proved over the past thirty years. The failures of the government's war on cancer have been matched by the unforeseen successes it led to in fighting other diseases; indeed, its greatest successes came from shattering its central premise—the belief in cancer viruses. As Moore predicted, the most promising results stemmed from basic biological inquiry. And yet both Congress and the public continue to view open-ended scientific investigations as nebulous, self-indulgent, and wasteful of taxpayers' money, and are reluctant to fund them. For this reason, oncologists talk in terms of imminent cures through directed research—both in their proposals for new projects and in their assessments of ongoing work. The media attention that results further misleads the public.

If Americans are unwilling to reject the national mythology of cancer, it may be because they fear that the only alternative is despair. That fear can be tempered by the rapid pace and diversity of new discoveries in science and technology that are influencing every dimension of cancer research. Of course, it is impossible to say which type of currently intractable cancer will be cured first. In the next ten years, the survival rate of people with a certain type of melanoma or lung tumor or lymphoma or breast cancer may not change. But it also might improve by fifty per cent, or ninety per cent. Because of the uncertainty inherent in scientific discovery, there is simply no way of knowing. Paradoxically, for cancer patients and their families this inability to predict the future becomes their sustaining hope.

GARY TAUBES

The Soft Science of Dietary Fat

FROM *SCIENCE*

Everyone knows that eating fat is bad for you. It raises cholesterol and promotes heart disease. The message is clear: Cut down on fat and live longer. There's just one problem: Nearly fifty years of research has failed to prove conclusively that restricting dietary fat confers any health benefits. Gary Taubes, a writer with a surgeon's skill in separating scientific myth from fact, shows how politics and wishful thinking joined to create the medical equivalent of an urban legend.

When the U.S. Surgeon General's Office set off in 1988 to write the definitive report on the dangers of dietary fat, the scientific task appeared straightforward. Four years earlier, the National Institutes of Health (NIH) had begun advising every American old enough to walk to restrict fat intake, and the president of the American Heart Association (AHA) had told *Time* magazine that if everyone went along, "we will have [atherosclerosis] conquered" by the year 2000. The Surgeon General's Office itself had just published its 700-page landmark "Report on Nutrition and Health," declaring fat the single most unwholesome component of the American diet.

All of this was apparently based on sound science. So the task before the project officer was merely to gather that science together in one volume, have it reviewed by a committee of experts, which had been promptly established, and publish it. The project did not go smoothly, however. Four project officers came and went over the next decade. "It consumed project officers," says Marion Nestle, who helped launch the project and now runs the nutrition and food studies department at New York University (NYU). Members of the oversight committee saw drafts of an early chapter or two, criticized them vigorously, and then saw little else.

Finally, in June 1999, 11 years after the project began, the Surgeon General's Office circulated a letter, authored by the last of the project officers, explaining that the report would be killed. There was no other public announcement and no press release. The letter explained that the relevant administrators "did not anticipate fully the magnitude of the additional external expertise and staff resources that would be needed." In other words, says Nestle, the subject matter "was too complicated." Bill Harlan, a member of the oversight committee and associate director of the Office of Disease Prevention at NIH, says "the report was initiated with a preconceived opinion of the conclusions," but the science behind those opinions was not holding up. "Clearly the thoughts of yesterday were not going to serve us very well."

During the past 30 years, the concept of eating healthy in America has become synonymous with avoiding dietary fat. The creation and marketing of reduced-fat food products has become big business; over 15,000 have appeared on supermarket shelves. Indeed, an entire research industry has arisen to create palatable nonfat fat substitutes, and the food industry now spends billions of dollars yearly selling the less-fat-is-good-health message. The government weighs in as well, with the U.S. Department of Agriculture's (USDA's) booklet on dietary guidelines, published every 5 years, and its ubiquitous Food Guide Pyramid, which recommends that fats and oils be eaten "sparingly." The low-fat gospel spreads farther by a kind of societal osmosis, continuously reinforced by physicians, nutritionists, journalists, health organizations, and consumer advocacy groups such as the Center for Science in the Public Interest, which refers to fat as this "greasy killer." "In America, we no longer fear God or the communists, but we fear fat," says David Kritchevsky of the Wistar Institute in Philadelphia, who in 1958 wrote the first textbook on cholesterol.

As the Surgeon General's Office discovered, however, the science of dietary fat is not nearly as simple as it once appeared. The proposition, now 50 years old, that dietary fat is a bane to health is based chiefly on the fact that fat, specifically the hard, saturated fat found primarily in meat and dairy products, elevates blood cholesterol levels. This in turn raises the likelihood that cholesterol will clog arteries, a condition known as atherosclerosis, which then increases risk of coronary artery disease, heart attack, and untimely death. By the 1970s, each individual step of this chain from fat to cholesterol to heart disease had been demonstrated beyond reasonable doubt, but the veracity of the chain *as a whole* has never been proven. In other words, despite decades of research, it is still a debatable proposition whether the consumption of saturated fats above recommended levels (step one in the chain) by anyone who's not already at high risk of heart disease will increase the likelihood of untimely death (outcome three). Nor have hundreds of millions of dollars in trials managed to generate compelling evidence that healthy individuals can extend their lives by more than a few weeks, if that, by eating less fat. To put it simply, the data remain ambiguous as to whether low-fat diets will benefit healthy Americans. Worse, the ubiquitous admonishments to reduce total fat intake have encouraged a shift to high-carbohydrate diets, which may be no better—and may even be worse—than high-fat diets.

Since the early 1970s, for instance, Americans' average fat intake has dropped from over 40% of total calories to 34%; average serum cholesterol levels have dropped as well. But no compelling evidence suggests that these decreases have improved health. Although heart disease death rates have dropped—and public health officials insist low-fat diets are partly responsible—the *incidence* of heart disease does not seem to be declining, as would be expected if lower fat diets made a difference. This was the conclusion, for instance, of a 10-year study of heart disease mortality published in *The New England Journal of Medicine* in 1998, which suggested that death rates are declining largely because doctors are treating the disease more successfully. AHA statistics agree: Between 1979 and 1996, the number of medical procedures for heart disease increased from 1.2 million to 5.4 million a year. "I don't consider that this disease category has disappeared or anything close to it," says one AHA statistician.

Meanwhile, obesity in America, which remained constant from the early 1960s through 1980, has surged upward since then—from 14% of the population to over 22%. Diabetes has increased apace. Both obesity and diabetes increase heart disease risk, which could explain why heart disease incidence is not decreasing. That this obesity epidemic occurred just as the government began bombarding Americans with the low-fat message suggests the possibility, however distant, that low-fat diets might have unintended consequences—among them, weight gain. "Most of us would have predicted that if we can get the population to change its fat intake, with its dense calories, we would see a reduction in weight," admits Harlan. "Instead, we see the exact opposite."

In the face of this uncertainty, skeptics and apostates have come along repeatedly, only to see their work almost religiously ignored as the mainstream medical community sought consensus on the evils of dietary fat. For 20 years, for instance, the Harvard School of Public Health has run the Nurses' Health Study and its two sequelae—the Health Professionals Follow-Up Study and the Nurses' Health Study II—accumulating over a decade of data on the diet and health of almost 300,000 Americans. The results suggest that total fat consumed has no relation to heart disease risk; that monounsaturated fats like olive oil lower risk; and that saturated fats are little worse, if at all, than the pasta and other carbohydrates that the Food Guide Pyramid suggests be eaten copiously. (The studies also suggest that trans fatty acids are unhealthful. These are the fats in margarine, for instance, and are what many Americans started eating when they were told that the saturated fats in butter might kill them.) Harvard epidemiologist Walter Willett, spokesperson for the Nurses' Health Study, points out that NIH has spent over $100 million on the three studies and yet not one government agency has changed its primary guidelines to fit these particular data. "Scandalous," says Willett. "They say, 'You really need a high level of proof to change the recommendations,' which is ironic, because they never had a high level of proof to set them."

Indeed, the history of the national conviction that dietary fat is deadly, and its evolution from hypothesis to dogma, is one in which politicians, bureaucrats, the media, and the public have played as large a role as the scientists and the science. It's a story of what can happen when the demands of public health policy—and the demands of the public for simple advice—run up against the confusing ambiguity of real science.

Fear of Fat

DURING THE FIRST HALF of the 20th century, nutritionists were more concerned about malnutrition than about the sins of dietary excess. After World War II, however, a coronary heart disease epidemic seemed to sweep the country. "Middle-aged men, seemingly healthy, were dropping dead," wrote biochemist Ancel Keys of the University of Minnesota, Twin Cities, who was among the first to suggest that dietary fats might be the cause. By 1952, Keys was arguing that Americans should reduce their fat intake to less than 30% of total calories, although he simultaneously recognized that "direct evidence on the effect of the diet on human arteriosclerosis is very little and likely to remain so for some time." In the famous and very controversial Seven Countries Study, for instance, Keys and his colleagues reported that the amount of fat consumed seemed to be the salient difference between populations such as those in Japan and Crete that had little heart disease and those, as in Finland, that were plagued by it. In 1961, the Framingham Heart Study linked cholesterol levels to heart disease, Keys made the cover of *Time* magazine, and the AHA, under his influence, began advocating low-fat diets as a palliative for men with high cholesterol levels. Keys had also become one of the first Americans to consciously adopt a heart-healthy diet: He and his wife, *Time* reported, "do not eat 'carving meat'—steaks, chops, roasts—more than three times a week."

Nonetheless, by 1969 the state of the science could still be summarized by a single sentence from a report of the Diet-Heart Review Panel of the National Heart Institute (now the National Heart, Lung, and Blood Institute, or NHLBI): "It is not known whether dietary manipulation has any effect whatsoever on coronary heart disease." The chair of the panel was E. H. "Pete" Ahrens, whose laboratory at Rockefeller University in New York City did much of the seminal research on fat and cholesterol metabolism.

Whereas proponents of low-fat diets were concerned primarily about the effects of dietary fat on cholesterol levels and heart disease, Ahrens and his panel—10 experts in clinical medicine, epidemiology, biostatistics, human nutrition, and metabolism—were equally concerned that eating less fat could have profound effects throughout the body, many of which could be harmful. The brain, for instance, is 70% fat, which chiefly serves to insulate neurons. Fat is also the primary component of cell membranes. Changing the proportion of saturated to unsaturated fats in the diet changes the fat composition in these membranes. This could conceivably change the membrane permeability, which controls the transport of everything from glucose, signaling proteins, and hormones to bacteria, viruses, and tumor-causing agents into and out of the cell. The relative saturation of fats in the diet could also influence cellular aging as well as the clotting ability of blood cells.

Whether the potential benefits of low-fat diets would exceed the potential risks could be settled by testing whether low-fat diets actually prolong life, but such a test would have to be enormous. The effect of diet on cholesterol levels is subtle for most individuals—especially those living in the real world rather than the metabolic wards of nutrition researchers—and the effect of cholesterol levels on heart disease is also subtle. As a result, tens of thousands of individuals would have to switch to low-fat diets and their subsequent health compared to that of equal numbers who continued eating fat to alleged excess. And all these people would have to be followed for years until enough deaths accumulated to provide statistically significant results. Ahrens and his colleagues were pessimistic about whether such a massive and expensive trial could ever be done. In 1971, an NIH task force estimated such a trial would cost $1 billion, considerably more than NIH was willing to spend. Instead, NIH administrators opted for a handful of smaller studies, two of which alone would cost $255 million. Perhaps more important, these studies would take a decade. Neither the public, the press, nor the U.S. Congress was willing to wait that long.

Science by Committee

LIKE THE FLOURISHING American affinity for alternative medicine, an antifat movement evolved independently of science in the 1960s. It was fed by distrust of the establishment—in this case, both the medical establishment and the food industry—and by counterculture attacks on excessive consumption, whether manifested in gas-guzzling cars or the classic American cuisine of bacon and eggs and marbled steaks. And while the data on fat and health remained ambiguous and the scientific community polarized, the deadlock was broken not by any new science, but by politicians. It was Senator George McGovern's bipartisan, nonlegislative Select Committee on Nutrition and Human Needs—and, to be precise, a handful of McGovern's staff members—that almost single-handedly changed nutritional policy in this country and initiated the process of turning the dietary fat hypothesis into dogma.

McGovern's committee was founded in 1968 with a mandate to eradicate malnutrition in America, and it instituted a series of landmark federal food assistance programs. As the malnutrition work began to peter out in the mid-1970s, however, the committee didn't disband. Rather, its general counsel, Marshall Matz, and staff director, Alan Stone, both young lawyers, decided that the committee would address "overnutrition," the dietary excesses of Americans. It was a "casual endeavor," says Matz. "We really were totally naïve, a bunch of kids, who just thought, 'Hell, we should say something on this subject before we go out of business.' " McGovern and his fellow senators—all middle-aged men worried about their girth and their health—signed on; McGovern and his wife had both gone through diet-guru Nathan Pritikin's very low-fat diet and exercise program. McGovern quit the program early, but Pritikin remained a major influence on his thinking.

McGovern's committee listened to 2 days of testimony on diet and disease in July 1976. Then resident wordsmith Nick Mottern, a former labor reporter for *The Providence Journal,* was assigned the task of researching and writing the first "Dietary Goals for the United States." Mottern, who had no scientific background and no experience writing about science, nutrition, or health, believed his Dietary Goals would launch a "revolution in diet and agriculture in this country." He avoided the scientific and medical controversy by relying almost exclusively on Harvard School of Public Health nutritionist Mark Hegsted for input on dietary fat. Hegsted had studied fat and cholesterol metabolism in the early 1960s, and he believed unconditionally in the benefits of restricting fat intake, although he says he was aware that his was an extreme opinion. With Hegsted as his muse, Mottern saw dietary fat as the nutritional equivalent of cigarettes, and the food industry as akin to the tobacco industry in its willingness to suppress scientific truth in the interests of profits. To Mottern, those scientists who spoke out against fat were those willing to take on the industry. "It took a certain amount of guts," he says, "to speak about this because of the financial interests involved."

Mottern's report suggested that Americans cut their total fat intake to 30% of the calories they consume and saturated fat intake to 10%, in accord with AHA recommendations for men at high risk of heart disease. The report acknowledged the existence of controversy but insisted Americans had nothing to lose by following its advice. "The question to be asked is not why should we change our diet but why not?" wrote Hegsted in the introduction. "There are [no risks] that can be identified and important benefits can be expected." This was an optimistic but still debatable position, and when Dietary Goals was released in January 1977, "all hell broke loose," recalls Hegsted. "Practically nobody was in favor of the McGovern recommendations. Damn few people."

McGovern responded with three follow-up hearings, which aptly foreshadowed the next 7 years of controversy. Among those testifying, for instance, was NHLBI director Robert Levy, who explained that no one knew if eating less fat or lowering blood cholesterol levels would prevent heart attacks, which was why NHLBI was spending $300 million to study the question. Levy's position was awkward, he recalls, because "the good senators came out with the guidelines and then called us in to get advice." He was joined by prominent scientists, including Ahrens, who testified that advising Americans to eat less fat on the strength of such marginal evidence was equivalent to conducting a nutritional experiment with the American public as subjects. Even the American Medical Association protested, suggesting that the diet proposed by the guidelines raised the "potential for harmful effects." But as these scientists testified, so did representatives from the dairy, egg, and cattle industries, who also vigorously opposed the guidelines for obvious reasons. This juxtaposition served to taint the scientific criticisms: Any scientists arguing against the committee's guidelines appeared to be either hopelessly behind the paradigm, which was Hegsted's view, or industry apologists, which was Mottern's, if not both.

Although the committee published a revised edition of the Dietary Goals later in the year, the thrust of the recommendations remained unchanged. It did give in to industry pressure by softening the suggestion that Americans eat less meat. Motterns says he considered even that a "disservice to the public," refused to do the revisions, and quit the committee. (Mottern became a vegetarian while writing the Dietary Goals and now runs a food co-op in Peekskill, New York.)

The guidelines might have then died a quiet death when McGovern's committee came to an end in late 1977 if two federal agencies had not felt it imperative to respond. Although they took contradictory points of view, one message—with media assistance—won out.

The first was the USDA, where consumer-activist Carol Tucker Foreman had recently been appointed an assistant secretary. Foreman believed it was incumbent on USDA to turn McGovern's recommendations into official policy, and, like Mottern, she was not deterred by the existence of scientific controversy. "Tell us what you know and tell us it's not the final answer," she would tell scientists. "I have to eat and feed my children three times a day, and I want you to tell me what your best sense of the data is right now."

Of course, given the controversy, the "best sense of the data" would depend on which scientists were asked. The Food and Nutrition Board of the National Academy of Sciences (NAS), which decides the Recommended Dietary Allowances, would have been a natural choice, but NAS president Philip Handler, an expert on metabolism, had told Foreman that Mottern's Dietary Goals were "nonsense." Foreman then turned to McGovern's staffers for advice and they recommended she hire Hegsted, which she did. Hegsted, in turn, relied on a state-of-the-science report published by an expert but very divergent committee of the American Society for Clinical Nutrition. "They were nowhere near unanimous on anything," says Hegsted, "but the majority supported something like the McGovern committee report."

The resulting document became the first edition of "Using the Dietary Guidelines for Americans." Although it acknowledged the existence of controversy and suggested that a single dietary recommendation might not suit an entire diverse population, the advice to avoid fat and saturated fat was, indeed, virtually identical to McGovern's Dietary Goals.

Three months later, the NAS Food and Nutrition Board released its own guidelines: "Toward Healthful Diets." The board, consisting of a dozen nutrition experts, concluded that the only reliable advice for healthy Americans was to watch their weight; everything else, dietary fat included, would take care of itself. The advice was not taken kindly, however, at least not by the media. The first reports—"rather incredulously," said Handler at the time—criticized the NAS advice for conflicting with the USDA's and McGovern's and thus somehow being irresponsible. Follow-up reports suggested that the board members, in the words of Jane Brody, who covered the story for *The New York Times,* were "all in the pocket of the industries being hurt." To be precise, the board chair and one of its members consulted for food industries, and funding for the board itself came from industry donations. These industry connections were leaked to the press from the USDA.

Hegsted now defends the NAS board, although he didn't at the time, and calls this kind of conflict of interest "a hell of an issue." "Everybody used to complain that industry didn't do anything on nutrition," he told *Science,* "yet anybody who got involved was blackballed because their positions were presumably influenced by the industry." (In 1981, Hegsted returned to Harvard, where his research was funded by Frito-Lay.) The press had mixed feelings, claiming that the connections "soiled" the academy's reputation "for tendering careful scientific advice" *(The Washington Post),* demonstrated that the board's "objectivity and aptitude are in doubt" *(The New York Times),* or represented in the board's guidelines a "blow against the food faddists who hold the public in thrall" *(Science).* In any case, the NAS board had been publicly discredited. Hegsted's Dietary Guidelines for Americans became the official U.S. policy on dietary fat: Eat less fat. Live longer.

Creating "Consensus"

Once politicians, the press, and the public had decided dietary fat policy, the science was left to catch up. In the early 1970s, when NIH opted to forgo a $1 billion trial that might be definitive and instead fund a half-dozen studies at one-third the cost, everyone hoped these smaller trials would be sufficiently persuasive to conclude that low-fat diets prolong lives. The results were published between 1980 and 1984. Four of these trials—comparing heart disease rates and diet within Honolulu, Puerto Rico, Chicago, and Framingham—showed no evidence that men who ate less fat lived longer or had fewer heart attacks. A fifth trial, the Multiple Risk Factor Intervention Trial (MRFIT), cost $115 million and tried to amplify the subtle influences of diet on health by persuading subjects to avoid fat while simultaneously quitting smoking *and* taking medication for high blood pressure. That trial suggested, if anything, that eating less fat might shorten life. In each study, however, the investigators concluded that methodological flaws had led to the negative results. They did not, at least publicly, consider their results reason to lessen their belief in the evils of fat.

The sixth study was the $140 million Lipid Research Clinics (LRC) Coronary Primary Prevention Trial, led by NHLBI administrator Basil Rifkind and biochemist Daniel Steinberg of the University of California, San Diego. The LRC trial was a drug trial, not a diet trial, but the NHLBI heralded its outcome as the end of the dietary fat debate. In January 1984, LRC investigators reported that a medication called cholestyramine reduced cholesterol levels in men with abnormally high cholesterol levels and modestly reduced heart disease rates in the process. (The probability of suffering a heart attack during the seven-plus years of the study was reduced from 8.6% in the placebo group to 7.0%; the probability of dying from a heart attack dropped from 2.0% to 1.6%.) The investigators then concluded, without benefit of dietary data, that cholestyramine's benefits could be extended to diet as well. And although the trial tested only middle-aged men with cholesterol levels higher than those of 95% of the population, they concluded that those benefits "could and should be extended to other age groups and women and . . . other more modest elevations of cholesterol levels."

Why go so far? Rifkind says their logic was simple: For 20 years, he and his colleagues had argued that lowering cholesterol levels prevented heart attacks. They had spent enormous sums trying to prove it. They felt they could never actually demonstrate that low-fat diets prolonged lives—that would be too expensive, and MRFIT had failed—but now they had established a fundamental link in the causal chain, from lower cholesterol levels to cardiovascular health. With that, they could take the leap of faith from cholesterol-lowering drugs and health to cholesterol-lowering diet and health. And after all their effort, they were eager—not to mention urged by Congress—to render helpful advice. "There comes a point when, if you don't make a decision, the consequences can be great as well," says Rifkind. "If you just allow Americans to keep on consuming 40% of calories from fat, there's an outcome to that as well."

With the LRC results in press, the NHLBI launched what Levy called "a massive public health campaign." The media obligingly went along. *Time,* for instance, reported the LRC findings under the headline "Sorry, It's True. Cholesterol Really Is a Killer." The article about a drug trial began: "No whole milk. No butter. No fatty meats . . ." *Time* followed up 3 months later with a cover story: "And Cholesterol and Now the Bad News. . . ." The cover photo was a frowning face: a breakfast plate with two fried eggs as the eyes and a bacon strip for the mouth. Rifkind was quoted saying that their results "strongly indicate that the more you lower cholesterol and fat in your diet, the more you reduce your risk of heart disease," a statement that still lacked direct scientific support.

The following December, NIH effectively ended the debate with a "Consensus Conference." The idea of such a conference is that an expert panel, ideally unbiased, listens to 2 days of testimony and arrives at a conclusion with which everyone agrees. In this case, Rifkind chaired the planning committee, which chose his LRC co-investigator Steinberg to lead the expert panel. The 20 speakers did include a handful of skeptics—including Ahrens, for instance, and cardiologist Michael Oliver of Imperial College in London—who argued that it was unscientific to equate the effects of a drug with the effects of a diet. Steinberg's panel members, however, as Oliver later complained in *The Lancet,* "were selected to include only experts who would, predictably, say that all levels of blood cholesterol in the United States are too high and should be lowered. And, of course, this is exactly what was said." Indeed, the conference report, written by Steinberg and his panel, revealed no evidence of discord. There was "no doubt," it concluded, that low-fat diets "will afford significant protection against coronary heart disease" to every American over 2 years old. The Consensus Conference officially gave the appearance of unanimity where none existed. After all, if there had been a true consensus, as Steinberg himself told *Science,* "you wouldn't have had to have a consensus conference."

The Test of Time

TO THE OUTSIDE OBSERVER, the challenge in making sense of any such long-running scientific controversy is to establish whether the skeptics are simply on the wrong side of the new paradigm, or whether their skepticism is well-founded. In other words, is the science at issue based on sound scientific thinking and unambiguous data, or is it what Sir Francis Bacon, for instance, would have called "wishful science," based on fancies, opinions, and the exclusion of contrary evidence? Bacon offered one viable suggestion for differentiating the two: the test of time. Good science is rooted in reality, so it grows and develops and the evidence gets increasingly more compelling, whereas wishful science flourishes most under its first authors before "going downhill."

Such is the case, for instance, with the proposition that dietary fat causes cancer, which was an integral part of dietary fat anxiety in the late 1970s. By 1982, the evidence supporting this idea was thought to be so undeniable that a landmark NAS report on nutrition and cancer equated those researchers who remained skeptical with "certain interested parties [who] formerly argued that the association between lung cancer and smoking was not causational." Fifteen years and hundreds of millions of research dollars later, a similarly massive expert report by the World Cancer Research Fund and the American Institute for Cancer Research could find neither "convincing" nor even "probable" reason to believe that dietary fat caused cancer.

The hypothesis that low-fat diets are the requisite route to weight loss has taken a similar downward path. This was the ultimate fallback position in all low-fat recommendations: Fat has nine calories per gram compared to four calories for carbohydrates and protein, and so cutting fat from the diet surely would cut pounds. "This is held almost to be a religious truth," says Harvard's Willett. Considerable data, however, now suggest otherwise. The results of well-controlled clinical trials are consistent: People on low-fat diets initially lose a couple of kilograms, as they would on any diet, and then the weight tends to return. After 1 to 2 years, little has been achieved. Consider, for instance, the 50,000 women enrolled in the ongoing $100 million Women's Health Initiative (WHI). Half of these women have been extensively counseled to consume only 20% of their calories from fat. After 3 years on this near-draconian regime, say WHI sources, the women had lost, on average, a kilogram each.

The link between dietary fat and heart disease is more complicated, because the hypothesis has diverged into two distinct propositions: first, that lowering cholesterol prevents heart disease; second, that eating less fat not only lowers cholesterol and prevents heart disease but *prolongs* life. Since 1984, the evidence that cholesterol-lowering drugs are beneficial—proposition number one—has indeed blossomed, at least for those at high risk of heart attack. These drugs reduce serum cholesterol levels dramatically, and they prevent heart attacks, perhaps by other means as well. Their market has now reached $4 billion a year in the United States alone, and every new trial seems to confirm their benefits.

The evidence supporting the second proposition, that eating less fat makes for a healthier and longer life, however, has remained stubbornly ambiguous. If anything, it has only become less compelling over time. Indeed, since Ancel Keys started advocating low-fat diets almost 50 years ago, the science of fat and cholesterol has evolved from a simple story into a very complicated one. The catch has been that few involved in this business were prepared to deal with a complicated story. Researchers initially preferred to believe it was simple—that a single unwholesome nutrient, in effect, could be isolated from the diverse richness of human diets; public health administrators required a simple story to give to Congress and the public; and the press needed a simple story—at least on any particular day—to give to editors and readers in 30 column inches. But as contrarian data continued to accumulate, the complications became increasingly more difficult to ignore or exclude, and the press began waffling or adding caveats. The scientists then got the blame for not sticking to the original simple story, which had, regrettably, never existed.

More Fats, Fewer Answers

THE ORIGINAL SIMPLE STORY in the 1950s was that high cholesterol levels increase heart disease risk. The seminal Framingham Heart Study, for instance, which revealed the association between cholesterol and heart disease, originally measured only total serum cholesterol. But cholesterol shuttles through the blood in an array of packages. Low-density lipoprotein particles (LDL, the "bad" cholesterol) deliver fat and cholesterol from the liver to tissues that need it, including the arterial cells, where it can lead to atherosclerotic plaques. High-density lipoproteins (HDLs, the "good" cholesterol) return cholesterol to the liver. The higher the HDL, the lower the heart disease risk. Then there are triglycerides, which contain fatty acids, and very low density lipoproteins (VLDLs), which transport triglycerides.

All of these particles have some effect on heart disease risk, while the fats, carbohydrates, and protein in the diet have varying effects on all these particles. The 1950s story was that saturated fats increase total cholesterol, polyunsaturated fats decrease it, and monounsaturated fats are neutral. By the late

1970s—when researchers accepted the benefits of HDL—they realized that monounsaturated fats are not neutral. Rather, they raise HDL, at least compared to carbohydrates, and lower LDL. This makes them an ideal nutrient as far as cholesterol goes. Furthermore, saturated fats cannot be quite so evil because, while they elevate LDL, which is bad, they also elevate HDL, which is good. And some saturated fats—stearic acid, in particular, the fat in chocolate—are at worst neutral. Stearic acid raises HDL levels but does little or nothing to LDL. And then there are trans fatty acids, which raise LDL, just like saturated fat, but also lower HDL. Today, none of this is controversial, although it has yet to be reflected in any Food Guide Pyramid.

To understand where this complexity can lead in a simple example, consider a steak—to be precise, a porterhouse, select cut, with a half-centimeter layer of fat, the nutritional constituents of which can be found in the Nutrient Database for Standard Reference at the USDA Web site. After broiling, this porterhouse reduces to a serving of almost equal parts fat and protein. Fifty-one percent of the fat is monounsaturated, of which virtually all (90%) is oleic acid, the same healthy fat that's in olive oil. Saturated fat constitutes 45% of the total fat, but a third of that is stearic acid, which is, at the very least, harmless. The remaining 4% of the fat is polyunsaturated, which also improves cholesterol levels. In sum, well over half—and perhaps as much as 70%—of the fat content of a porterhouse will improve cholesterol levels compared to what they would be if bread, potatoes, or pasta were consumed instead. The remaining 30% will raise LDL but will also raise HDL. All of this suggests that eating a porterhouse steak rather than carbohydrates might actually improve heart disease risk, although no nutritional authority who hasn't written a high-fat diet book will say this publicly.

As for the scientific studies, in the years since the 1984 consensus conference, the one thing they have not done is pile up evidence in support of the low-fat-for-all approach to the public good. If anything, they have added weight to Ahren's fears that there may be a downside to populationwide low-fat recommendations. In 1986, for instance, just 1 year after NIH launched the National Cholesterol Education Program, also advising low-fat diets for everyone over 2 years old, epidemiologist David Jacobs of the University of Minnesota, Twin Cities, visited Japan. There he learned that Japanese physicians were advising patients to raise their cholesterol levels, because low cholesterol levels were linked to hemorrhagic stroke. At the time, Japanese men were dying from stroke almost as frequently as American men were succumbing to heart disease. Back in Minnesota, Jacobs looked for this low-cholesterol–stroke relationship in the MRFIT data and found it there, too. And the relationship transcended stroke: Men with very low cholesterol levels seemed prone to premature death; below 160 milligrams per deciliter (mg/dl), the lower the cholesterol level, the shorter the life.

Jacobs reported his results to NHLBI, which in 1990 hosted a conference to discuss the issue, bringing together researchers from 19 studies around the world. The data were consistent: When investigators tracked all deaths, instead of just heart disease deaths, the cholesterol curves were U-shaped for men and flat for women. In other words, men with cholesterol levels above 240 mg/dl tended to die prematurely from heart disease. But below 160 mg/dl, the men tended to die prematurely from cancer, respiratory and digestive diseases, and trauma. As for women, if anything, the higher their cholesterol, the longer they lived.

These mortality data can be interpreted in two ways. One, preferred by low-fat advocates, is that they cannot be meaningful. Rifkind, for instance, told *Science* that the excess deaths at low cholesterol levels *must* be due to preexisting conditions. In other words, chronic illness leads to low cholesterol levels, not vice versa. He pointed to the 1990 conference report as the definitive document on the issue and as support for his argument, although the report states unequivocally that this interpretation is not supported by the data.

The other interpretation is that what a low-fat diet does to serum cholesterol levels, and what that in turn does to arteries, may be only one component of the diet's effect on health. In other words, while low-fat diets might help prevent heart disease, they might also raise susceptibility to other conditions. This is what always worried Ahrens. It's also one reason why the American College of Physicians, for instance, now suggests that cholesterol reduction is certainly worthwhile for those at high, short-term risk of dying of coronary heart disease but of "much smaller or . . . uncertain" benefit for everyone else.

This interpretation—that the connection between diet and health far transcends cholesterol—is also supported by the single most dramatic diet-heart trial ever conducted: the Lyon Diet Heart Study, led by Michel de Lorgeril of the French National Institute of Health and Medical Research (INSERM) and published in *Circulation* in February 1999. The investigators randomized 605 heart attack survivors, all on cholesterol-lowering drugs, into two groups. They counseled one to eat an AHA "prudent diet," very similar to that recommended for all Americans. They counseled the other to eat a Mediterranean-type diet, with more bread, cereals, legumes, beans, vegetables, fruits, and fish and less meat. Total fat and types of fat differed markedly in the two diets, but the HDL, LDL, and total cholesterol levels in the two groups remained virtually identical. Nonetheless, over 4 years of follow-up, the Mediterranean-diet group had only 14 cardiac deaths and nonfatal heart attacks compared to 44 for the "Western-type" diet group. The likely explanation, wrote de Lorgeril and his colleagues, is that the "protective effects [of the Mediterranean diet] were not related to serum concentrations of total, LDL or HDL cholesterol."

Many researchers find the Lyon data so perplexing that they're left questioning the methodology of the trial. Nonetheless, says NIH's Harlan, the data "are very provocative. They do bring up the issue of whether if we look only at cholesterol levels we aren't going to miss something very important." De Lorgeril believes the diet's protective effect comes primarily from omega-3 fatty acids, found in seed oils, meat, cereals, green leafy vegetables, and fish, and from antioxidant compounds, including vitamins, trace elements, and flavonoids. He told *Science* that most researchers and journalists in the field are prisoners of the "cholesterol paradigm." Although dietary fat and serum cholesterol "are obviously connected," he says, "the connection is not a robust one" when it comes to heart disease.

Dietary Trade-offs

One inescapable reality is that death is a trade-off, and so is diet. "You have to eat something," says epidemiologist Hugh Tunstall Pedoe of the University of Dundee, U.K., spokesperson for the 21-nation Monitoring Cardiovascular Disease Project run by the World Health Organization. "If you eat more of one thing, you eat a lot less of something else. So for every theory saying this disease is caused by an excess in *x*, you can produce an alternative theory saying it's a deficiency in *y*." It would be simple if, say, saturated fats could be cut from the diet and the calories with it, but that's not the case. Despite all expectations to the contrary, people tend to consume the same number of calories despite whatever diet they try. If they eat less total fat, for instance, they will eat more carbohydrates and probably less protein, because most protein comes in foods like meat that also have considerable amounts of fat.

This plus-minus problem suggests a different interpretation for virtually every diet study ever done, including, for instance, the kind of metabolic-ward studies that originally demonstrated the ability of saturated fats to raise cholesterol. If researchers reduce the amount of saturated fat in the test diet, they have to make up the calories elsewhere. Do they add polyunsaturated fats, for instance, or add carbohydrates? A single carbohydrate or mixed carbohydrates? Do they add green leafy vegetables, or do they add pasta? And so it goes. "The sky's the limit," says nutritionist Alice Lichtenstein of Tufts University in Boston. "There are a million perturbations."

These trade-offs also confound the kind of epidemiological studies that demonized saturated fat from the 1950s onward. In particular, individuals who eat copious amounts of meat and dairy products, and plenty of saturated fats in the process, tend not to eat copious amounts of vegetables and fruits. The same holds for entire populations. The eastern Finns, for instance, whose lofty heart disease rates convinced Ancel Keys and a generation of researchers of the evils of fat, live within 500 kilometers of the Arctic Circle and rarely see fresh produce or a green vegetable. The Scots, infamous for eating perhaps the least wholesome diet in the developed world, are in a similar fix. Basil Rifkind recalls being laughed at once on this point when he lectured to Scottish physicians on healthy diets: "One said, 'You talk about increasing fruits and vegetable consumption, but in the area I work in there's not a single grocery store.' " In both cases, researchers joke that the only green leafy vegetable these populations consume regularly is tobacco. As for the purported benefits of the widely hailed Mediterranean diet, is it the fish, the olive oil, or the fresh vegetables? After all, says Harvard epidemiologist Dimitrios Trichopoulos, a native of Greece, the olive oil is used either to cook vegetables or as dressing over salads. "The quantity of vegetables consumed is almost a pound [half a kilogram] a day," he says, "and you cannot eat it without olive oil. And we eat a lot of legumes, and we cannot eat legumes without olive oil."

Indeed, recent data on heart disease trends in Europe suggest that a likely explanation for the differences between countries and over time is the availability of fresh produce year-round rather than differences in fat intake. While the press often plays up the French paradox—the French have little heart disease despite seemingly high saturated fat consumption—the real paradox is throughout Southern Europe, where heart disease death rates have steadily dropped while animal fat consumption has steadily risen, says University of Cambridge epidemiologist John Powles, who studies national disease trends. The same trend appears in Japan. "We have this idea that it's the Arcadian past, the life in the village, the utopia that we've lost," Powles says; "that the really protective Mediterranean diet is what people ate in the 1950s." But that notion isn't supported by the data: As these Mediterranean nations became more affluent, says Powles, they began to eat proportionally more meat and with it more animal fat. Their heart disease rates, however, continued to improve compared to populations that consumed as much animal fat but had less access to fresh vegetables throughout the year. To Powles, the antifat movement was founded on the Puritan notion that "something bad had to have an evil cause, and you got a heart attack because you did something wrong, which was eating too much of a bad thing, rather than not having enough of a good thing."

The other salient trade-off in the plus-minus problem of human diets is carbohydrates. When the federal government began pushing low-fat diets, the scientists and administrators, and virtually everyone else involved, hoped that Americans would replace fat calories with fruits and vegetables and legumes, but it didn't happen. If nothing else, economics worked against it. The food industry has little incentive to advertise nonproprietary items: broccoli, for instance. Instead, says NYU's Nestle, the great bulk of the $30-billion-plus spent yearly on food advertising goes to selling carbohydrates in the guise of fast food, sodas, snacks, and candy bars. And carbohydrates are all too often what Americans eat.

Carbohydrates are what Harvard's Willett calls the flip side of the calorie trade-off problem. Because it is exceedingly difficult to add pure protein to a diet in any quantity, a low-fat diet is, by definition, a high-carbohydrate diet—just as a low-fat cookie or low-fat yogurt are, by definition, high in carbohydrates. Numerous studies now suggest that high-carbohydrate diets can raise triglyceride levels, create small, dense LDL particles, and reduce HDL—a combination, along with a condition known as "insulin resistance," that Stanford endocrinologist Gerald Reaven has labeled "syndrome X." Thirty percent of adult males and 10% to 15% of post-menopausal women have this particular syndrome X profile, which is associated with a several-fold increase in heart disease risk, says Reaven, even among those patients whose LDL levels appear otherwise normal. Reaven and Ron Krauss, who studies fats and lipids at Lawrence Berkeley National Laboratory in California, have shown that when men eat high-carbohydrate diets their cholesterol profiles may shift from normal to syndrome X. In other words, the more carbohydrates replace saturated fats, the more likely the end result will be syndrome X and an increased heart disease risk. "The problem is so clear right now it's almost a joke," says Reaven. How this balances out is the unknown. "It's a bitch of a question," says Marc Hellerstein, a nutritional biochemist at the University of California, Berkeley, "maybe the great public health nutrition question of our era."

The other worrisome aspect of the carbohydrate trade-off is the possibility that, for some individuals, at least, it might actually be easier to gain weight on low-fat/high-carbohydrate regimens than on higher fat diets. One of the many factors that influence hunger is the glycemic index, which measures how fast carbohydrates are broken down into simple sugars and moved into the bloodstream. Foods with the highest glycemic index are simple sugars and processed grain products like pasta and white rice, which cause a rapid rise in blood sugar after a meal. Fruits, vegetables, legumes, and even unprocessed starches—pasta *al dente,* for instance—cause a much slower rise in blood sugar. Researchers have hypothesized that eating high–glycemic index foods increases hunger later because insulin overreacts to the spike in blood sugar. "The high insulin levels cause the nutrients from the meal to get absorbed and very avidly stored away, and once they are, the body can't access them," says David Ludwig, director of the obesity clinic at Children's Hospital Boston. "The body appears to run out of fuel." A few hours after eating, hunger returns.

If the theory is correct, calories from the kind of processed carbohydrates that have become the staple of the American diet are not the same as calories from fat, protein, or complex carbohydrates when it comes to controlling weight. "They may cause a hormonal change that stimulates hunger and leads to overeating," says Ludwig, "especially in environments where food is abundant. . . ."

In 1979, 2 years after McGovern's committee released its Dietary Goals, Ahrens wrote to *The Lancet* describing what he had learned over 30 years of studying fat and cholesterol metabolism: "It is absolutely certain that no one can reliably predict whether a change in dietary regimens will have any effect whatsoever on the incidence of new events of [coronary heart disease], nor in whom." Today, many nutrition researchers, acknowledging the complexity of the situation, find themselves siding with Ahrens. Krauss, for instance, who chairs the AHA Dietary Guidelines Committee, now calls it "scientifically naïve" to expect that a single dietary regime can be beneficial for everybody: "The 'goodness' or 'badness' of anything as complex as dietary fat and its subtypes will ultimately depend on the context of the individual."

Given the proven success and low cost of cholesterol-lowering drugs, most physicians now prescribe drug treatment for patients at high risk of heart disease. The drugs reduce LDL cholesterol levels by as much as 30%. Diet rarely drops LDL by more than 10%, which is effectively trivial for healthy individuals, although it may be worth the effort for those at high risk of heart disease whose cholesterol levels respond well to it.

The logic underlying population-wide recommendations such as the latest USDA Dietary Guidelines is that limiting saturated fat intake—even if it does little or nothing to extend the lives of healthy individuals and even if not all saturated fats are equally bad—might still delay tens of thousands of deaths each year throughout the entire country. Limiting total fat consumption is considered reasonable advice because it's simple and easy to understand, and it may limit calorie intake. Whether it's scientifically justifiable may simply not be relevant. "When you don't have any real good answers in this business," says Krauss, "you have to accept a few not so good ones as the next best thing."

JOSEPH D'AGNESE

Brothers with Heart

FROM *DISCOVER*

In the market for donor organs, demand far exceeds supply. Every year hundreds of thousands of people around the world die for lack of suitable organs—or because their bodies reject the organs that are available. The journalist Joseph D'Agnese has followed four visionary doctors who are working on a high-tech solution: using a patient's own tissue to grow custom-made body parts. Surprisingly, they're making progress. Perhaps their success lies in another surprising fact: they're brothers.

The doctor they call Chuck stands over the body with an electric blade, ready to make the first incision. The knife whirs, peeling crisp brown skin off the breast and digging into the firm white flesh below. The doctor wields the knife confidently, humming to himself, as if he finds pleasure in severing muscle from bone. His two brothers, Marty the pathologist and Frank the anesthesiologist, stand nearby, ready to offer advice. (Jay, the oldest brother, is on call.)

It is the last Thursday in November and the patient is dead, a 23-pound turkey roasted to feed the assembled Vacanti family. As Chuck buzzes away, slicing white meat and dark, the grown-ups sip Riesling, the kids and Uncle Frank gulp cola. The table is groaning with side dishes. All in all, the scene is about average for Thanksgiving Day.

Still, it does not take long to discover that there is little or nothing average about the Vacanti brothers. Joseph (Jay), Charles, Martin, and Francis Vacanti work together as researchers in the new field of tissue engineering, a discipline they practically invented. What they are trying to create is nothing less than lab-grown human organs, produced from a patient's own tissue. Their work is urgently needed—roughly 100,000 patients in this country die each year because not enough people donate organs, and many of those who are saved by transplantation ultimately die because donor organs are rejected.

In the new world the Vacantis hope to build, an infant born without intestines and destined to die will get a new gut grown from a clutch of her own cells. As the fruit of the child, that new intestine will never be rejected. Imagine a world, say the Vacantis, in which diseased pancreases, lungs, and spinal cords can be replaced as easily as the transmission in an old Chevy. Imagine a world in which salvation grows in an incubator. Imagine a world in which hope is a given.

So far, artificial skin and cartilage are the only lab-grown tissues available to surgeons. But that is about to change so quickly that even doctors familiar with the research will find it difficult to comprehend the possibilities. It was only in 1996 that Chuck and Jay Vacanti held the first conference of their fledgling Tissue Engineering Society. Today two of their former colleagues, Anthony Atala, a pediatric urologist at Children's Hospital in Boston, and Laura Niklason, a Duke University researcher, have already performed what seem like miracles—Atala successfully implanting lab-grown bladders in beagles and Niklason growing fresh pig arteries in her lab. This year there are more than 50 laboratories in the United States alone racing to create people-made people parts. And the researchers in all of those labs are indebted to five breakthroughs made by the four Vacanti brothers.

Breakthrough #1: How Do We Make a Scaffold?

THE BROTHERS GREW UP in Omaha, born to a dentist father and a mother who stopped six credit hours shy of a premed degree to marry. There were eight children, and money was always tight. Jay, the oldest son, longed to go to Harvard. His father pulled him aside one day and told him that he could work his way through the Ivy League school or go to Creighton, where Dad was on the faculty, for free. Jay reluctantly chose Creighton but soon made it to Harvard as a surgical intern. His brothers got the same speech, also attended Creighton, and eventually followed Jay to Massachusetts.

Meanwhile, working at Massachusetts General Hospital, the largest teaching hospital for Harvard Medical School, Jay witnessed an endless parade of gurneys delivering children who could not be saved. "I have always thought that being a pediatric surgeon was the most gratifying kind of surgery," he says. "You start with the most helpless and vulnerable of humans, diagnose a potentially harmful condition, definitively manage it, oftentimes with surgical intervention, and return the child to his family and to possibly another 80 years of life." But too many children Jay saw needed new livers, bladders, intestines. There weren't enough organs to go around; there wasn't nearly enough hope. Jay Vacanti determined to fix that.

He knew that in 1979 Eugene Bell, an engineer at the Massachusetts Institute of Technology, had grown skin cells in flat sheets of tissue. Jay, along with his colleague Robert Langer, a chemical engineer at MIT, became obsessed with figuring out how to extend Bell's work beyond two dimensions.

The number one challenge in tissue engineering is to get the specimen cells to grow at all. Freshly plucked from a human, a batch of cells hasn't got long to live. They need oxygen, a temperature of about 98 degrees Fahrenheit, and nutrients. So a tissue engineer doesn't waste time. He places the cells in a petri dish with liquid nutrients—carbohydrates and amino acids do nicely—then tucks them away in an incubator. With a little luck they multiply and in a few days produce enough cells to be considered tissue. As amazing as the process is up to this point, it's relatively useless, because clumps of tissue are of little use to a patient. The tissue engineer must convince cells to morph from a meaningless jumble of flesh into a functioning organ.

Jay and Langer's vision was radical—build a scaffold out of plastic on which the cells could build a three-dimensional organ. "The original organ we had in mind was the liver," Jay recalls. "I thought degradable plastics would make an ideal scaffold. I knew from my work in cell biology that cells adhere to plastic dishes for in vitro culture, secreting their own scaffold as they settle to the bottom of the plate. I also knew you could treat the surfaces of plastics so they would be more likely to cause cell adherence."

But Eugene Bell was already on the record: Plastic polymers simply wouldn't work as a scaffold. He dismissed the idea as fatuous. That word stuck in Jay's craw.

In the summer of 1986 Jay took his family to Cape Cod. As his children played in the surf, he perched on a jetty, mulling the problem. You could get cells to populate the exterior of a hunk of polymer, where they had easy access to oxygen and nutrition, but the interior was another story—you might as well try to grow a houseplant inside a basketball. Out of the corner of his eye, Jay saw a sheet of seaweed bobbing in the water. Then it hit him. Nature, the original tissue engineer, had already solved the problem. Under the seaweed's rubbery skin lies a network of fine, hollow branches that pipe fresh oxygen into the organism while pumping out expended gases. Jay shot an entire roll of film of that seaweed and phoned Langer before the day was out. Their scaffold interior had to be light and airy, like spun candy, they decided. Build the branches, and the cells would do the rest.

Today that's the game plan in labs everywhere. Tissue engineers build a porous, three-dimensional polymer model of an organ, squirt a soup of nutrients and living cells over the structure, incubate, and wait. "It makes sense," says Jay, "that cells thrive best in this environment because they are designed to live and function in three-dimensional space." In time the biodegradable framework dissolves in the body's water through the process of hydrolysis, until what is left is wholly alive.

Now when Jay talks to young tissue engineers, he sometimes slaps a transparency on the overhead projector that displays *Webster's* definition of the word *fatuous:* "smugly or foolishly stupid." His intention is to encourage his successors to trust their instincts and go forward, ignoring criticism that might hold them back.

Breakthrough #2: The First Human Experiments

IN 1989 CHUCK and his team, then also based at Mass General, submitted a paper to a top journal announcing they'd grown a piece of human cartilage. It was rejected outright: "No practical implications," the reviewers wrote.

Chuck was stunned. No practical implications? He saw it as a challenge. So he quizzed plastic surgeons: What's the toughest cartilaginous structure to repair? The human ear, they answered, no question. Every day infants are born with underdeveloped ears; children and adults lose ears in car accidents. Somewhere between bone and skin, cartilage is a tricky substance to work with, and the ear is the body's most intricately shaped and visible piece of it.

So Chuck and his team decided to build one. They needed a living host that wasn't human, so they implanted an ear scaffold under the wrinkled skin of a hairless lab mouse they nicknamed *Auriculosaurus.* The mouse grew an ear on its back. The image was beamed into newsrooms all over the world. Facing hard questions from the public and colleagues about his motives,

Chuck had to explain that he'd only intended to show the medical world what could be done.

In April 1994 Chuck and Jay got a chance to prove what could be done in a human. They met Sean McCormack, 12, who had been born with a protruding sternum and no cartilage or bone under the skin of his left torso. Unprotected, his heart could be seen beating just below the skin. As a Little League pitcher, he badly needed a chest wall. Boston's Children's Hospital let the Vacantis conduct a procedure so bizarre the Food and Drug Administration had no regulations to cover it. The doctors harvested cartilage cells from Sean's sternum to seed a flat, round scaffold about the size of a compact disc. Awash in nutrients, the cells multiplied and permeated the polymer. Weeks later the construct was inserted in Sean's chest. As his body grew, it incorporated the shield for his heart; seven years later, he's a star BMX bicycle racer.

In 1998, Raul Murcia, a machinist, crushed and severed his left thumb in a cargo elevator. Chuck and his team, now at the University of Massachusetts in Worcester, leaped to the challenge: Think we can grow this guy a new distal phalanx? Chuck carved a piece of surgical coral into the shape of a thumb, seeded Murcia's bone cells onto it, and in a few weeks he had a thumb digit ready for implanting. By then newspapers were screeching "test-tube thumb," and Chuck was fielding a call from the FDA: What are you guys doing up there?

Since Sean McCormack's operation, the FDA had developed a set of regulations governing cultured cells. The surgical coral was FDA approved, and the FDA had decided that one could implant autologous cells—those grown from a patient's own cells—without approval. However, if a doctor combined two FDA-approved technologies, the technique required separate approval. You can implant the new bone, they told Chuck, but you absolutely cannot implant any tissue-engineered cartilage. "I asked what to do, and they suggested I apply for retroactive approval, which I did," Chuck says.

In the end Murcia got a new thumb-tip, but Chuck was not allowed to attach cartilage or tendons to it. In a paper published in May 2001 in *The New England Journal of Medicine,* Chuck and his colleagues report the case, including the happy results of Murcia's recent biopsy: Much of the coral still exists, but its pores have filled in with Murcia's own cells. More surprising, the new cells are intelligently transforming the coral, remodeling it to look more like a human thumb bone.

Today researchers in tissue labs all over the world have moved beyond thumbs and ears, as they struggle to grow more important and complex tissue structures. In the search for fool-proof methods, they play with variables: What's the best polymer for arteries? For tracheae? They jet to symposia, deliver papers, send e-mails to find out what other labs are up to. Such encounters are civil, even genteel, but an undercurrent of rivalry often fills the air.

Breakthrough #3: The Miracle Cells

THE VACANTI BROTHERS have raised sibling rivalry to an art form. Jay knows, for example, that if he's courting an applicant for residency at Mass General, chances are the would-be tissue engineer has already visited Chuck and Marty at the University of Massachusetts, who told him: "If you're offered a position at Harvard, you should consider it just because it's Harvard. If you're offered a position in tissue engineering at UMass, you should accept it because we're better."

Chuck says the brothers are "very competitive" with each other but not destructive. They needle each other in the same delighted way they always have, but they know they are safe. "It's absolutely better to work with your own brother. You trust your own brother in a way you can never trust someone else," says Chuck, who adds, laughing, "If one of us does something to break that trust, we can always go tell our mom."

Their rivalry is balanced by an intimacy few scientists will ever enjoy. "You don't have to go into the dance that you would if you were sharing an idea with a colleague," says Jay, "where both of you are trying to be polite. If you're dealing with your brother, he can say, 'You know, that's really stupid.' It's an efficient way to problem-solve. On the other hand, if he says, 'That's really smart,' you know it's genuine."

That dynamic engendered their most startling breakthrough, which was announced last fall.

In 1996 Chuck had convinced Marty, the pathologist, to leave Nebraska and join him in Worcester. Chuck had grown increasingly frustrated with the fragile adult-tissue cells he had been working with. Most cannot last more than 30 minutes without an oxygen supply. Fetal stem cells are hardier, but harvesting them is controversial.

Chuck told Marty to find an alternative: "Look for stem cells in adult tissue."

He instantly replied: "They don't exist."

"They have to exist," Chuck insisted, intent on driving his point home. "If the human body is constantly trying to repair itself, it must have immature cells somewhere. Find them."

"You're nuts," Marty told him.

"Just do it."

"If I had talked like that to anyone other than a family member," Chuck says, "he would have gone home and told his wife I had an attitude problem."

Instead, Marty decided to give it a try. For 15 months he drew cells from living animals, only to watch them die. He scrounged lab animals other researchers had sacrificed for their work. He scraped flesh with scalpels and dissolved it in enzymes. He peered into the resulting broth, magnified 200 times, to no avail. At every staff meeting, Marty had nothing to report. It became embarrassing.

Then one day, peering through the microscope, he spotted tiny circular shapes. Adult-tissue cells are about 15 micrometers wide. Marty saw cells only 3 micrometers wide. He began showing them around. They're too small to be stem cells, everyone said. Just debris. Junk.

Tired and depressed, Marty stood in his lab staring at flasks of the cell soup, thinking, "Wastebasket or incubator?" For reasons he does not comprehend, he stuck them in the incubator. Three days later, those little specks of junk had multiplied. What's more, they had gone without oxygen for more than an hour before he put them in the incubator, an ordeal adult–tissue cells could not have survived.

At staff meetings Marty took center stage. Eventually someone asked: What do you call these cells? Privately, Marty had begun to call them "sporelike cells." They had a faintly prokaryotic, sporelike look about them. Until 2.5 billion years ago, life on this planet was limited to bacteria and algae that reproduced through the agency of single-celled bodies called spores, which lay dormant until called upon to create new life. In time prokaryotes morphed into eukaryotes, multicelled creatures. Marty's mind reeled when he thought about it: What if the most primitive process of evolution and self-repair was still going on inside our bodies? At one large meeting, on Jay's turf in Boston, his concentration was broken by Jay chanting sotto voce: "Fungus, fungus, yeast, yeast!"

Weeks later, Chuck phoned with a suggestion, but Marty cut him off. Obsessed now, he had been examining every scrap of tissue he could lay his hands on and had isolated sporelike cells in every one. He'd bought a tray of chicken livers at the grocery. Even there, he found them.

Chuck was agog but, being Chuck, couldn't wait to up the ante. Freeze 'em and cook 'em, he said. Marty took them down to -121 degrees Fahrenheit. The cells survived. He left them at 187 degrees Fahrenheit for 30 minutes. They were still alive.

Marty tried to keep a lid on his excitement. He'd learned early that it was prudent to get the data in the bag before you crowed over a new discovery. His confidence soared the day he showed his work to Guido Manjo, an eminent Italian-born pathologist who lectures at UMass. Manjo's advice: Test those cells for DNA—and publish as soon as possible. Then came the ultimate compliment: "Dr. Vacanti," said the senior scientist, "you may have discovered a fundamental process of nature that has not yet been described."

Manjo was correct. DNA was present in the cells, and no one in the history of biology had ever identified such minuscule formations living in mammalian tissue. They were the kind of cells that the Vacantis had been dreaming about: They could live in the body without oxygen for days until blood vessels grew to supply them. Marty's most recent research shows the cells may actually be able to differentiate into tissues other than those of the organs from which they originated.

Properly incubated, they grow like grass on a prairie. The team in the lab at Worcester has used them to grow everything from retinal rods and cones to liver, bone, fascia, skin, and heart tissue. They have pulled sporelike cells out of a diabetic pancreas and grown insulin-producing islets in 12 weeks. They have cut a golf-ball-sized section from a living sheep's lung, stuffed the wound with a scaffold seeded with pulmonary sporelike cells, and watched as the lung incorporated the new tissue in eight weeks. Everyone was in awe: A lung is perhaps the most complex organ in the body, possessing at least half a dozen different types of tissue.

Breakthrough #4: Seeking to Heal the Spine

MARTY'S DISCOVERY paved the way for the most astounding experiment at Worcester to date. In the late summer of 1998, the Vacanti team inserted scaffolds seeded with sporelike cells into the severed spinal cords of eight lab rats. They hoped new tissue would bridge the gap. But first, the team cut themselves a big break. Scarring of nerve ends in severed spinal cords interferes with healing. So they put the scaffolds in place immediately after the cords were cut and before scarring set in. The cells quickly stitched themselves into the fibers of the existing cords, and the paralyzed rats regained a significant degree of feeling and movement in their previously paralyzed limbs.

"After 10 days," Chuck recalls, "you could see little twitches in their toes . . . In three months, some rats could stand on their hind legs and eat what you fed them." After several months a few of them walked.

Meanwhile, in Jay's lab at Mass General, Frank, an anesthesiologist and the youngest of the brothers, was building an interest in the same problem. Frank had started out in stroke research, where he made several breakthroughs. He was, for example, the first to realize that slightly lowering the body temperature of a patient at risk for a stroke, such as during an operation, could minimize complications. That interest in neurology developed into a fascination with spinal-cord repair.

Frank suspected the severed fibers in a spinal cord wouldn't be able to resist an easy pathway along which they could grow back toward each other. So he used a laser to drill tiny tunnels through his scaffolds, which he implanted into the cords of rats immediately after he severed them.

The procedure fits his personality. Frank loves building things, so much so that he almost didn't become a doctor like his brothers. As a teenager he wanted to be an engineer but noticed that engineers were having a hard time finding work after Nixon decimated the space program. He craved a job where he could work with gadgetry. A medical lab appealed to him, but shifting gears from a technician's mind-set to that of a biologist would be a challenge. Nonetheless, he harnessed his inner Vacanti and plowed through academia, taking the Hippocratic oath at 23.

Now a seasoned mechanic, he labors away in the lab on weekends, drilling his 2-millimeter holes in solitude, which he prefers. At first, it looked as if his idea was headed for success. After just six weeks spinal-cord tissue appeared to have nearly replaced the scaffold. Sadly, the effects on the rats ranged from mixed to negligible. Some of them died; others lived to barely wiggle their toes. "Not enough," says Frank. "I wanna see them jump rope."

Analyzing his failure, he saw that he could improve life-support systems for the recovering rats. Redesigning the study got him thinking about salamanders: "If a salamander damages his spinal cord, he can repair it. They don't scar. But in mammals the cords form scars . . . At some point, the ability to scar must have been an evolutionary advantage."

As organisms became more complex, tissues required more oxygen to function. And that, Frank thinks, hurts us in the regeneration department. If a human spine is damaged, the cells cannot tolerate being torn away from blood and oxygen. Scar tissue sweeps in to obscure the damage. If we once had the ability to heal, Frank reasons, we must be able to restore it, and he intends to find out how.

His imagination and curiosity have taken him beyond medicine. He has begun writing physics essays, one of which has been published in a prestigious journal. "You can't advance science unless you take a risk," he says. "It's like fixing a spinal cord. Most people think you're crazy. But to have any success you have to let your mind wander. You have to look for relationships, see how things fit. It's beautiful. But it's not the scientific method."

Breakthrough #5: A Heart of Foam

SOONER OR LATER, all tissue engineers are haunted by the body's need for oxygen. It was no accident that the earliest successes in tissue engineering came in hatching skin, bone, and cartilage, which can survive for hours in the body until blood vessels mosey over to attach themselves. "Now we're getting into organs like livers and hearts that are too thick to work that way," Jay says. "To survive, they need oxygen, they need nutrition, they need to dump waste." They need a circulatory system. But how does one coax cells to grow into something as complex as a network of capillaries?

These days Jay, along with researchers at the Massachusetts Institute of Technology and the Draper Laboratory, has embarked on a radical protocol: They are etching blood vessels onto silicon wafers. Cells are then deposited onto the wafers, where they grow into circuits just micrometers in diameter and shaped like branching vessels. In time the entire fragile sheet can be lifted from the wafer and the vessels rolled like cigars or stacked like checkers to build a circulatory structure. They work: Jay and his team enjoy showing visitors a video of blood cells coursing through the man-made vessels like water in a stream. This project has taken them one step closer to meeting their greatest challenge yet.

In a corner of the lab lies a polymer cast of a sheep's heart. To build it, Jay and his team took a real sheep's heart, pumped its vessels full of liquid plastic, then dissolved the tissue in a bath of flesh-eating enzymes. When Jay saw the first cast, it looked like a ball of Styrofoam. "I thought they had made a mistake," he says. "Then we looked at it under a microscope and saw: These are all capillaries. It showed us where we're headed. In organs like the heart, circulation is structure." MIT engineers have used the specs from the cast to design a scaffold, another big step down this road.

Someday, Jay and his team will implant cells on a scaffold and try to grow them into a heart. But that is still a good while off. In the meantime, Jay's lab is working on smaller projects such as building an esophagus for children who are born without one or with a portion missing. Before 1938 this condition, called esophageal atresia, was always fatal. Then doctors began using skin to fashion a replacement. In the 1950s they hit upon the technique used today of replacing the missing part with a section of the colon. However, a high incidence of esophageal cancer is associated with this technique. A grown-to-order organ would be a godsend.

The Vacantis can't rest.

"It was a driving motivation for me to specialize in the surgical care of children," Jay says. "I'm certain that our Sicilian-American cultural tradition imprinted on us the primacy of caring for and nurturing children. We were reared in a family of eight children with a large extended family, including grandparents, great-grandparents, and many cousins."

GROWING UP IN OMAHA, the four brothers could not help but pick up that lesson—and more. From their parents, Joanne and Charles, they learned that no theory, no question, no design of theirs was too ridiculous to explore. They grew to manhood knowing in their bones that they did not labor solely for themselves. They've lost their father, who taught them this lesson through a life devoted to his family, students, patients, and science. He died of a heart attack in 1994, and those who gather each year at the table on Thanksgiving miss him. But their mother is still around and does not hesitate to remind them they were put on this Earth to work for the good of others.

Chuck's own heart has led him to an even deeper understanding of the values his parents instilled. Nine years ago chest pains led to an angiogram for Chuck. It showed that his left coronary artery had completely occluded. He exercised and shed 30 pounds. Life looked good. Then one day five years ago his heart sounded another alarm as he sat in his office. Despite crushing pain, he calmly instructed his secretary to cancel appointments for six weeks and call the emergency room while he paged his cardiologist.

"I think I need to come in," Chuck told him.

"I can see you in two or three days."

"Maybe I haven't made clear the severity of the problem," Chuck said. "I believe I'm infarcting as we speak."

He wasn't scared. Denial chased danger from his mind. After another angiogram, doctors told him that three coronary arteries had occluded. Interesting, thought Chuck: My diagnosis was correct. You need a bypass immediately, they said. He stalled, enumerating other procedures. Can you try this or that? Uh, yeah, we can try all those things, but you'll be dead. We're doing this now.

So it came down to the healer's own cells rebelling, his sternum sawed and spread, his heart exposed to the glare of lights. Before he went under, he was still curious: I wonder whether I'm going to die.

That's when it hit him. Even in your most desperate hour, it is not just about you. No life ever is.

"When you're gone," he says now, "what you really have is what you're remembered for. I realized I would like people to say, 'He did something good for man. For mankind.' "

CHRISTOPHER DICKEY

I Love My Glow Bunny

FROM *WIRED*

For a molecular biologist at a French laboratory, Rabbit Number 5256 is just one of a group of genetically modified animals that serve as important tools for research. For a conceptual artist named Eduardo Kac, that rabbit (whom he calls Alba) is a work of "transgenic art" that he claims as his creation. Christopher Dickey, who in his distinguished career as a foreign correspondent has covered many battlegrounds, is on hand as art and science collide.

Eduardo Kac has come to Paris to get his bunny back. The Brazilian-born, Chicago-based conceptual artist put himself on the map of global curiosity in the summer of 2000 when he announced that he'd created—in the name of art—a transgenic rabbit that glowed green under blue light. Of course, he hadn't done the technical work. As he described it, he'd "commissioned" the bunny from top researchers at France's Institut National de la Recherche Agronomique, an organization of 8,600 people scattered among agricultural research centers around the country. INRA technicians took green fluorescent protein from the little Pacific jellyfish *Aequorea victoria,* and spliced it into the genes of a rabbit zygote. The resulting bunny, which Kac called Alba, was to be shown at an exhibition of digital art in the Provençal city of Avignon in June 2000. Kac (pronounced Katz) was going to create a small living room in one corner of the show and demonstrate what it would be like to have a green-glowing rabbit as a pet. And then he was going to take Alba home to his wife and child in the Windy City.

But on the eve of the exhibition, the then-director of INRA, Paul Vial, suddenly said no. He didn't really say why. Didn't have to. His people made the rabbit. His people funded it. His people would keep it. The organizers of the Avignon show denounced what they called disguised censorship. Kac returned, bunnyless, to Chicago. And the legend began to grow. "Cross Hare: Hop and Glow Mutant Bunny at Heart of Controversy Over DNA Tampering," read the headline in *The Boston Globe* that September. The wires picked up the story. So did ABC News and *The Washington Post.* Half-amused, half-horrified, the general public was presented with the implications of what Kac called "transgenic art." What would be next? Blue horses? Vermilion cows? Prada poodles? Vuitton pugs?

Like many a strange tale, Alba's story was widely read for a day or two, then widely forgotten. But Kac is determined to keep Alba alive in the public imagination, and eventually to take *his* rabbit home. She is a political victim yearning for liberation, a slave to bureaucracy awaiting emancipation.

When I hooked up with Kac in late 2000, he had returned to Paris to mobilize support among his peers in the field of high-tech art. He was exhibiting recent work, speaking at conferences at the Sorbonne and the École des Beaux-Arts, appearing on local TV programs, and giving interviews to the press. He was also planning to rally the populace.

Kac is a Ph.D. research fellow at the Centre for Advanced Inquiry in Interactive Arts at the University of Wales. He's an assistant professor of art and technology at the School of the Art Institute of Chicago, he has won numerous awards and grants, and his work is in the permanent collections at MoMA in New York and the Museum of Modern Art in Rio, among others. But, he assured me, "I'm not a geek who spends all his life in front of a computer."

Time to take the Free Alba campaign to the streets. And the next thing I know, Kac is putting up posters all over the Left Bank and Montmartre. Each shows a photograph of him holding Alba in his arms. At the top, they have different words on them: *Religion, Média, Éthique, Art, Science, Famille.*

One of the first posters goes up on the boarded window of a restaurant just across from the entrance to the École des Beaux-Arts. A street sweeper wearing the uniform of the Paris sanitation department—Day-Glo-green coat and trousers, green plastic broom, green handcart—stops to see what's going on. Kac explains. The sweeper nods sagely. *"C'est bien, c'est bien, c'est bien,"* he says. *"Transgenique, c'est l'avenir!"* Transgenics, it's the future! Kac, oblivious to blue-collar irony, is overjoyed by this initial reaction.

A couple of days later, Kac is still at it. We wander up the rue du Départ, past the Gare Montparnasse, and into a street fair of schlock art. Kac puts up an Alba FAMILLE flyer among a collection of graffiti-covered political posters. He's watched by a bunch of guys who've been drinking beer in a stall that features driftwood sculptures.

One of the men, who identifies himself as "the artist Christophe," asks Kac whether he'd mind if he drew a little graffiti on the poster. Kac says sure, and stands to one side, explaining where he's from and what he's doing, and that this is the transgenic rabbit he created and wants to take home. *"Ah, c'est magnifique,"* says Christophe, concentrating on his graffiti: a rough scar across Kac's forehead, a twisted smile on his face, and bolts protruding from his body. The reference to *Frankenstein* is unmistakable—but Kac doesn't get it. Taking the spindly bolts for arrows, he concludes that Christophe "made me into Saint Sebastian," the oft-painted saint who was skewered by the shafts of his enemies. But Christophe isn't finished. He's drawing garters and fishnet stockings on Alba. And a dialog balloon, with a lone question mark inside, coming out of Kac's mouth.

"Why the question mark?" asks Kac.

"It's 'why,' " says Christophe.

"It's 'why'? Why not?" says Kac.

"Exactly," says Christophe. He steps back to examine his work as if it were on an easel, then takes another swig of his Kronenbourg 1664. "In fact, I didn't understand anything you said," he says, taking another look at his opus and glancing at his drinking buddies for approval. "It's nice, no?" There's a hint of belligerence. "It's Dr. Frankenstein, isn't it? It's Brazilian? It's not one of the descendants of that old Nazi who cobbled together clones, is it?" Alba as one of Josef Mengele's boys from Brazil. Now there's a thought.

But Kac is missing this. Christophe raises his voice. "And its prick, is that fluorescent, too?" Transgenic transgendering. "I mean, what the fuck do you do with that rabbit there? What about the suffering it feels? And where's the art in this? You say you're an artist. Where's the art? You create life. You are God. There! You are Dr. Frankenstein."

Kac starts to talk about the long history of breeding animals, saying it took 50,000 years to create the domestic dog of today. But Christophe clearly doesn't give a shit. He's looking around for another can of Kronenbourg.

"Well," says Kac, "that's enough discussion with the man on the street."

"A fluorescent dog! That would be great!" shouts Christophe as we walk away up the hill.

When I started emailing Kac to arrange our meeting, I was thinking maybe we'd spend dawns in the French countryside, casing the hutches where Alba is confined. I vaguely imagined us wearing black balaclavas and gloves as we crept among the cages in search of Kac's long-eared opus glowing in the night like electric lettuce. Then I saw the flashed snapshot that appears on Kac's Web site and his posters. The artist is standing in front of some hideous brown geometric wallpaper, his face as radiant as a proud father's in the delivery room. Alba looks a little puzzled, but placid. And white. In the one bit of documentary evidence Kac has for his claim on the luminous rabbit, she's quite white. With pink eyes. No hint of green; no suggestion of a glow.

Deep in the Alba literature on the Web site, there's an explanation. "Alba is not green all the time. She only glows when illuminated with the correct light. When (and only when) illuminated with blue light (maximum excitation at 488 nanometers), she glows with a bright green light (maximum emission at 509 nanometers)." But why was another online photo of Alba so obviously tinted green, conveying the *idea* of a glowing rabbit instead of the actual glow? More disturbing, it seemed that INRA had a slightly different take on her creation. For them, the bunny wasn't art—she was just one among many green fluorescent protein specimens. The organization's chief researcher had been quoted as saying that " 'Alba' doesn't exist. For me, it's rabbit number 5,256 or so."

Kac's glow-bunny project felt at once wildly imaginative and vaguely fraudulent. Who was this bunnyman from Brazil? And what did this transgenic rabbit really represent? Did she *mean* anything at all?

At the University of Paris' *Archiving as Art* show, where we first met, Kac was dressed entirely in black: rumply turtleneck, rumply jeans. His hair was short and curly, cut high on his forehead like a Roman emperor's. He was watching a video of himself projected over a framed syringe that bore the most lethal-looking needle I'd ever seen. Beside it was a little plastic pill. The display was part of an installation called *Time Capsule.* The pill was a microchip, and one day in 1997, live on Brazilian TV, Kac had injected the thing into the flesh above his ankle. "The syringe is the one I used," he said, his voice a mix of self-satisfaction and surprise. "The microchip on the wall is of course a replica, because the original is still in my leg."

In a short book on Kac's work, *Eduardo Kac: Telepresence, Biotelematics,*

Transgenic Art (on the cover of which the bunny-hugging snapshot has been retouched in Day-Glo green), art critics offer their opinions of his installations and performances. This is what Christiane Paul, a curator at the Whitney Museum of American Art, had to say about the monstrous syringe and the microchip: "Kac's radical approach to the creation and presentation of the body as a wet host"—yuk!—"for artificial memory and 'site-specific' work raises a variety of important questions that range from the status of memory in digital culture to the ethical dilemmas we are facing in the age of bioengineering and tracking technology."

The questions that interested me were a little less transcendental. "Where'd you get the chip, Eduardo?" I was thinking he had the thing elaborately constructed in partnership with some high-tech lab, and that there was a message of far-reaching significance buried deep within. You know, like a time capsule.

"I bought a pack of five," he said.

"Oh," I said, feeling a shiver of apprehension about misapprehensions. "A pack of five?"

Sure. There was nothing really special about the time-capsule chip. It's an ID made for tagging pets, livestock, and endangered species. It can be scanned electronically, like an embedded barcode. Kac just latched onto the technology—bought a pack of five (for about $50)—and made it part of his art by making it, well, part of him. He even registered himself online with Identlchip as the owner of his body. The form is reproduced in his book:

Animal Identification: Chip Number 026109532
Call Name: Eduardo Kac
Date of Birth: July 03, 1962
Registered Name: Eduardo Kac
Species: Other
Breed: Human
Sex: Male
Spay/Neutered: NO

And so on.

This world of Kac's is about making connections—some logical, many not—in the name of art. It's supposed to be high concept, about metaphor. It is not supposed to be about hanging things on a wall.

Except that there *are* things hanging on the wall in Kac's installation. There's the framed syringe. There are also sepia photographs—the kind taken with a Brownie camera between the two world wars—of a woman in her late twenties or thirties. "These are actual photographs of my grandmother," he says. Her name was Perla Cukier (Pearl Sugar). She left Poland in 1939, a Jew fleeing the Nazis, and joined Kac's grandfather, Perec Przytyk, in Brazil. Their daughter—Kac's mother—married early and divorced quickly, then remarried and left Eduardo with her folks. "I was raised by my grandparents," says Kac, looking at Perla's picture. "She was particularly influential in my life. . . . As a child, the world that you know is obviously circumscribed. That is, grown-ups are always telling you no. The contrast between other adults and the world they let me develop, and her, and the world she let me develop, was so great."

One of the little sepia squares shows Perla astride a motorcycle on a tree-lined lane in 1930s Poland. Her confident smile is the most arresting image in the installation. "Her part of the family did not survive," says Kac. "These are analog memories that I internalized," he says. "This celebrates the lives that could have been."

"She was from Warsaw," he adds. "So she would have gone to Auschwitz." When it's scanned, he tells me, the ID chip in his leg displays a number. Like the tattoos at Auschwitz. The elegant literalness of this connection is almost an embarrassment to Kac in his role as a professional spinner of conceptual jargon. But there is something here that sheds light on the rabbit that glows in the dark. It is Kac's obsession with "otherness," with the way we treat the strange and the alien. Could it be this bunny is really a Maus?

Kac has got this thing about animals. He wants to get inside of them, he wants, in his way, to *be* them and to share the experience via his Web site. He draws heavily on the work of Chilean biologist and philosopher Humberto Maturana, who talks about creating "consensual domains." In one of Kac's installations, *Rara Avis,* a bright-red robotic macaw sits on a branch in an aviary amid dozens of drab sparrows. Using a headset, you see through the macaw's camera eyes, and when you turn your head, so does the macaw, so that you're interacting with the other birds. In a cave in the Rotterdam zoo, Kac constructed what he calls his batbot, which makes bat sounds while surrounded by 300 real Egyptian fruit bats. "When you put on the [batbot] headset," he tells me, "what you saw was a visualization of dots moving in and out of a circle representing those fleeting moments of mutual awareness" between the dangling robotic bat, which was you, and the living bats in the cave. "So there was a moment when we got closer to that only mammal that flies."

"Don't psychoanalyze the work," says Kac. But why not? The more closely you look at the technology itself, the less impressive it is. The bird and bat robots are probably less sophisticated than a Furby. The time capsule is nothing but an Identlchip. Even *Aequorea victoria*'s green fluorescent protein is nothing very extraordinary. GFP is widely marketed through scientific supply catalogs for use in various kinds of biochemical research projects. (This January, researchers in Oregon used it to create the first genetically engineered primate, a baby rhesus monkey—minus the glow.)

Kac may quote the impenetrable Maturana and cite Hungarian artist Lászlò Moholy-Nagy as his inspiration, but it's probably more relevant, and certainly more revealing, to know that Kac—whose greatest theme is mutation and whose greatest works are mutants—grew up reading *X-Men* comics.

The next day, as we talk in Café Le Select on Boulevard Montparnasse, a couple of small dogs saunter by. Kac watches them closely, almost covetously. One is a black-and-white wirehaired Jack Russell with star-quality cuteness. "A good potential GFP dog," I suggest. In 1998, Kac announced a plan to make a GFP K-9, but he couldn't find anyone to do it. Now, he thinks he might have found a lab in California that can take on the job.

"I'm looking for dogs," says Kac. "But I'm looking for a hairless dog."

A GFP Mexican hairless?

Kac nods. The Aztecs used them on cold nights around Tenochtitlán as living, panting, licking hot-water bottles. A very weird breed—they're newborn-like, with almost all the skin exposed. But, since only living cells express GFP, hair doesn't glow, so the less fur the better. How perfect: the progeny of a bizarre canine race, bred by the earthly but alien culture of the Aztecs, combined with a jellyfish protein to make a mutant pet for the Chicago family of an American, Brazilian, Polish, Jewish, atheist artist-professor.

And then I remember something Kac said the night before, when we were talking about his indulgent grandmother. Was there anything she ever denied him? "I did ask for a dog," he says. But she'd never agreed to it.

Eduardo, have you *ever* owned a dog?

"Well, no, but I am looking forward to . . . I mean, my wife grew up with four dogs."

But you don't actually have any pets? Kac shakes his head. "My last name is Kac and the three of us are allergic to cats. But my daughter loves dogs."

For the moment, however, there's Alba.

What Kac keeps saying is that the work of art is not the rabbit; the work of art lies in creating the debate around the rabbit, and bringing the bunny home to live with Ruth and 5-year-old Mimi. "I didn't want it to be an experiment," he says. "The kind of interaction that would be real was important to me."

Kac holds up a piece of paper with a list of appointments on it. "This is my intervention plan," he says. "I'm developing this action on many fronts: to tell my story directly, to gather support from people who are influential, to mobilize opinion to liberate Alba." But he's not planning any visits to INRA, or any meetings with people there. "I can't drop in from America and tell the French how to operate," he says.

He is right about that, certainly. But Kac seems almost entirely unaware of the political and social environment he's dropped into. Europe is in a panic over mad cow disease, spread by the practice of feeding living cattle the processed remains of dead ones. Scientists said that inducing cow cannibalism was OK. Now they're saying bovine spongiform encephalopathy in hamburgers could lead to brain-rotting Creuzfeldt-Jacob disease in humans. Both BSE and CJD have appeared on the Continent. All this is feeding, as it were, a deep current of anti-scientific sentiment in Europe. France, in particular, is still reeling from the 1980s scandal over HIV-tainted blood that was pumped into thousands of hemophiliacs because the government refused to use an American test to screen the blood supply. The French people blamed their politicians and their scientists (only three or four minor officials paid any penalty), and they blamed the United States because, well, because it was there. Spite is more important than right when it comes to anti-American reactions among the Gauls. This special bias is also reflected in the raging debate over genetically modified foods, which are seen as part of a plot to globalize American big-money science.

Into the middle of all this comes Eduardo Kac with his GFP bunny. No wonder the reception he's getting isn't altogether friendly.

Though the French hold artists in higher esteem than politicians, scientists, or CEOs, there is clear suspicion that the bunny is a front—a kind of cuddly Trojan horse—for multinationals trying to foist genetically modified organisms (GMOs) on the European public. When Kac gives his lecture at the Sorbonne, a ponytailed young man in an orange turtleneck wonders about what he calls "transference" from the scientific to the artistic realm: "Isn't it bothersome that an artist can promote genetic manipulation, given that scientists are under much more control?"

Again and again Kac is asked to comment on a recent paper, "Notes on the Human Zoo," by the German philosopher Peter Sloterdijk, which defends the idea that the human race can be improved by scientific means. In other words, by eugenics—or transeugenics. It's not a great technological or even ethical leap from trying to prevent hereditary diseases to trying to build better immune systems, and from there to making people stronger, or more intelligent. Dr. Frankenstein had an idea like that. So did Dr. Mengele. And if Kac can create a special rabbit in the name of art, why not create a special human?

"No," Kac says emphatically. "If I would do something with a human being, then I would be the father. And I do not believe it's all the same. It's an issue of difference." He pauses for a moment and a mischievous smile lights his face. "But people do often come to me in these lectures and say, 'You know, I would like to glow.' "

After days of listening to Kac expound, I start to think that Alba is fewer than six degrees of separation from every major issue of our time. No small achievement in a work of art. But one of the criticisms leveled at the GFP bunny by Kac's colleagues in Paris and by academics visiting his Web site is that the rabbit wasn't *created* by the artist, but *appropriated* by him.

When I ask whether this is true, Kac expounds on the history of appropriation. "Artists have always borrowed from each other's work," he says. "But when you say appropriation in art, that's not quite what you mean." He points to Marcel Duchamp's 1917 presentation of a urinal as a work of art. *Fountain,* Duchamp called it. And appropriation was born, affecting everything from pop art to musical sampling.

Kac doesn't smile at the idea of Alba as a urinal. "It's very important to distinguish my work from that. My work is not about appropriation; it's about creation."

Kac has never said that Alba was the first transgenic rabbit, or even the first to receive a GFP graft. But he doesn't think any other rabbits glow like his does. And he is sure he commissioned this one.

Is he saying this was the first rabbit created in the name of art?

"Oh, there's no doubt about it," he answers.

Kac left Paris on December 13, no closer to winning Alba's release than when he arrived. And I was no closer to meeting either the elusive bunny or Louis-Marie Houdebine, the man credited with her physical creation. As director of research into biological development and biotechnology at INRA's center in Jouy-en-Josas, Houdebine is one of the country's leading proponents of GMOs. After several phone calls to him went unreturned, I tried an end run: I went to visit one of Houdebine's colleagues, Brother Jacques Arnould—author of the anti-creationist book *God, the Ape, and the Big Bang*—who had appeared on a brief television debate with Kac.

An influential figure in the French scientific establishment, Arnould is a Dominican friar who works at the National Center for Space Studies. He explained that Houdebine and his higher-ups at INRA wanted to control the message at a time when they're under enormous pressure—politically, financially, and emotionally.

"They are very sensitive," he told me. "They are all a little crazy, the researchers. You know, they say a researcher's occupational disease is to become mad. And it's not only a joke." Now there was the risk that their world could fall apart. "You spend 20 years in research trying to improve agriculture and suddenly you discover that all your work is condemned, that you're blamed for mad cow, that genetically modified organisms are feared." Lifelong projects can die, said Arnould, "because they're not loved anymore." The passion to pursue them, and the funding, disappears. Arnould continued: "I think the researchers of INRA are afraid that they wouldn't be loved anymore. There's a need to be loved, a fear of not being loved. We think of that as natural with an artist, but it's the same with a scientist."

I said I understood and sympathized.

That same afternoon, whether prompted by Arnould or not, Houdebine returned my phone calls. A few days later, I was at the gates of the INRA center in Jouy-en-Josas, about 30 minutes southwest of Paris.

The setting was beautiful and bucolic, on the outskirts of the little village that gave the world the famous fabric, printed with beautiful and bucolic scenes, known as toile de Jouy. A stream flows through the narrow vale beneath weeping willows near a riding ring. A large, 19th-century chateau dominates the hillside. INRA looks like the campus of a small university; dozens of buildings are scattered around the grounds, but with rather uncollegiate names: Genetic Fish, Freezing Rooms, Abattoirs. The phrase I couldn't get out of my mind was "Soylent Green."

I found Houdebine, who has worked at INRA for almost 33 years, in an office overlooking the valley, his desk crowded with papers and burdened with a large potted plant. He pointed to a collection of three long, low structures outside his window. "The famous green rabbits are in those buildings," he said. And he began to tell their tale.

Much of his account matched Kac's. Yes, Louis Bec, the director of the Digital Avignon festival, had approached his friend Patrick Prunet of INRA in 1999 about getting a GFP bunny for Kac's exhibit. Prunet got in touch with Houdebine, and Houdebine agreed. Then the INRA director said no, and that was it. But there was no question, said Houdebine, of creating a rabbit expressly for Kac. There was no "commission." As for Kac taking the rabbit home to Chicago, there may have been a misunderstanding, because that was never really an option.

Don't think I wasn't suspicious of Houdebine's motives. He recently signed on as scientific adviser to Bio-Protein Technologies, a company created to "prepare pharmaceuticals in the milk of transgenic animals"—including rabbits. He thinks bunnies could do an excellent job of producing erythroprotein, for instance. Though this human protein is much sought after by athletes because it increases red blood cells and is hard to detect in drug tests, a little EPO goes a long way. The worldwide market, according to Houdebine, is less than one kilo a year. "You need a few hundred rabbits, and that's enough."

Transgenic rabbits have been used to provide the enzyme alpha glucosidase, vital to the treatment of a rare and fatal ailment in children called Pompe's disease. Rabbits are also useful as research models for human diseases, specifically for the vascular system. "Rabbits are exquisitely sensitive to cholesterol, just as we are," he said.

To confuse this picture of rabbits as profitable pharmaceutical producers with a frivolous image of green fluorescent bunnies as objets d'art might alienate BioProtein's financial backers in Holland and France. So Houdebine had an incentive to play down Kac's inspiration for Alba's creation, and his own role in the project. Yet the logic and detail of his account made sense.

Houdebine said he had nothing against either bio-art or Kac: "I think he's a good guy, a nice person. But he cheated, in a way."

Back in 1998, Houdebine's team took some commercial GFP, combined it with the human promoter EF1Alpha—which expresses the protein in all cell types—and injected it into the eggs of three albino rabbits. When the rabbits reached maturity, the researchers bred them and raised the offspring that showed signs of GFP; the non-GFP offspring were destroyed. Eventually the GFP bunnies numbered about 150, which was more than the hutches could handle, so they were culled back. There is a more or less permanent population on hand of six or seven, but few will live to enjoy "their full rabbitness," as Kac would say. Most of the GFP bunnies are wanted for their glowing ovaries. When these are removed for research, the donor is killed. That could be Alba's fate as well. "She's an experimental animal," said Houdebine. "If we need her, we'll use her."

When Louis Bec called from Avignon in mid-1999, Houdebine found his request for a GFP rabbit tempting. The proselytizing author of *GMOs: The True and the False,* Houdebine liked the idea of presenting a cuddly glow bunny in a mocked-up living room. "The rabbit would be observed by a few people; there would be a debate. We planned to show a number of photos of [GFP] cells, showing that we have created a number of [scientific] tools, and they can be beautiful." Kac and Bec would help Houdebine make his case for genetically modified organisms. "So we said we *have* green rabbits—or we suppose they're green," said Houdebine. But nobody had ever checked to see. The researchers were interested in rabbit genes that would glow green through a microscope down through the generations. They assumed the whole animal might look green under the right light, viewed through the right filters, but they didn't have the equipment to test it.

One day in April of 2000, Kac and Bec brought in the lights and the goggles they needed to examine the rabbits: Three were taken from their cages. One, a control, had no GFP; under the blue light, its eyes were bright red. But the other two had the gene, and their eyes shone dazzling green. Their skin glowed, too, though it was hard to say how much, because of the fur. Kac picked up one GFP bunny to hold in his arms, but it was fretful and hard to handle. He picked up the other, and she was very quiet, Houdebine recalled, "very kind and nice." Later, Kac named her Alba.

As Houdebine and I talked, we toured the main biotechnology center. It was constructed about 10 years ago with a butterfly-shaped floor plan. "France still believed in biotechnology then, so we got this nice building," said Houdebine. "Now, they still believe, but more . . . moderately." In fact, there is almost no public forum left in which to defend genetic engineering. Even the thoughtful daily, *Le Monde,* seems to have turned against it. "This is recent," said Houdebine. "After the contaminated blood affair and mad cow, people changed."

In a cluttered lab, Houdebine showed me a rack of little plastic tubes on ice. Each contained a quantity of liquid about equal to a teardrop, which contained specifically chosen genetic fragments. "When the gene is prepared, we put it in bacteria, and the tubes are placed in a generator to amplify the genes." The device looks like an old fax machine with a cribbage board on top to hold the tubes. "You take just one cell and you have millions of copies of a gene in just a few hours."

Up the hill in a little outbuilding is the lab where technician Céline Viglietta implanted the GFP-laced DNA in the rabbit eggs almost three years ago. Like many techniques, this one seems remarkably simple once it's mastered. The zygote is positioned under a microscope and held in place by a tiny glass pipette. The first bit of the operation is accomplished by the technician sucking lightly on the other end of the tube. The gene is put in place using another pipette that's been brought to an egg-penetrating microscopic point on a precision glass-cutting machine.

The abattoir, or slaughterhouse, on the slope below Viglietta's lab is where unwanted animals meet their doom. And below the abattoir are the rabbit hutches, which, Houdebine was quick to point out, are in deplorable condition. He blamed the local Green Party for blocking funding because it was hostile to his GMO research. "You will see when we go to see Alba," he said, "our facilities are very old."

It was the first time he'd said we might actually visit her—and the first time he'd called her by name. But before that close encounter, we'd have lunch. Houdebine drove us to the nearby Robin des Bois restaurant in his beat-up old Citroën, which had little potted plants in the dusty console. "I'm a biologist! I love living things!" he said. "I'm surprised more people don't have plants like this in their cars." At the restaurant, I ordered steak, having long since given up worrying about mad cow. Houdebine had duck. "You know, we have tests now for CJD," he said. "But what's the use? There's nothing we can do for the disease."

Back at the INRA campus, Houdebine opened the door to the rabbit barn. I was hit with a smell that reminded me of that hamster cage I kept forgetting to clean when I was 11. To the right of the entrance was a little washroom with a large sink and a rabbit-milking machine. "Notice the tiles on the wall," said Houdebine. They were the unforgettably ugly brown tiles that appeared in the photograph of Kac holding the bunny.

Inside, there were perhaps 100 albino bunnies, each in a little wire cage about two hops long and no hops wide. "Ahhh, well," said Houdebine. "Where is Alba?" He was looking at the little manila cards on each of the cages. "I don't know," he said, gesturing toward a row of six rabbits. "One of these," he suggested. On the cards identifying the rabbits—by race, in effect—were the letters OGM, which is French for GMO, along with the rabbits' GFP numbers: GFP.081, GFP.082, GFP.011, GFP.016, GFP.015, GFP.014. The 8s and the 1s indicated that the rabbits were from different founders, said Houdebine. "I remember I put a cross on the label . . ." But there were no crosses, and I think that for a minute both of us were wondering if Alba had been sent to the abattoir. Houdebine turned to the rabbit keeper, Georges Oxaran, a slow-talking man in blue work clothes that were wet from rabbit washing and speckled white by rabbit hair. Which rabbit was the one all the fuss had been about? Oxaran pointed to GFP.014, and as we looked more closely, we saw the word *Reserved* on the card.

"Ah, yes," said Houdebine. "There she is."

Alba, I presume.

"It's especially this one that's more green," said Oxaran. "It's especially the eyes."

She sat crosswise at the back of her cage, her nose against the bars on one side, her tail against them on the other. Glutinous turds hung like grapes on the bars beneath her. In the next cage, GFP.015 was restless, looking us up and down. Not Alba. She barely seemed to notice her visitors. In the normal light of the barn, there was no green glow to be seen. Her eyes were placid and pink.

Fidgety GFP.015 let loose a torrent of rabbit piss that splattered on the concrete floor, forcing us to step back.

"Notice," said Houdebine, "that part of their ears have been cut off to show us they're transgenic." I noticed, too, tattooed inside one ear of each GFP bunny, a designation and number. I thought of Kac's *Time Capsule* exhibition, the one that "celebrates the lives that could have been."

I copied down the details from the card on Alba's cage. GFP.014 had been born, along with her three siblings in the adjacent cages, on December 31, 1999.

I called Eduardo Kac in Chicago and told him I'd seen Alba. There was a pause on the other end of the line.

"Eduardo?"

"Yes."

"Do you know when Alba was born?"

"No. Do you?"

"Yes," I said, and then told him what I'd seen and what I'd heard from Houdebine.

"I was afraid his discourse would change after there was a scandal."

I said I wasn't sure that's what had happened.

Kac said there was no question that he had proposed such a project as early as 1998, and that Bec had told him it was on course in 1999. Bec had told him the rabbit was born in February 2000 (and, indeed, GFP.081 and .084 were, on the 6th).

But Kac was scrambling to account for how little he really knew about the rabbit he'd supposedly caused to be created. Houdebine "was probably in the process, scientifically speaking, of starting the pregnancy. . . . Perhaps he started it specifically for me, or it was just in the pipeline."

Perhaps, I suggested, it was all just a misunderstanding.

"Another thing that's very important," said Kac, "is the difference between the language of science and the language of art. In science, say you make a drug that will cure people; the focus of the research is the drug, everything is focused on this physical object. But art doesn't work that way. Is the urinal to be peed on? No. Obviously not. It's this symbolic role that a lot of scientists don't get. For Houdebine, he may not fully grasp what this represents to me."

"I'm just trying to nail down the facts, Eduardo."

"There *are* facts and they need to be right. But it's not just that."

"No?"

"It's my hope that she'll come here."

"It doesn't sound like that's going to happen."

"We can do things that science is not interested in—this contextualization in the symbolic realm."

"Ah," I said. "Houdebine seems to have liked you, Eduardo."

"A visionary," said Kac. "He took a chance with an artist he didn't know and I am going to do everything I can to express my recognition."

"But, you know, Eduardo," I told him, "Houdebine says if they need Alba's zygotes, they'll take them and kill her, just like they do with the other GFP bunnies."

"Oh maaaaan, this really breaks my heart. Shit! I can see why scientists don't want to get personal."

"And the future of transgenic art?" I asked.

Kac's tone was businesslike again. "In terms of getting the work done, this is something artists are learning, can learn, and will learn."

Michael Specter

Rethinking the Brain

FROM *THE NEW YORKER*

That overused term, "paradigm shift," was originally coined by the philosopher Thomas Kuhn to describe what happens when an established scientific theory is overturned by a new one. Such a paradigm shift has occurred recently in the study of the brain. For years, it was widely accepted that neurogenesis—the creation of new cells in the adult brain—was impossible. But new research in birds and mammals is showing otherwise. Focusing on a brilliant young woman whose pathbreaking work is central to the debate, the distinguished journalist Michael Specter guides us through the shock waves that a paradigm shift can cause—not just for a theory, but for egos and reputations as well.

Fernando Nottebohm has lived transfixed by the melodies of songbirds. He is sixty now, and it has been decades since he left the plains of Argentina—first to study agriculture in Nebraska, then zoology at Berkeley, before coming to rest, in 1967, at Rockefeller University, in New York. But his interest in birds has sustained him since his earliest childhood. "Finding out how birds sing and why they would bother and what it means has been the puzzle of my life," he told me when we met for the first time, early in 2001. Nottebohm is a courtly man, and though he has spent the bulk of his career at Rockefeller, his restrained demeanor seems out of place at the giant biotechnology mill on York Avenue. After all, this is the era of genomes and molecular biology; mice crafted from engineered genes and bred to live in Plexiglas boxes have come to dominate medical research. To Nottebohm's colleagues, his preoccupation with the song systems of zebra finches and canaries and with how black-capped chickadees remember where they hide their food has always seemed quaint, even touching—if perhaps beside the point. Yet, over the past three decades, in dozens of elegant experiments that produced results nobody had envisioned (and for years very few believed), Nottebohm's obsession with how birds learn to sing set of a chain of discoveries that have fundamentally altered the way scientists think about the brain. It has also opened a tantalizing, if tentative, new route toward treating degenerative conditions that are often considered beyond hope—from Parkinson's disease and multiple sclerosis to spinal-cord injuries, strokes, and Alzheimer's disease.

The bird brain has an undeservedly bad reputation. It's not easy to fly or to learn meaningful music. To do both is an anatomical triumph. Nottebohm was certainly not the first man to be beguiled by birdsong. Beethoven, Bach, and Vivaldi all transformed avian music into instrumental works; Mozart turned a starling's song into the closing variations of one of his best-known piano compositions, the Concerto in G. Nottebohm believed that if he could understand how birds acquire their songs it would make a wonderful model of the way the brain learns. Many birds produce just one tune and sing it until they die. Nottebohm was more interested in those birds, like canaries, which can learn new melodies each year. Canaries live, on average, for ten years, cover a wide octave range, and sing for several reasons: to announce themselves, to claim territory, and to scare away other males when they look for a mate. (Females rarely sing.) As Charles Darwin noted, a songbird's early, rudimentary attempts at vocalization—called subsong—have a lot in common with the babbling of a human infant. By the time canaries are eight months old, though, they sing like adults, and their habits never vary: they sing throughout the breeding season, in the spring, and then, during the summer molting season, they shed the songs as if they were feathers. The next spring, the same birds will turn up with an entirely new repertoire. Who was teaching the birds these new songs, Nottebohm wondered. And what was happening in their brains to let them learn?

"It's not that I was uninterested in human health, but I really cared most about birdsong as a model for the brain," Nottebohm told me when we met at his Rockefeller laboratory. He doesn't come to the lab often; most days, he can be found in the rolling fields of the university's ethological-research center, in upstate New York, among thousands of carefully tended canaries and zebra finches. "As it happens, there are some obvious connections between birds and humans. It was just a practical example of the ways in which scientific discovery is totally unpredictable. And the complexity of the brain—well, I have never stopped being amazed by it.

"I have always been intrigued by religious questions," he went on. "To what extent were people special? What is this thing called the mind, and how is it different from the brain?" Whether the brain was simply the sum of its molecules—"You're nothing but a pack of neurons" was how the Nobel laureate Francis Crick put it—or whether all that biology added up to something more has been debated for centuries. "We have some close relatives," Nottebohm said. "Chimps, even monkeys. But they can't speak. No primate can speak. It's only humans who do it. When you look around the animal kingdom, birds are one animal that attempts vocally to do anything like what we do."

By the early seventies, Nottebohm had begun to publish a series of remarkable observations that traced the genesis of birdsong to specific clusters of neurons—the cells into which memories are wired and through which complex actions are processed. First, almost by accident, he demolished the notion that handedness—the idea that one is born either right-handed or left-handed—was the exclusive province of humanity. The syrinx, the songbird's voice box, turns out to have two sources of sound, which originate on different sides of the trachea. In an attempt to establish their role in singing, Nottebohm cut the nerves leading to one side or the other. The results astonished him. Cutting the left nerve mostly silenced the birds; cutting the right had practically no effect. "Some property of their brain induced canaries to be left-handed singers," he told me. "With other birds the right side is dominant." If birds demonstrated such a uniquely human quality, Nottebohm reasoned, maybe the patterns of avian behavior would be relevant in other ways, too.

Next, he tried to figure out why male canaries sing and females almost never do. To his surprise, Nottebohm noticed that certain parts of the brains in the songbirds were as much as four times larger in males than in females. He also found that if you give testosterone to a female canary its song nuclei will double in size, and it will sing more like a male. "That was a real shock, because we had all been taught that an adult brain was supposed to stay the same size, with the same cells, forever," Nottebohm said. "It was one of the few uncontested facts about the brain. So how could it get bigger? That contradicted everything I had ever learned."

To study the environmental effects, Nottebohm compared the brains of birds kept in cages with those of birds that lived in the wild. Again, the differences were striking: a free-ranging chickadee, which has to avoid predators and forage for its food, produced larger numbers of new neurons in the hippocampus—the part of the brain that plays an essential role in the storage of memories—than a caged chickadee. In cold weather, a chickadee becomes desperate for calories; it must eat before it sleeps or it will die. So remembering the many places where it stashes seeds is of urgent importance.

At first, Nottebohm had wondered if neurons grew in bulk to accommodate these challenges. In 1981, he wrote a paper, called "A Brain for All Seasons," in which he speculated that the cells swell and shrink at different times of year. But even as he wrote, he told me, he wasn't sure that he was right. "Damn it, I said, this is strange. It's not supposed to happen. We all know that brains in adult animals don't change. Cells die as you get older, and that's it. What was going on here?"

For many years, it had been held as one of neuroscience's basic principles that sophisticated animals—and certainly humans—are born with essentially every brain cell they will ever have. Throughout the twentieth century, attempts to suggest otherwise were dismissed, largely because neurons are not like other cells. After infancy, they don't divide and they don't grow. Although the process was not fully understood, brain researchers assumed that adding new memories and knowledge required us somehow to rewire the circuitry of cells that have been in place from the beginning of our lives. But one day in 1981, while Nottebohm was in the shower, he had the type of insight that happens in books far more often than in life. "I think I actually said the word 'eureka,' " he told me as we sat in the Rockefeller faculty cafeteria one snowy day last winter. "I dried myself off and went to my wife and said, 'Do you know what could explain all these changes we are seeing? What if every day new cells are born in the brain and others die? Wouldn't that explain why some birds learn new songs and forget old ones? The neurons filled with old memories could be exchanged for new ones.' I thought, Maybe the dogma of our lifetime was just completely wrong. My wife, Marta, was very excited and urged me to test this idea at once."

The more he thought about his idea, the more sense it made. If, in order to survive the winter, a black-capped chickadee had to remember hundreds of places where it had hidden food, or if a canary needed to keep the exact melody of a forty-note song in its brain in order to attract a mate, it might require more neurons than birds that didn't have such demands. "The idea that neurons in the adult brain come and go was considered the view of a lunatic," he said. "If you cut your arm, new cells will grow. If you cut your brain, it's going to stay cut. That's one reason strokes are so devastating and why brain injuries rarely heal. Neurons don't come back. But I decided to look again at that assumption. I have always been seen as one of those scientists with good intuition, but one who is maybe simple in his approach. Now people were saying my intuition had dried up. People in my own lab begged me to stop. I saw the pity in their eyes. They were saying, 'Fernando has lost it completely.' "

Nottebohm needed to prove that neurons were replaced in the adult brain. By the early sixties, technology had been developed to help. When a cell is about to divide, it starts making DNA. A radioactive hydrogen molecule attached to the thymidine needed for cell division could be injected into a brain cell and become a permanent part of the cell; if the cell divided, the resulting cells would all be marked as well. In that way, it would be possible to determine the time and place of any cell born after the injection. Nottebohm and one of his doctoral students, Steven Goldman, injected birds with the radioactive molecule every day for a week. Then they waited a month, killed the birds, and examined neurons from various parts of their brains. "What we found," Nottebohm told me, still shaking his head in surprise nearly twenty years later, "was a huge pool of labelled cells—and many of the cells were new neurons. Every bird, young or old, was producing thousands of them each day."

The discovery that new nerve cells are generated in an adult brain—the process is called neurogenesis—overturned a century of scientific theory. And it has the potential to do much more: if neurons are continually born in the brain of a human adult, as Nottebohm discovered they were with canaries, researchers might be able to influence how those neurons develop and to replace dying and failing cells with new ones. That would allow advances in the treatment of brain injuries and many types of degenerative disease. "That is the Holy Grail for us," said one of Nottebohm's former students, Arturo Alvarez-Buylla, who is now a professor of neurosurgery at the University of California, San Francisco. "What we are talking about is teaching the brain to repair itself with its own cells. It's not going to be a simple task. It's a type of magic, really, but eventually I think it's going to be possible. And for that we should thank Fernando and his birds."

FERNANDO NOTTEBOHM HOLDS A CHAIR at one of the nation's most prestigious universities, and his research is considered beyond reproach.

"Intellectually, Fernando is a free spirit, which is what I admire most about him," Eric Kandel, of Columbia University, said when I called to ask about Nottebohm's work. In 2001, Kandel won the Nobel Prize for research into how synapses in the brain affect learning and memory. "He turned out to be absolutely right about neurogenesis, and it has led to one of the great paradigm shifts of modern biology."

Nevertheless, Nottebohm's discovery that adult birds give birth to a steady stream of new brain cells was hardly greeted with jubilation; Kandel himself was highly skeptical. In 1984, Nottebohm presented his most important findings to a conference in New York sponsored by the Institute for Child Development Research. He demonstrated not only how canaries produce new neurons but also how those neurons function at times when memory was particularly essential. He also mentioned, in an offhand way, that if new neurons could intergrate themselves so successfully into the brains of adult canaries, perhaps that would be the case with humans. Many in the audience were hostile to the idea; others laughed. Skepticism is the prime currency of science, and challenging a basic belief about how the brain works brought much of that attitude to the surface. Researchers wanted to know how Nottebohm could be sure these new cells were neurons. The brain is composed mostly of glial cells—often seen as the glue that binds neurons together. But there are many types of neurons, and it is not always easy for a professional to distinguish between them and glia, even under a microscope or after using sophisticated labelling techniques. Nottebohm's colleagues also wanted to know how he could be certain that the cells were new, and how they had managed to migrate from one part of the brain to function in another.

There was another, largely unspoken, response to Nottebohm's research. "People basically said, 'Even if this is true, big deal. It's just birds. All they do is fly around,' " Charles G. Gross told me. Gross, a professor at Princeton for thirty years, knows about the skepticism of colleagues. He withstood a wall of disbelief in the late sixties after discovering the neurons that the brain uses to recognize faces. "First, people said Fernando must be wrong," Gross told me. "He suggested from the start this could have important implications for learning and memory in humans. But when they saw how convincing his work was . . . people smiled and said, 'Old Fernando found a cute thing about birds.' "

One important reason for the doubts about Nottebohm's work was that questions had been raised by Pasko Rakic, who is perhaps the foremost student of the primate brain in America. Rakic, who for many years has been the chairman of the neurobiology department at the Yale University School of Medicine, has spent much of his life looking at the brains of rhesus monkeys, which are closely related to humans; and although few believed that primates could generate new neurons, the proposition had never been tested when Nottebohm released his findings on canaries. It didn't take long for Rakic to recognize the significance of the studies, though. The implications for humans "of even a minute turnover" of neurons would be "enormous," he wrote in a widely read paper called "Limits of Neurogenesis in Primates," which he published in 1985. New brain cells would mean new approaches to even the most terrible neurological problems and diseases.

Rakic's paper described his study of the brains of twelve rhesus monkeys ranging in age from six months to eleven years. He injected each of the monkeys with specially labelled thymidine and then killed them after intervals that varied from three days to more than six years. The labelled thymidine allowed Rakic to trace the development of neurons in the brain of every monkey he studied; by following the labels, he was able to examine more than a hundred thousand individual cells in each of them. The results were not ambiguous. "Not a single" cell with the physical characteristics of a neuron born after infancy "was observed in the brain of any adult animal," Rakic wrote. Although he acknowledged that no biological finding is ever final, he concluded that the dogma should stand: by the time a monkey—and, by inference, a human baby—is a few months old, it has all the neurons it is going to get.

Not long ago, I went to New Haven to visit Rakic. He is sixty-seven years old, a nearly bald, dapper man with a wry sense of humor. Rakic has been in America for years and his English is flawless, although he has retained the accent of his native Yugoslavia. Rakic showed me his slides; cells from monkey brains were stored in boxes scattered around his office. There were thousands, all neatly labelled. "You know, I am often considered as the bad guy in this discussion of neurogenesis," Rakic said. "People want the new cells because they think it offers new hope. And they think I am the guy who always says, 'Read my lips—no new neurons.' But that was never really my position. I did not object to Fernando's birds. I only objected when he said that what he saw in canaries could be applied to human beings."

Rakic says that it makes no biological or evolutionary sense for human adults to replace the building blocks that provide their memories. "We learn our memories and store them in synaptic circuits and in neurons," he said. "If you replaced them, you would not have those memories anymore. I speak with this accent because I use the neurons that were wired into my brain when

I learned how to talk. Then, unfortunately, when I learned English as an adult, those neurons were still in control of my vocal cords. If I were somehow able to replace them, as canaries do, I would speak perfect English. But if I then went back to Europe I wouldn't recognize my own mother, because the new neurons in my brain would never have seen her."

Rakic argues that gradually, over millions of years, humans traded the ability to make new neurons for the ability to keep them. For an adult human to shed thousands of neurons and slip a few thousand new ones into the same space would be a bit like trying to rip out two floors of the Empire State Building and replace them brick by brick without affecting the rest of the building. "Even if you could do it, it would be a Faustian bargain," Rakic said. "Perhaps you would get rid of the neurons that gave you problems and get new ones that worked right. And the price for that could be that you—as a unique person with a unique group of memories—would no longer exist.

"You could take a canary from Northern California, put it in Southern California, and the next year it might even sing with a Southern California accent. That's a hell of a trick, particularly since after all these years I still speak with a Croatian accent. But, when Fernando stood up and said that even while we are talking you are making a bunch of new neurons in your brain, I simply said no, you don't. We have never seen that. It just doesn't make sense."

THE ISSUE DISAPPEARED after Rakic published his paper, in 1985. He is a persuasive man, and those who believed that adult neurogenesis mattered decided that it mattered only in lower animals, where the complexities of human memory did not exist. By chance, however, in 1989, in another laboratory at Rockefeller University, a young postdoctoral researcher in behavioral neuroscience named Elizabeth Gould, who was investigating the action of specific hormones in the brains of rats, stumbled onto something in her research that didn't add up. Gould had arrived at Rockefeller that year to work with Bruce McEwen, one of the world's leading experts on how stress affects the brain. "We noticed that if we took the adrenal glands out of a rat, many cells in the hippocampus rapidly began to die," she told me not long ago. People with Addison's disease, which is caused by a severe deficiency of the hormones normally created in the adrenal glands, suffer similar cell destruction. "The effect is massive," Gould said. "You don't even need statistics to see it." Yet, when she counted the cells that remained, she could detect no decrease in the number of neurons. She was stunned. Gould asked herself, "Were our accounting methods completely screwed up? How could thousands of cells disappear and there still be the same number as there were before?"

Gould, who was then twenty-six, went to the Rockefeller library in search of some precedent for the bizarre effect she had noticed. (This was before the Internet provided the most efficient way for a scientist to review what had previously been published in her field.) "I have strong memories of sitting in this ancient room, looking through the Index Medicus, and going back a long, long time until I finally found evidence of adult neurogenesis," she said. She found what she was looking for in a series of reports published beginning in 1962—the year Gould was born—by a researcher at MIT named Joseph Altman. At the time, the new technique of labelling a cell with thymidine to determine the birth date of neurons was used in newborns, since adult animals were not thought to create new neurons. But Altman decided to try the technique with adults. He published several papers in the most reputable scientific journals, claiming that new neurons are formed in the brains of adult rats, cats, and guinea pigs—a discovery that Nottebohm later made with canaries. Because the techniques Altman used were primitive, however, they were open to reasonable doubt. It was a classic example of a discovery made ahead of its time. At first, Altman was ignored, then he was ridiculed, and finally, after failing to receive tenure at MIT, he moved to Purdue. With no recognition, he was quickly forgotten. The field almost dried up. A decade later, Michael Kaplan, a researcher at Boston University and later at the University of New Mexico, used an electron microscope to supply more compelling evidence that several parts of the adult brain, including the cortex, also produced neurons. He, too, met resistance from researchers who did not find his work convincing. ("Those may look like neurons in New Mexico," Kaplan remembers Rakic saying at the time. "But they don't in New Haven.") Kaplan had published his findings in important journals and even suggested a novel way to test the phenomenon in humans, but he, too, was ignored, and he left the field.

Gould barely knew Nottebohm in 1989, although their labs were only a few hundred yards from each other. But she also came across his work in the library, and suddenly it all clicked. "I realized what had to be going on," she told me. "The brain was making new neurons to compensate for the ones that died. That is why the numbers didn't change. It was so simple, but it was one of these things you were trained not to think about." With McEwen's support, Gould shifted the focus of her research from hormones to neurogenesis. "For a long time, although nobody was interested in what we were doing and we couldn't get our papers into fancy journals, there was a sustaining excitement to it," she said. "I felt that if I don't study this no one else will. It was interesting and it was potentially very important. But I have to tell you I also enjoyed it because the field was so small."

For eight years, Gould carried out her work on neurogenesis in McEwen's lab. In 1997, she moved to Princeton. She was thirty-four, with many publications to her name; but neurogenesis had only sporadic scientific support and she was as far out on a limb as a researcher can go. In little more than three years, however, she had been given tenure and a full professorship—a previously unimaginable leap in her department at Princeton—having demonstrated with new and more convincing techniques that cells are born in the brains of adult rats. She pushed the field further than anyone else had. The year Gould arrived in Princeton, she reported that neurons were produced in the hippocampus of adult tree shrews (which are similar to early primates). In humans, the hippocampus is the principal area where Alzheimer's disease develops.

The next year, she published a paper demonstrating that a New World monkey, the marmoset, also makes neurons as an adult. (Primates that live in Africa and Asia—places that Europeans had explored before Columbus—are known as Old World monkeys; New World monkeys live in Central and South America.) Then she repeated the work using macaques, Old World primates that are more closely related to humans. Finally, in 1999, she and her colleagues discovered that not only are cells produced in the adult primate brain but they even appear in the neocortex—the most sophisticated region of the brain, which is responsible for language and complex thought. It was her most controversial work and it has yet to be repeated, but Gould reported that these new cells had migrated to the cortex from another part of the brain, had quickly developed into mature neurons, and had integrated themselves into the circuitry there.

Gould, whose background was in behavioral psychology, also undertook a series of experiments that suggested a strong relationship between the number of neurons an animal generates and the challenges it faces. Certain types of events seem to require the adult brain to make more neurons, and others appear to prevent that from happening. She found, for instance, that a brain needs to "use it or lose it"—if new cells are not put to work, they will die more rapidly than if they have a purpose. Also, through several studies in which she examined the effects of stress on the brain, Gould demonstrated the adverse effects that social subordination or fear can have: expose a rodent to the scent of a predator (in this case a fox) and it will become so anxious that its production of new neurons will quickly fall away. The studies, when combined with results from others, echoed Nottebohm's earlier research with birds and showed not only that new neurons were generated by adults but that active animals appeared to generate more of them.

There was one problem with Gould's work, though. The results seemed to contradict the theories of Pasko Rakic, and he has not been reticent about suggesting that Gould's methods were flawed. Disagreement and debate require scientists to repeat their studies; it's a fundamental precept that if you can't repeat something it cannot be taken seriously. Yet this was debate of a different order. Rakic, a former president of the Society for Neuroscience, is one of the seminal researchers in his field. In order to avoid a clash with someone so eminent, Gould would have had to permit herself to become marginalized, like Altman, or follow her predecessor Kaplan out of the business. She had no intention of doing that.

GOULD, AN ANIMATED WOMAN with long dark hair, is the youngest tenured member of the Princeton psychology department, and among the most prominent. She is sought after by other universities, teaches what she wants, and in 2001 was able to persuade Princeton to buy a four-hundred-thousand-dollar confocal microscope for the exclusive use of her lab. She is tenacious; her third child was born in November of 2000, and four days later she was standing in Peyton Hall, lecturing to a roomful of students. Gould grew up on Long Island, went to college at St. John's and graduate school at UCLA, and then married her high-school sweetheart. (He is a vascular radiologist at a Philadelphia hospital.) When I asked her how she came to select UCLA for graduate school, she replied, not completely in jest, "Good weather." Not until she received a Ph.D., in 1988, did she think seriously about an academic career. "I was not one of these people who knew when they were a little kid that they wanted to be a scientist," she told me. "I was not a person who had some quest or problem in my destiny to solve. Basically, I wanted to have a good time. I hung around the beach, and I thought psychology was reasonably interesting. It wasn't until I came back to New York and I was doing my postdoc at Rockefeller that I became so consumed by it."

In most ways, Gould is a typical bench scientist: driven, perfectionist, aggressively interested in teasing out the most inexplicable elements of a complex story. Yet her rise has not been without complications. Gould told me that when her first child, a girl, was born, just as her career was taking off, in 1991, she didn't see how she could continue to teach. "I had decided to put her in day care and go right back to work," she said. "Then she was born and I fell in love with her and I thought she couldn't possibly survive without me. I was at this weird point of moving from a postdoc to the junior faculty, and I had to write big grants to keep moving up, and for a while there I was just falling apart. My husband was really great. He said, 'You know, you worked so hard to get to this point, if you give up you are going to be miserable. You will feel like a failure.' If he had said something else—if he had said, 'Oh, it's terrible, I can see how you feel, why don't you just take a year off'—well, that would have been bad in the end. Bad for me, for my children, and for my work. I would have never been happy in my life if I had taken that turn. So I bit the bullet. It was hard, but I went back to work."

Gould's controversial successes do not always thrill her. "You can find yourself thinking about what you should do next to satisfy your critics, instead of what is the most interesting thing you could do as a scientist," she told me. "That is the route right to death. When you make decisions about your life based on what the scientific community is saying, you should quit. I think about that a lot these days. I mean, if you are doing your research for some other scientist, why even bother?"

Gould has a compelling air of distracted urgency. She manages to be both completely focussed and endearingly forgetful at the same time. ("I left my slide carrousel once in a cab in Boston, once at a conference in Greece, once in Maryland, and once in D.C. I got it back from the cab in Boston. But losing a carrousel with all your work in it four times is not a good record. I took it as a sign.") With the demands of a family, a full teaching schedule, and many experiments constantly in progress, Gould turns away speaking engagements by the dozen. She often finds conference a useless distraction, and acknowledges that the politics of such events make her queasy—mostly because she is not one of the boys. It is an attitude that worries her mentor, Bruce McEwen. "There is a danger," he told me. "There is the green eye of jealousy, and Liz has to face that. If she were a pretender on her way up, she would be dead. But she is a full professor and already widely recognized in our field. So I think she can ride out the opposition. . . . But I have intended to have a talk with her, at least when her youngest kid is a little bit older. Because, frankly, I think she could be hurting herself. You don't have to love it or focus on it, but you have to play the game you are in. It just gets misinterpreted if you withdraw, and I don't want that to happen to her."

Gould told me that she isn't even certain that she wants to continue with neurogenesis, and although she would not say it directly, Pasko Rakic is clearly part of the reason. When I went to see Rakic at Yale, he spent a long time disputing some of her latest findings; he could find no evidence of neurons in the cortex, and he is convinced that Gould (together with her Princeton colleague Charles Gross) made a mistake in labelling the cells. (They, on the other hand, wonder whether Rakic is fully comfortable with the complicated new molecular-labelling techniques needed to do this research.) A week after we met, Rakic telephoned me. "I wanted to tell you about something, but I didn't know if it was appropriate," he said. He went on to say that after consulting with a Yale University ethicist he had decided he could go ahead. "We examined the slides from that Gould study and photographed them and we did not find new neurons. I asked for permission to use the slides in a paper I am writing," Rakic told me. "And they refused." It was an extraordinary accusation, so I asked Gould about it. She said that Rakic had asked to visit her lab, but that she was about to deliver her third child, had preeclampsia, and thought she might have to be induced into labor at any moment. So she sent Rakic the slides, and, as is not uncommon in scientific disputes, he interpreted the data differently. (Gould and Gross intend to use the slides in an article they are writing and don't want Rakic to publish them first.)

That was only the latest skirmish between Gould and Rakic. In the fall of 1998, they came close to a very public showdown at a scientific forum in Los Angeles. Earlier that year, Gould had found neurogenesis in adult macaques; Rakic had not. Before a scheduled press conference, at the annual meeting of the Society for Neuroscience, Rakic suddenly announced that he, too, had discovered neurogenesis in Old World monkeys. Rakic told me that evidence was hard to find because the brain produced so few of the new cells. Gould disagrees, noting that she has found the phenomenon in rats, mice, tree shrews, marmosets, and two species of macaques and has never noticed a significant difference in the quantity of new cells.

At Princeton, Gould shook off several direct questions about Rakic. After I visited Yale, however, I asked again whether she thought his continued skepticism about her research was fair. In replying, she finally permitted herself to look back on this steady opposition to her work. "When I was studying adult neurogenesis in the hippocampus of the rat," she said, "the rat was unimportant [to Rakic]. When we found adult neurogenesis in the hippocampus of the marmoset, a New World monkey, the New World monkey was unimportant. Then, when we studied adult neurogenesis in the hippocampus of the macaque, an Old World monkey that Rakic has studied throughout his career, our methods were faulty. Then he used these same methods to demonstrate the identical finding. Now that we have found adult neurogenesis in the neocortex of the macaque, it is our methods again."

Fernando Nottebohm, who admires Rakic and considers him one of the most insightful people in neuroscience, was more direct. "Pasko has taken on the role of hard-nosed defender of standards," he said. "And that's fine—it's even warranted. But we have to keep in mind that he missed this discovery altogether. It's something he should have seen, and he just blew it. And, frankly, as much as I hate to say this, I think Pasko Rakic single-handedly held the field of neurogenesis back by at least a decade."

AT FIRST GLANCE, San Diego seems a strange place to claim as the capital of American brain research. It is filled with seals sunning themselves on the beaches and tourists in search of aquatic adventures. People seem constantly to be hovering in the air, hang gliding from promontories above the Pacific Ocean. When I was there, during the Buick Invitational golf tournament, the conversation almost everywhere, as absurd as it now seems, centered on whether Tiger Woods would get his groove back. Yet, if you ride around La Jolla for long, you will almost certainly drive past the Scripps Research Institute or the Salk Institute for Biological Studies. Neither is far from Nobel Drive or, for that matter, from the Burnham Institute or the University of California, San Diego, which has one of the world's foremost centers of brain research. In fact, San Diego has far more than its demographic share of members of the National Academy of Sciences, not to mention Nobel laureates. Francis Crick, the eighty-five-year-old president emeritus of Salk, still shows up at his office. There are also dozens of private companies spread along the sun-drenched coast with names like Advanced Tissue Sciences and Neurome.

Scores of laboratories at universities and in private industry are now in on the search for the origins, mechanism, and meaning of neurogenesis. But if Elizabeth Gould has one genuine competitor—and a complete antithesis—it is Fred Gage, who is co-director of the Laboratory of Genetics at the Salk Institute. Where Gould guards her privacy and declines invitations to most meetings, Gage is one of America's most public scientists. Gage, who is fifty, holds one of three endowed chairs at Salk. He is the chairman of the scientific advisory council of the Christopher Reeve Paralysis Foundation, a member of the scientific steering committee of the Michael J. Fox Foundation for Parkinson's Research, and president-elect of the Society for Neuroscience, among many similar positions. He is on the editorial board of scientific publications ranging from the *Journal of Comparative Neurology* to *Research and Perspectives in Alzheimer's Disease.* Gage's curriculum vitae lists two hundred and ninety-seven scientific articles, and it isn't even up to date. In comparison with Gould's lab, where just five or six scientists work closely together, Gage's laboratory at Salk is a vast scientific field house, with revolving teams of researchers pursuing dozens of projects.

Gage is accomplished, but he is also well known for being well known. When I was in La Jolla one day and was introduced as a reporter to a Salk researcher, she said, "Oh, then you must be here to see Fred Gage." A rangy man with thinning sandy hair and a mustache on the verge of drooping, he has the manner of a mellow Californian. Gage grew up in Rome, and he is a descendant of Phineas Gage, who, in 1848, was a foreman on a railway-construction crew in Cavendish, Vermont. One day, an explosion shot a thirteen-pound tamping spike into his skull and out again, ending up twenty-five yards away after running through his brain. Gage didn't die or even become permanently incapacitated, but his personality changed completely (and not for the better). The accident turned him into the most famous brain patient in American history.

In the nineteen-eighties, Fred Gage lived for several years in Lund, Sweden, where he worked with the scientist Anders Björklund on some of the earliest fetal-cell-transplant approaches to treating Parkinson's. Fetal cells are flexible because they have not yet fully developed, and it was widely hoped that, once implanted in the brain, they would be able to "train themselves" to become the type of neurons that fail in Parkinson's patients. By the beginning of the eighties, experiments at Lund and at the Karolinska Institute, in Stockholm, had demonstrated that fetal-tissue grafts could replace cells that were destroyed by Parkinson's and other diseases, like juvenile diabetes, and that in many cases the grafts could restore the lost functions, at least temporarily. Yet there have always been doubts that placing the cells into such highly organized and established circuitry would work. One recent study has been particularly discouraging, suggesting that transplanted cells, while capable of surviving, and even adapting to their new surroundings, may actually be able to hijack the brain, becoming uncontrolled and malevolent.

Fetal-tissue research had obvious implications, though, and the work set Gage, and scores of other scientists, on a quest: How could you program cells in the brain so that they develop normally when other cells start to fail? Stem cells, which are created at the earliest stages of embryonic development, seem to provide an answer. (Stem-cell science often employs frozen embryos left over from in-vitro fertilization, and the field has become the most recent battlefield in the war over abortion.) A week doesn't pass without encouraging reports of the potential for stem cells to treat any number of diseases. Stem cells can mature into almost every type of cell a human needs, and the most promising results have come with cells taken from the brain. If neuroscientists can make cells, particularly new neurons, grow in adult brains, they should, in theory, be able to find ways of getting them to emerge at the right time in the right places. That has already proved possible in animals. One Italian researcher, Angelo Vescovi, after extracting just a few stem cells from the brain of a healthy mouse, can now routinely grow the equivalent of several brains' worth of tissue in laboratory dishes.

Gage performed experiments that demonstrated that age affects the production of new neurons in rats, and he also showed that if a mouse has regular exercise—something as simple as running on a device that looks like a miniature Ferris wheel—the number of new neurons will increase. Rodents are not humans, though, and Rakic's theory that adult neurogenesis is likely to play a diminished role in advanced animals was not encouraging. It was hard to know how to test humans, since researchers cannot sacrifice them or take slices from their brains to study under a microscope. In 1998, though, Gage and his colleagues at Salk, along with a team from Sahlgrenska University Hospital, in Sweden, managed to use an approach that had initially been suggested by Michael Kaplan in 1982. It was the last piece of the neurogenesis puzzle, and in many ways the most vital.

Gage's team knew that many cancer patients receive injections of a chemical marker, bromodeoxyuridine, or BrdU, which allows cancer specialists to assess how many new cells are being born. Since BrdU attaches itself to every new dividing cell, and not just to those with cancer, Gage's team realized that it could also reveal whether new neurons are being formed. Gage and his group studied five people between the ages of fifty-seven and seventy-two who had cancer of the throat or the larynx. After the patients died, the researchers looked for BrdU in several sections of their brains, and found that primitive neural stem cells had divided and created from five hundred to a thousand new cells each day. "All of the patients showed evidence of recent cell division," Gage said at the time. "It's interesting to note this was not a particularly young or healthy group of people, so new cell growth may usually be even more prominent than we observed."

Gage's study had just five patients, a number that could not support definitive conclusions. But the Gage paper, when added to the earlier primate work of Gould (and also to that of Rakic, who in 1998 reported seeing new neurons in rhesus monkeys), unleashed a flood of research, political maneuvering, and idle speculation. The implications were too promising to ignore. Neither Gage nor Gould is a clinician; their job is to figure out the fundamental principles of science. Still, each receives scores of messages a month from people who wonder whether there is a magical elixir that can reverse a stroke or save somebody they love from a deadly neurological condition. At least one medical group has promoted its ability to grow human stem cells in laboratory dishes and transplant them into the brain of a sick person.

"It's absolutely heartbreaking," Gage told me. "I get these E-mails asking whether people should spend fifty thousand dollars on this stuff. And it's just theft. We are a long way from that kind of treatment, and I can't give anybody any reason to hope for what may never happen. On the other hand, I am not frightened to admit that I believe this information is going to be useful to sick human beings. How soon, or for what specific conditions, I cannot say. But I really do believe that this will eventually work."

We had been sitting in Gage's study, above his lab at Salk. His phone there only dials out; it's the one place where he can escape the frenzied research he directs. Outside the window, across the hills, the sky above the Pacific was filled with the Mylar sails of hang gliders. "It's astounding, and as we learn more about basic biology we are going to be able to take these stem cells and reproduce the steps inside them and make them behave in a specific way," he told me. "It's very complicated, but you have to remember one thing: the embryo does it. It develops the whole system. So if we can learn how the embryo does it we can make something fairly similar to what is lost in certain illnesses. And when we do that we are in business."

One morning in the winter of 2001, I drove up to Rockefeller University's Center for Field Research to see Fernando Nottebohm and his birds. The center—a cluster of austere farm buildings not far from Poughkeepsie—is an estate that was bequeathed to the university in 1971. For many years, Nottebohm shared it with two senior colleagues. These days, behavioral science is not in vogue, and nobody uses the place except him and his lab mates. A foot of snow had fallen the night before I arrived, and the place was silent. As I left my car, however, I heard a muted whirring in the distance;

it sounded like an electrical appliance. By the time I reached the main house, the whirring had turned into the rising crescendo of birdsong.

This is where Nottebohm and Ofer Tchernichovski, who is an assistant professor at Rockefeller, and their colleague Thierry Lints are trying to create the first detailed molecular map of how a bird's brain changes as it learns to sing. Nottebohm and his team are now studying how the brain changes physically—including an analysis of which genes are affected and in what way—every time a young bird opens its mouth.

The lab is filled with sensitive recording equipment, thousands of gigabytes of computing power, stacks of compact disks onto which tens of thousands of birdsongs have been recorded, and a few dozen Igloo beer coolers, which Tchernichovski has transformed into soundproof booths for baby birds. There are also a thousand bright-yellow canaries, and fourteen hundred zebra finches each no bigger than a child's fist. Many of the birds live in room-size cages filled with trees and a long cuttlebone, on which they can sharpen their beaks.

There is no other research facility in America like the field center. "People are not using birds in scientific research now," Nottebohm said, as we stomped through snowdrifts between his office and one of the main houses. "Behaviorists love rats. They can watch them run the mazes; it gives them lots of numbers. That's the American approach, because Americans believe, above all, in statistics. There is also this feeling that mice and rats are like little people.

"But I look at it in a different way," he continued. "What kinds of things do animals do in their natural circumstances, what kind of problems do they have, and how do they solve them? For a brain scientist like me, that is a much nicer approach, because brains are not all-purpose machines. They have evolved to deal with specific existential problems: How do you make it through a year with all kinds of different seasons? How do you claim and defend a territory? How do you find a mate? How do you look after your offspring? How do you remember where you hid your seeds?"

That, of course, is the leitmotiv of Nottebohm's career: you can understand how animals behave, and how their brains function, only if you watch them live normally. It has been Nottebohm's singular perception that behavioral analysis alone would never explain how birds learn to sing, and that just examining the molecular basis of the cells won't do it, either. "Unless you understand the needs, the habits, the problems of an animal in nature, you will not understand it at all," he said. "Put rats and mice into little plastic boxes and you will never fully comprehend why they do what they do. Take nature away and all your insight is in a biological vacuum."

With help from Bell Laboratories, Tchernichovski, a transplanted Israeli with a first-rate ear and a deep knowledge of computers, designed a program that takes control of what a bird can learn and traces it by the second. First, he built a sophisticated sound system into a five-dollar plastic model of a bird, "the type you put on a Christmas tree," he told me. When the baby birds are thirty days old, the researchers place them in a cooler with the plastic father, which is perched in the center of what is essentially a tiny recording studio. The chicks respond immediately to its songs. They quickly get used to the plastic bird. Two big red keys are at the back of the cooler, and it doesn't take long for a young bird to realize that it can make the plastic model sing by pecking on the keys. The computer registers every move the bird makes, recording how many notes it sings, how often it pecks the key, the exact composition of each song, and the vocal register the bird uses. The system then analyzes every note.

Tchernichovski whistled a bit of Gershwin. The computer immediately recorded his version of "Rhapsody in Blue," analyzed the vocal patterns, the notes, the syllables, and the timing. Suddenly, the plastic bird in the middle of the cage is singing Gershwin. "If we wanted to, we could then have the young bird learn that song," he told me with a big smile, since Gershwin is a bit too complex for a songbird to master. Birds learn to sing by the time they are two months old, but it has never been possible to understand the process very well. What is learned and what is programmed from birth? The computer system has finally permitted the team to try to provide an answer.

"Now, if we want to say a certain note was learned at a certain instant, we can take the bird and sacrifice it the second we see him learn that note," Nottebohm said. "Then we can look at what genes are expressed and what cells are there in the brain. We will literally be able to pull those cells out of the brain and say, 'How have you changed the way your genes work?' That is something we will need to know if we are ever going to program the brain to make up for its problems."

Nottebohm is delighted—up to a point—to see that Gage and Gould, as well as experts at the National Institutes of Health and in every major center of science, are now fully engaged in the field that for so long was his alone. He told me more than once—never sourly—that he was surprised by how little publicity Rockefeller sought for his research. Neurogenesis is hardly ever mentioned in the university's brochures, and that also surprises him. "I have always had a passion for clinical relevance," Nottebohm told me as we strolled from one room filled with canaries to the next. "I wanted to discover lovely basic things and I wanted to listen to the music of the birds. But there is so much suffering out there, and it would be so nice to have a solution. Yet I have to admit it's not quite as exciting for me as it was. For so long, this field was my backwater, my sandbox. And I enjoyed it. I saw Eric Kandel"—his friend who had just won the Nobel Prize—"not long ago, and he said, 'You must be so happy that all the things you said turned out to be true,' and of course I am. But, honestly, it used to be much more fun when nobody believed it. In science, by the time everybody tells you it's true you have to scratch your head and look for another business."

MARY ROGAN

Penninger

FROM *ESQUIRE*

The biologist Josef Penninger is only in his late thirties but has already made several breakthrough discoveries that not only help us understand how the human immune system functions but may well lead to better treatments for a wide variety of serious health problems—heart disease, arthritis, osteoporosis, even cancer. Mary Rogan tries to get inside the astoundingly fertile, surprisingly intuitive mind of this scientific star.

The greatest scientist of our time worries about you every day. He worries about the diseases lurking in your genes, the ache in your bones, the viruses sneaking up on your heart. And he worries about your soul, too. He worries whether you have one, and he worries about what you're going to do with it. But he's working while he's worrying. He's finding the genes that will make your bones feel good again, he's flipping the switch for your immune system, and he's hunting down the virus that is going to murder your heart. And he's getting closer, every day, to discovering the gene that will rock your world.

Josef Penninger may well save your life someday, but he'll still wish he could have touched your soul. When he wins the Nobel prize for discovering God, he'll feel his own troubled soul pounding away inside him, nestled between his heart and his thymus, protected by an army of T cells. But that's a long bicycle ride from here. Right now, he's just getting started.

CD45 IS THE GENE that controls the on/off switch for your immune system. When CD45 is switched on, T cells are released to fight infection. After the T cells have done their job, CD45 is switched off and your T cells go home to wait for the next battle. It's a great system when it works. When it doesn't, the results are disastrous. Until recently, scientists weren't sure which gene told your T cells when it was time to come in swinging and when it was time to call it quits.

CD45 is huge. It could be the heart of autoimmune diseases. Think of it this way: There's trouble in your neighborhood; some guy is knocking over trash cans and keeping your daughter out late. So you call in the big guy who keeps order in the neighborhood—CD45—to reach out and set things straight. CD45 unleashes his heavy hitters—T cells—which is a good thing, unless these heavy hitters get hooked on the juice of being heavy and won't stop. Pretty soon, they're killing everyone in the neighborhood, including your daughter and your wife, and after a few beers and crazy talk, they'll be coming for you. That's what happens when your immune system overreacts.

If your T cells keep fighting, they'll eventually start fighting you, attacking healthy organs like, say, your pancreas. If they shred your pancreas, you'll be sucking on insulin for the rest of your life.

Or let's say your heart quits and you're lucky enough to get a new one. CD45 turns on the T cells because T cells will destroy anything foreign that enters your body, and they won't exactly welcome the poor motorcycle rider who was kind enough to sign his donor card. So you'll have to spend the rest of your life taking antirejection drugs that poison your liver but prevent your immune system from chewing up your new heart and spitting it out.

This is where Josef Penninger walks in, with his Sideshow Bob hair. He grows a few mice in his lab and fingers the little man working the machines behind the curtain. That's CD45. Now we're only a few steps away from Dorothy clicking her heels together and saying, There's no place like home, there's no place like home. You can wake up in Kansas, insulin-free, with that new heart splashing away inside you, as welcome as a Dunkin' Donuts on the highway between nowhere and home.

MEETING JOSEF PENNINGER is like a firecracker going off in your ear. For an instant, you can't hear a thing. Then all you hear is this whirring sound. That's Josef's brain working like a bicycle going downhill. You listen and think, What does that feel like? If you get close enough, rub against the gray stuff pounding away in his skull, you can see the asphalt whizzing by underneath his wheels. Josef Penninger is riding the smartest bicycle in the world, and he's not even halfway down the hill.

There are two kinds of scientists: the guys who don't hunt what they can't kill, and the real players, who come hard or don't come at all. If you're a real player, you publish in the top three magazines: *Nature, Cell,* and *Science.* Josef, he's a player. In the last three years, he's published seven articles in *Nature* alone. In the science world, this just doesn't happen.

At the end of each year, *Science* lists its "Hot Scientists," the players who lit the loudest firecrackers in our ears. It isn't just the number of papers you crank out, it's the impact of your work. To make the list, you have to be asking the kind of questions that will change the way we live or die. This year, number one on the list is Craig Venter, the human-genome wizard. Number two doesn't matter. Number three with a bullet is Josef Penninger, the thirty-six-year-old son of an Austrian farmer. CD45 is only one in a string of hits that just keep coming. Everybody is waiting to see what comes next.

There's nothing in your childhood town. This is Austria, the land of sheep and more sheep. Forty kilometers up the road, Julie Andrews sang her heart out over the roar of the helicopter trying to get that shot of her running up the hill. It took about fifty takes because she kept getting knocked over.

So you count sheep. And other things. You're going to be a great mathematician. Numbers are music that trips off your tongue. It doesn't matter that your village teacher is a bastard who figures you'll never be more than a farmer. You have your friends. And math. And soccer. That's enough.

HOW DO YOU CHART the trajectory of a genius? The path that takes you from counting sheep to changing the way we live and die? You can begin anywhere.

Josef's four, playing out behind the barn. There's a dung heap, a pile of cow shit, that he loves to sit in for hours by himself. One day, he's there and he can feel this space opening up in his head. Room for his bicycle to pick up speed. But then his mother comes around the corner and drags him off to school. There's a man there, that village teacher, who says, *Match the red with the red, put the squares with the squares,* and Josef falls for it. Does everything he's asked and gets kicked into the school system one year early.

Or he's ten and his mother doesn't have time for him anymore. Ships him off to a *Young Torless* boarding school where the monks beat the kids. They try to break him like a pony, but he won't even bend. He manages to stagger through, soul intact but with a new topography for his skin. It's thicker. Josef is the smartest kid in his class and he hates every minute of it.

Or he's twenty-four and kicking around Paris, getting drunk when he should be back in Innsbruck, studying. He meets a girl at a bus station and follows her to Canada. The girl thing goes belly-up, but in Toronto, Josef's soon-to-be-ex-girlfriend introduces him to Tak Mak, another pyromaniac with firecrackers in his pocket.

Tak Mak is the guy who unlocked the black box of immunology. Back in 1984, he discovered the gen for the T-cells receptor, which is like discovering what those cave drawings are all about T-cell receptors are the language of immunology. T cells fight infection, but how do they know what to fight? They take their orders from the T-cell receptor. Without the receptor, T cells would be blind, staggering around like the Keystone Kops looking for a fight.

In 1990, Tak Mak *is* immunology. He gets a look at Josef's bicycle and lets out a long whistle. Josef signs on as a postdoctoral fellow in Tak's lab. Three years later, Tak handpicks him for a team of six scientists out of six hundred contenders, and Amgen, the world's biggest biotech corporation, sets them up in Toronto with a whack of cash to chase whatever dragons they want for fifteen years. At Amgen, Josef's bicycle is at the top of the hill.

Amgen is located on Toronto's Avenue of Misery—one gray slab after another of hospitals, each specializing in some dreadful thing that hasn't happened to you yet. There's the Hospital for Sick Children for things you can't even think about, Toronto General for just about anything, Mount Sinai for delivering babies and losing parts of your colon, and, finally, the Princess Margaret for cancer. Amgen is in the Princess Margaret.

Security lets me up to the seventh floor. From there, a secretary uses an elevator pass card to get me to the tenth. The door opens, and there's Josef. And his hair is fucked up. It sort of floats this way and that above his head and occasionally comes to rest. It's not a color I've seen before except maybe on a Chesapeake Bay retriever. A mixture of rust and pink and gray. And he's taller than he looks. That's because he bends a little at the waist when he meets someone. Twists his torso around until it looks as if he's coming and going at the same time. The first word Josef says to me is "Oh." He'll say that every time we meet. He grabs at my hand and lets it drop like a live wire.

His office is the size of a bathroom. His desk looks as if it's just been tossed by the cops. The lab is around the corner, but I can smell it from here, a mixture of cleaning products and sewage. Josef keeps offering me vials of bacteria to smell. In another room, there's an oven where he grows mice. Little blobs of mouse DNA in petri dishes. Josef notes the look on my face and laughs. He says he finds it hard to believe, too.

You're ten and you don't even own a toothbrush. At the dinner table, you hold your fork like a spear. Boys like you aren't supposed to go to special boarding schools for the children of doctors and lawyers. They point at your shitty clothes and snicker at your low German. You're the "it" boy, the guy they come at every day. It's their lack of imagination that staggers you. How they never get bored with tormenting you. But you're not so imaginative, either. "Fuck you" is the only thing you can think to say. You can feel it coming, though. The ideas are on the tip of your tongue.

SOMETIMES, SOMETHING EATS AWAY at your bones and turns them into Necco Wafers. When that happens, you step off a curb and your leg snaps like a piece of kindling. Pretty soon, you can't even get out of bed because stuff keeps breaking. Don't be fooled by those milk-mustached models. Milk won't fix it. But Josef will.

In five years, anyone with osteoporosis or arthritis will get a drug to reverse the process. Not stop it. *Reverse* it. Grow your bones back like fingernails. Amgen stands to make about $5 billion a year from Josef's discovery. But the best part of this story is how he figured it all out, riding that bicycle around dead man's curve.

In 1997, Josef was thinking about bone loss. Another Amgen scientist had asked him to take a look at a gene called OPGL. People already knew OPGL had something to do with bone loss, but it was only one of about thirty proteins that kept showing up in experiments. In science, this is like being stuck in the middle of an answer. You're surrounded by clues, but you're seeing the same thing and the picture won't shift. You're staring so hard, you can't even blink.

Then someone like Josef comes along. He peers over your shoulder and says, *Hey, why not try it like this?* He reaches back to all that space in his head. He takes the glass dome full of snow and gives it a shake. When the snow settles, the village is moved fifty miles up the road. It's the wet dream of science. No extra steps, no more staring until you're asleep with your eyes open.

Scientists wait a lifetime for a moment like this. Josef has about two a year. When it's coming, the tips of his fingers tingle.

His osteoporosis moment happens at the corner of Gerrard and Jarvis in Toronto. It's the middle of the day and Josef's out walking, pushing against the edges, trying not to think. He's at an intersection that looks a lot like Times Square before Giuliani turned it into Disney World. Jarvis is hookers with bad legs, junkies twitching, and people talking to themselves. And then it hits him. The whole thing lies down in front of him like Tuesday's answers to Monday's crossword puzzle.

Back at his lab, Josef grows a mouse and leaves out the OPGL gene. He doesn't worry about the thirty other proteins he could be working with. He picks OPGL and rides it, and what he gets is a mouse you could use as a hockey puck. Bone density up the wazoo, with jaws so dense that teeth can't even break through. Josef's bicycle is humming now. Then something else happens. In the female mice without the OPGL gene, mammary glands fail to develop during pregnancy and the pups can't feed. This is astounding because, later, Josef will use it to explain why women are so susceptible to osteoporosis.

When we grow bones, some of the bone has to be eaten away and used to create fingers, toes, and teeth. Otherwise, we'd be like that concrete mouse. Think of your body as a piece of clay with a beautiful sculpture inside. In a healthy person, OPGL is turned off and on according to what the body needs. It's a well-regulated system that allows us to grow bones and shed cells throughout our lives. OPGL is the gene that sets us free.

Now that Josef knows what OPGL is for, he wants to know what's happening inside people with a hyperactive OPGL gene. He wants to know what causes their bones to chew away at themselves.

His first eureka moment comes when he discovers that T cells trigger OPGL. Josef could see that T cells and OPGL go hand in hand. People with immune diseases like arthritis, lupus, and leukemia are swimming in T cells. Their bodies are fighting all the time, and their bones are paying the price. With the T cells comes a signal for the OPGL gene.

The wheels of Josef's bicycle aren't even touching the ground now. On a hunch, he takes OPG, another gene implicated in the bone-loss mystery, and injects it into rats that can't stand up anymore because their arthritis is so bad. Within days, they're running around in their cages, doing whatever it is that rats do when they suddenly feel better. OPG has completely shut down the crippling effects of their arthritis.

The last part of this story is something even Josef couldn't have predicted. Through all of this, he couldn't stop thinking about the female mice without the OPGL gene. Scientists already knew that osteoporosis was a sex-hormone-related disease, but nobody could make the link. When Josef's female mice didn't develop functional mammary glands, he had the answer in his hands.

Turns out OPGL is one of those cruel Darwinian genes. To develop mammary glands, extra OPGL is released during pregnancy. Without extra OPGL, women can't feed their babies. In a Darwinian world, the one we live in, it's all about survival. Not ours, mind you, but our genes'. So feeding our babies is more important than what happens to our bones when we're seventy and can't get out of bed. With Josef's discovery, the OPGL gene in women could be turned off after pregnancy.

Amgen is in phase two of its clinical trials of OPG. Injecting OPG into patients with arthritis or cancer is just around the corner. From here, Josef is a short bike ride away from curing the two hundred million women who suffer from osteoporosis.

Maybe what hurts the most is that your boarding school is only six miles from your mother's house. But there's no time for you now. She wants more than the farm. So your parents sell off some land and buy a local bar. And all that means is more work. All she does is work. Cooking and cleaning and pouring drinks for drunks who sit there until your father screams at them to get out of the bar.

When you go home to visit, you sit and watch men drink in the middle of the day. It scares the hell out of you because you can feel the violence simmering. You're waiting for someone to explode.

SCIENTISTS, POCKET PROTECTORS notwithstanding, are tough bastards. They're ruthless competitors all scrambling to get into the same journals, all clawing for the cash to keep doing what they do. Labs are rigid and hierarchical but full of cowboys trying to imagine the unimaginable. The work is monotonous but wildly unpredictable. You can't do it by yourself, but nobody wants to help you. Josef says science is like a chickenshit ladder: You're trying to climb, and the chickens on top are shitting on your head and greasing the steps.

In science, it's not cool to jump from idea to idea. Better to throw on that white coat and stare at the same gene for twenty years. Josef's habit of jumping from bone loss to immune responses to heart disease to whatever catches his eye next gives his colleagues a rash. He's got what some scientists call a monkey brain. But right now, they can't touch him. Because he keeps crunching the ball, and, more important, because of the proof he offers. In science, it doesn't matter if you prove something once. What matters is whether any guy in a lab coat can duplicate your results. There has to be a series of steps to follow that will lead the next person to the same door. Josef's discoveries are rock solid because he's made them using what's become the gold standard in genetic science: knockout mice.

It's simple, really. With knockout mice, you start with a question: What does CD45 do? Instead of poking around with CD45, trying to change or influence what it does, you yank it out. Imagine your lunch keeps disappearing from the fridge at work. You're pretty sure it's Bob, the same guy who always steals the swimsuit issue from the lunchroom. But you're not sure. Maybe it's Sally. They're all thieving bastards, anyway. You could try and spy on Bob, but if he knows you're looking, what's the point? So you yank Bob out of the office. Still your lunch is gone. Put Bob back and take Sally out. There's your lunch, unmolested in the fridge. Now you know what Sally does.

Geneticists love knockout mice because the evidence they offer is unpolluted. But only when it works. This is high-risk science. Knockout mice are hard to grow. You take the DNA of a mouse, cut out the gene you're asking about, and then inject the DNA back into an embryo—in the case of autoimmune diseases, mouse DNA without CD45. But this new DNA still has to blend with what's already in the embryo. So Josef's version, the one with the missing gene, won't necessarily dominate. He also includes a color code, which will show up in the pups' fur. The darker the pup, the more of Josef's DNA there is. And on and on. He selects the darkest to breed until eventually he's got a bunch of mice with no CD45 gene. The whole process can take up to two years. That's a lot of work before you even get to ask your question. The risk is, you could spend two years and get absolutely nothing.

In Josef's hands, knockout mice are pure gold. Ask around and people will say it's because he asks the right questions. His hunches always pay off. Even his wrong turns are manna from heaven. He's got a horseshoe up his ass. But it's more than luck: People also know Josef chases his mistakes. He's not throwing mice in the wastepaper basket. If he gets something unexpected, he works that, too. Like CD45. When Josef was chasing CD45, other scientists thought he was wasting his time because it was a "dead" gene, insignificant. Over drinks, one scientist told Josef that if he wasn't careful, CD45 would ruin his career.

Sometimes you go home for the weekend. But it's lonely at the farm because nobody's there. When you're twelve, some idiot builds an illegal dump on the outskirts of your village. Pretty soon, you're sharing your kitchen with rats the size of Volvos. It's midnight and you wake up with King Rat sitting on the edge of your bed. You can't move. You can't even breathe. The house is empty and you have two choices: stay with the rats or get on your bicycle and ride the four miles through the forest to your parents' bar. There's a tree you have to ride by, the one your mother always told you eats bad children for breakfast. The tree or the rats. You choose. And you wonder if this is the way life always goes.

AFTER HIGH SCHOOL, Josef went to medical school in Innsbruck. He says it was an attack of holiness that drove him there, this Catholic thing about doing good in the world that he could never shake. So medical school it was. And he was bored silly. He read what he had to read and never went to class. He aced his exams and used his spare time to study for a second degree in fine arts. He traveled around Europe looking at beautiful art, imagining what it would be like to be a painter.

When he was twenty-two, he went to Nigeria with another medical student. They worked in a children's ward the size of your bedroom. The kids were stacked like cords of wood, four or five to a bed. They kept dying from really simple things like measles and pneumonia, all stuff Josef could fix if the army didn't steal every supply in the place. On the street, vendors lined up like hot-dog sellers, hawking needles and penicillin.

There was this one kid, and Josef was watching him die. All he needed was a needle to get the medicine in, so he sent the boy's father to go find one. The father came back with a nail. The poor bastard spent two weeks' salary on a fucking nail that someone told him was a needle. But Josef says that wasn't the worst thing. The worst thing was watching a kid die from tetanus. The whole body seizes like a busted lock, and the pain is unbearable. In a perfect world, you get a tetanus shot. In a less than perfect world, you get a dark, quiet room, because sound and light trigger the spasms. In hell, you get a nail.

Things were clearer for Josef after Nigeria. He continued to do some work in hospitals when he got back to Austria, but he knew he didn't want to practice medicine. He wanted to do research. He wanted to ask questions that mattered. He wanted to cure things.

At the end of medical school, Josef hooked up with Georg Wick, a professor of immunology at the University of Innsbruck. The way Josef tells it, they grabbed on to each other. Wick hadn't met anyone like Josef before, and Josef hadn't met a professor in five years who actually wanted to hear what he had to say.

Wick and I meet in a café in Vienna because, he explains, all great ideas are discussed in Viennese cafés. He seems disappointed when I won't order the apple strudel, and he lets out a little sigh before he begins to eat. Wick is maybe the most charming man I've ever met. Perfectly manicured, not a hair out of place. He does that thing that gracious men do with their ties, holding it gently against his chest every time he leans in to sip his espresso.

Wick comes from a prestigious Austrian family. He's the great-grandson of a famous architect and was educated at the finest schools in Switzerland. He is exactly where you might have predicted him to be. I can't for the life of me see what the son of a farmer and this man have in common. And then he tells me about how he came to study immunology.

It's the sixties and immunology is brand-new. Wick can't get over the idea of how the body can turn on itself, how a system designed to do so much good can turn so bad. Immune disease as Hamlet. But there's no support for what he wants to do. No labs, no money. Like Josef, Wick never stops to think about what's in his way. He sets himself up in Vienna and starts breeding chickens, because chickens' T cells are particularly good to work with. He rents a shack halfway across the city from his lab, and he and his wife spend hours cleaning up chickenshit and carting birds back and forth across town. For both men, it's about the science. The pure drive to know.

Wick tells me that he figures the thing he'll be remembered for, and the thing he's proudest of, is that he picked Josef. He saw what was in front of him and didn't shut his eyes. He says he understands the tingling Josef gets in his hands sometimes. He felt something like that when he first met Josef.

In his heart, he'd love for Josef to come back so they could work together again. But he knows Josef can't breathe in Austria. He never could.

You write this essay for your tenth-grade teacher. You're fifteen, so it's from the heart. Maybe the hip. You write about priests and how all churches should be burned. The language is clunky, but it's pure anarchy, and the school wants to fry your ass. It's the first time you realize that people have to listen. If you stick their heads in a vise and say, "Look," they'll punish you for that. The monks are waiting for you when you get back to the dorm. The school wants to throw you out on your farmer ear, but there's one guy who's got your back. Seitel, the physics teacher. Head the size of a basketball and hands to match. He fights for you and gives you the best advice of your life: Don't fuck yourself. Stay alive long enough to say the things you need to say.

THE NEXT THING Josef writes is now in a museum in Innsbruck. It's the thesis he wrote for Georg Wick when he was twenty-six. It's about T cells in the thymus. It has chapters like "The Purgatory of Immunity," which tries to answer why some people don't get sick. Or "Taming the Desires of the Soul," which wonders if it's our thymus pounding away inside us or our souls. It's the most beautiful piece of science writing you'll ever read, and it tells you why the world fell at Josef's feet at Jarvis and Gerrard a decade later.

Josef wanted to know about T cells in the thymus, but he also wanted to know how art and science stopped being friends. So he turned to Plato, the first scientist to understand that science begins with matters of the soul. Plato believed we have two souls—our immortal soul, which lives in our head and flies away to heaven when we die, and our mortal soul, the one we wrestle with every day. The monks had taught Josef about Plato, and they had taught him to worry about his own soul. (The Catholic thing, Josef says, just can't be helped.) Plato decided our mortal soul also has two parts. The evil part, which lives in our stomachs and causes bile to spill out of our livers, and our good part, which lives in our thymus, cooling the passions of the heart and beating away inside us until we die.

In his thesis, Josef wanted to explain how T cells go to school, how they learn to recognize their enemies and knock them down. He watched small T cells swim inside larger T cells and thought about the art he'd seen in Italy. In Caravaggio's *Medusa* he saw T cells hiding in a tangle of hair, learning how to ride their bicycles.

From there he traced our descent into the kind of science that doesn't worry about our souls anymore. He wrote about Frederick II, the Holy Roman emperor who figured the best way to find out if we had souls was to nail prisoners into barrels until they died, then take the lid off and see if a puff of smoke followed. No smoke, no souls. It was 1210, and souls no longer existed.

It is a heartbreaking work that tells you that while Josef's bicycle may be fast, it's a hard ride. When he won his award for the best thesis in Austria, he couldn't get into a reception held in his honor because he wasn't wearing a tie. The next year, Josef left Austria.

After the essay scandal, they leave you alone. That terrible kind of alone that makes you feel invisible. You've frightened everybody, even your classmates. But there's relief, too. Because the boot in your face you always knew was coming has finally arrived. The boot is the past no one talks about. It's the high and low

German that separate the dirty farm boy from the bourgeois. And when it finally arrives and hits you square in the teeth, it's not so bad. There's freedom in the pain and the loneliness. Room to find your soul. Watch it rise like a puff of smoke. You don't know it yet, but you'll leave this place.

I WENT TO AUSTRIA to find out how you build a bicycle like Josef's. The closest I came to an answer was figuring out that things happen not *because* but *in spite of.* Nasty monks don't make you smarter. Class distinction doesn't give you wings. A country that won't look at its past and serves up a new version of an old idea in the present doesn't produce geniuses. It produces emigrants.

The train ride from Vienna to Josef's village takes two and a half hours. We roll past one perfect green field and farmhouse after another and every few miles, a lonesome-looking church. I feel as if I could reach out and rearrange the pieces and they'd reassemble themselves before I reach the next town.

Josef's brother Norbert meets my train, and I see it right away—the cleverness in him that Josef spoke about. But it's muffled somehow. I hear Josef, but the volume is turned down. In his family, Josef is the youngest of three boys, the only one to go past high school. His mother and father made it as far as grades eight and four, respectively. Norbert has worked for the railways since he was seventeen.

Norbert looks like Josef, only thinner and strangely fragile. He seems surprised when I get off the train. It will be like that the whole time I'm in Austria. There's a whisper behind everyone I meet that tells me they can't believe I'd travel this far just to hear about Josef.

Their village is beautiful in a stifling sort of way. Gurten lies in a valley, and from the crest of the hill, you can see the tidy farmhouses and green fields that sweep down into the center of town and yet another church. Norbert takes me to meet Josef's first teacher. We eat apple strudel, and Herr Teacher tells me he always saw something special in Josef. He keeps clearing his throat and leaning into my little tape machine. I don't believe a word he's saying. Later, back in Toronto, Josef laughs when I tell him this story. This is the teacher who told his parents not to bother sending him to school. He'd never be anything more than a farmer.

The farmhouse Josef grew up in is shaped like a horseshoe, with the living quarters on the left and the barn on the right. Nothing is out of place. All the doors are painted green. Max, Josef's father, is waiting for us in the courtyard at the center of the horseshoe. He doesn't speak any English, but he knows why I'm here. When I get out of the car, he hugs me and kisses both sides of my face. He smiles, and it seems to me he'll never stop smiling. Norbert stands back, poking at something on the ground with his foot.

At Josef's old boarding school, I meet Seitel, the physics teacher with balls of stone. The vice-principal, a fussy little man, translates for me. Seitel tells me that this place almost killed Josef. That there's no one like him in the universe. And the best thing he could tell Josef was that he should keep his head down until he could get away. The vice-principal spits the words out like razors. Seitel stares straight ahead, working his hands like worry beads. When I leave, I want to take him with me.

Later, we stop for a drink and Norbert talks a little about Josef. But he's guarded, and it feels as if he's trying to guess what I want to hear. He settles on this: The last time Josef came to Austria, he passed through Norbert's hometown, Wels, but couldn't stop. It seems the rental car Josef picked up in Vienna ran on diesel. By mistake, Josef filled it to the brim with gasoline. His car bucked and lurched its way through Wels on the way to Gurten without stopping.

Norbert and I make a quick stop at the hospital to visit Josef's mother. Seven years ago, she had a stroke that hit her like a train. All that work, all that ceaseless motion, just ground to a halt. Max is there, wiping the spittle from her mouth. One eye sags, and her right arm keeps dropping to her side. Max keeps smiling and putting her arm back in her lap. Norbert strokes her arm and says hello over and over. I'm looking as hard as I can, trying to find her soul.

ANTONIO OLIVEIRA-DOS-SANTOS works for Josef in Toronto. Like his boss, Antonio studied medicine before becoming a research scientist. He believes Josef's lab is the chance to bring the power of genetics to the bedside. Antonio possesses the phenotype for narcissism—a ponytail. He's the young Turk.

He loves working for Josef. The high is incredible because the risks are insane. As a young student, you have to decide: Do you want to play it safe, look at a few things and publish a couple of decent papers? Or do you want to get on the back of a bicycle going downhill?

Antonio did his time under the boot. He worked at a lab in Austria and learned all about ass-kissing and the feudal system that is science. He saw people running the same experiments over and over again because the professor wearing the tie told them to. Biting his own tongue got tiresome, so he jumped at the chance to come to Josef's lab four years ago. To be so close to the bicycle that you feel your own gears kicking in. Antonio is typical of the people Josef attracts to his lab. Not just the best and the brightest, but the most fearless, as well.

Then there's Mike Crackower, a postdoctoral fellow in Josef's lab. Mike tells me that Josef is one of those old souls who's been here before. That's why he has all the answers. Or maybe he's a genius. Either way, Mike's just happy to be along for the ride. He came on board just as the party was cranking up in Josef's lab. Before that, he was grinding out a thesis on lobster-claw syndrome, an unfortunate single-gene defect that gets you a free pass into the circus. But he wanted more. He wanted to take a look at the stuff that isn't so black-and-white. The little miseries that feel more like real life and leave us gasping like trout on a deck.

When he came to Josef's lab in 1999, Mike had no idea he'd be looking at heart disease. He knew Josef liked chasing a million ideas at once, but when they met, Josef told Mike heart disease was the ball he wanted to knock out of the park. His mother had just suffered the stroke that jammed her bicycle wheel. He picked Mike as the front man in his lab looking into heart disease.

Mike sees the difference between a guy like Josef and himself. The work, he tells me, is great. But for Mike, the passion stems from his own success, not the work itself. Josef, he says, doesn't care who's watching or what they think. He's pure drive. Mike loves being next to this kind of courage but admits it's quieter in his own head. He figures the thing that gives Josef his vision is that he never doubts for a second that he could be the guy who cures cancer. Or gives his mother her life back.

It was heart disease that first put Josef on the map in 1996. After he finished his postdoctoral work with Tak Mak, Josef was in a bind. He had his own lab, and he had money from Amgen. What he didn't have was credibility. He wanted to chase immunology, but that was the big dog's bone. Competing directly with Tak wasn't wise. If he tackled immunology, people would say he was just riding Tak's success. If he branched out, they'd say he didn't have the expertise. Tak was kind enough to throw Josef a piece of his own bone.

A Toronto cardiologist named Peter Liu came to Tak with a mystery. In the early nineties, Peter had grown sick of watching healthy people die of heart attacks. People who never had their own booth at McDonald's, who didn't smoke or eat fast food or have a family history of heart disease, or the teenager who played on the football team, caught the flu, and died two weeks later from heart failure. Peter found that what all these folks had in common was Coxsackie, a virus from the same family as the common cold.

Tak told Peter that Josef was his man because Josef had the bicycle to ask the question, Can you *catch* a heart attack?

Josef and Peter spent the next few years following a hunch Josef still can't explain. He created a knockout mouse that was missing a gene called p56lck. Like OPGL, it's just one of many genes banging around our immune systems. There was no guarantee he would get anything by knocking it out. What he got was one of the biggest breakthroughs in heart-disease research in the last two decades.

What Josef discovered was that, yes, you can catch a heart attack. Before Josef, all we'd heard about was diet and lifestyle, but they didn't account for the 50 percent of heart attacks in seemingly healthy people. The idea that the guy sneezing all over you on the subway can give you a heart attack seemed outrageous.

At his lab, Josef discovered that if you knock out p56lck, you can throw all the Coxsackie virus you want at a mouse and it won't get heart disease. But why?

It works like this: Coxsackie uses p56lck to reproduce. As the virus multiplies, it works its way into the T cells sweeping through your body. Now your T cells are infected with Coxsackie and become a Trojan horse landing on your heart. When the T cells reach your heart, Coxsackie jumps out and starts devouring healthy heart cells. Now your T cells go into overdrive. They don't know they're carrying deadly cargo; all they know is your heart is under attack. So more and more T cells arrive to help, only they're infected, too. You can catch a cold, catch a heart attack, and be dead inside of a week.

After meeting Mike and Antonio, I get together with Tak Mak, Josef's mentor at Amgen. Tak's office is down the hall from Josef's. As soon as I sit down, he starts busting my chops. Why am I doing this article on Josef? There are lots of good scientists I could be looking at. I wonder if this is jealousy I'm hearing, and then Tak switches gears.

Journalists, he tells me, can be devious people. Reporters have a viewpoint they want to prove, and they'll twist the facts to do it. He knows because he's been there. When he was in the spotlight, he had to beat them away. I'm not sure where this shit storm is coming from, but it's on my head, and there's nothing to do but sit it out. And he keeps talking.

A few years ago, when Tak was still the golden boy of cancer research in Canada, his wife got breast cancer. A colleague was doing experimental treatment out in California and offered to help. Tak took his wife to California and sat with her while she was nuked with the most potent chemotherapy on the planet. She died anyway. At home, the press crucified him for taking his wife out of Canada for treatment. For crying uncle and running to big brother down south. Somehow, Tak Mak's loss became Canada's shame. Even after she died, the press called, looking for quotes. All he wanted to do was talk about his science. All they wanted was a quote about his dead wife.

Now all I want to talk about is science, but that's not going to happen. Eventually, Tak tells me that Josef is doing excellent work. Amazing, even. Everything he touches turns to gold.

I leave wondering what it feels like to chase a dragon for twenty years, turn around, and find out it ate your wife.

When someone has a stroke, she looks like a dead person. The same outline, but all the stuff in the middle is missing. So you recognize her when you get to the hospital. The hair and the housedress. That's your mother, only stopped. No more work. You sit and wonder if she even knows you're there. You wonder about the cascade of events that brought her here. The science you can't get out of your head lays it out for you until you can imagine the breakdown, cell by cell.

So you try to think of something else. You try to string together enough moments from those first few years on the farm. The way she'd come around the corner and find you in your dung heap, sitting by yourself. Maybe take you by the hand. It's enough.

There's one more thing Josef discovered recently. Something that might change the way we live and die. When I get back from Austria, Josef's bicycle has left the planet.

He's discovered AIF, the gene that controls why every living cell on this planet dies. In all the searching through T cells, bones, and broken hearts, Josef was looking for our souls.

Programmed cell death is what happens inside so you can be here. The cells inside you die so you can go from looking like something in your fish tank to looking like the babe at the end of the bar. AIF is the master gene that sets everything in motion. It tells which cells to die, and when. We start off as this blob, and inside us, dying to get out, are our eyes and ears, mouth and nose. Our fingertips and toes. All this other stuff falls away, and we emerge. AIF is probably the oldest gene in our body. It is the missing link scientists knew existed but didn't have the wheels to find. Now it belongs to Josef.

AIF may be his greatest discovery, and also the scariest. Put a drop of AIF on a bone tumor and it disintegrates in front of your eyes. A drop in your body and you're dead in twenty-four hours.

Cell-death guys, he says, are the real cowboys. All elbows and high sticks and, until Josef, all stuck at a dead end.

A decade ago, scientists isolated caspase, the family of proteins they thought was responsible for cell death. But when they tested it on knockout mice, the results were disappointing. If caspase had been it, the gene creating cell growth and death, knocking it out in a mouse embryo would have stopped the clock. The embryo would never have grown into a mouse. Instead, mice with severe brain damage were born. Caspase wasn't the gene solely responsible for cell death.

In 1998, Josef got a call from an old friend in Paris. Guido Kroemer was looking at AIF and was sure this could be it, the gene that controlled cell death, but he couldn't figure out how to prove it. He needed Josef to come in and peer over his shoulder, see what he was missing. Josef got to work and cloned the AIF gene. Then, with hands tingling, he knocked it out. The results were staggering. There is no life without AIF. Nothing grows. It's a prehistoric gene, and it's present everywhere. It's in you, and it's in worms and flowers and even slime mold. It's God.

AIF gives Josef the willies. It would give any sane person the willies. But he wants to harness it. Figure out a way to drop it on the virus that's attacking your heart. Or turn it off in the brain that's traveling back in time, devouring memories. Put a little splash on the bone cancer that's eating your nine-year-old son alive.

The possibilities are mind-boggling. Pick anything that scares you: cancer, Alzheimer's, Parkinson's. Cancer is cell overgrowth. Cells that are too stupid or mean to die. Alzheimer's and Parkinson's are about cell death. Ideas that disintegrate in mid-sentence.

If Josef is right and AIF is calling the shots, then we're a heartbeat away from telling AIF what *we* want it to do. Genetics is all about who's boss. And we always want to be the boss. But AIF, like God, is a monster. Josef shakes a little when he talks about what could go wrong if AIF is mishandled. A drop of AIF on the cancer cells eating away at your brain will produce a miracle, but if you miss by the tiniest fraction and AIF touches healthy cells, your brain will be destroyed. With AIF, we could blow our souls right out of the water.

JOSEF SAYS SCIENCE is better than the best Greek tragedy. It's got everything. Blood and guts. Kids dying and people melting in front of your eyes. And the hero waiting in the wings to take you by the hand.

I think this might be another attack of holiness, but I know what he means. It's a great bicycle.

Sarah Blaffer Hrdy

Mothers and Others

FROM *NATURAL HISTORY*

Our human ancestors relied on more than just a mother to raise a child. Like many other species, they were practitioners of "cooperative breeding"—in which several members of the tribe played an active role in rearing young. Or so argues the anthropologist Sarah Blaffer Hrdy, who has often overturned conventional ideas about sex roles. She believes that not understanding our evolutionary history may have dangerous consequences for how we raise our children today—and for the future of our species.

Mother apes—chimpanzees, gorillas, orangutans, humans—dote on their babies. And why not? They give birth to an infant after a long gestation and, in most cases, suckle it for years. With humans, however, the job of providing for a juvenile goes on and on. Unlike all other ape babies, ours mature slowly and reach independence late. A mother in a foraging society may give birth every four years or so, and her first few children remain dependent long after each new baby arrives; among nomadic foragers, grown-ups may provide food to children for eighteen or more years. To come up with the 10–13 million calories that anthropologists such as Hillard Kaplan calculate are needed to rear a young human to independence, a mother needs help.

So how did our prehuman and early human ancestresses living in the

Pleistocene Epoch (from 1.6 million until roughly 10,000 years ago) manage to get those calories? And under what conditions would natural selection allow a female ape to produce babies so large and slow to develop that they are beyond her means to rear on her own?

The old answer was that fathers helped out by hunting. And so they do. But hunting is a risky occupation, and fathers may die or defect or take up with other females. And when they do, what then? New evidence from surviving traditional cultures suggests that mothers in the Pleistocene may have had a significant degree of help—from men who thought they just might have been the fathers, from grandmothers and great-aunts, from older children.

These helpers other than the mother, called allomothers by sociobiologists, do not just protect and provision youngsters. In groups such as the Efe and Aka Pygmies of central Africa, allomothers actually hold children and carry them about. In these tight-knit communities of communal foragers—within which men, women, and children still hunt with nets, much as humans are thought to have done tens of thousands of years ago—siblings, aunts, uncles, fathers, and grandmothers hold newborns on the first day of life. When University of New Mexico anthropologist Paula Ivey asked an Efe woman, "Who cares for babies?" the immediate answer was, "We all do!" By three weeks of age, the babies are in contact with allomothers 40 percent of the time. By eighteen weeks, infants actually spend more time with allomothers than with their gestational mothers. On average, Efe babies have fourteen different caretakers, most of whom are close kin. According to Washington State University anthropologist Barry Hewlett, Aka babies are within arm's reach of their fathers for more than half of every day.

Accustomed to celebrating the antiquity and naturalness of mother-centered models of child care, as well as the nuclear family in which the mother nurtures while the father provides, we Westerners tend to regard the practices of the Efe and the Aka as exotic. But to sociobiologists, whose stock in trade is comparisons across species, all this helping has a familiar ring. It's called cooperative breeding. During the past quarter century, as anthropologists and sociobiologists started to compare notes, one of the spectacular surprises has been how much allomaternal care goes on, not just within various human societies but among animals generally. Evidently, diverse organisms have converged on cooperative breeding for the best of evolutionary reasons.

A broad look at the most recent evidence has convinced me that cooperative breeding was the strategy that permitted our own ancestors to produce costly, slow-maturing infants at shorter intervals, to take advantage of new kinds of resources in habitats other than the mixed savanna-woodland of tropical Africa, and to spread more widely and swiftly than any primate had before. We already know that animal mothers who delegate some of the costs of infant care to others are thereby freed to produce more or larger young or to breed more frequently. Consider the case of silver-backed jackals. Patricia Moehlman, of the World Conservation Union, has shown that for every extra helper bringing back food, jackal parents rear one extra pup per litter. Cooperative breeding also helps various species expand into habitats in which they would normally not be able to rear any young at all. Florida scrub-jays, for example, breed in an exposed landscape where unrelenting predation from hawks and snakes usually precludes the fledging of young; survival in this habitat is possible only because older siblings help guard and feed the young. Such cooperative arrangements permit animals as different as naked mole rats (the social insects of the mammal world) and wolves to move into new habitats and sometimes to spread over vast areas.

What does it take to become a cooperative breeder? Obviously, this lifestyle is an option only for creatures capable of living in groups. It is facilitated when young but fully mature individuals (such as young Florida scrub-jays) do not or cannot immediately leave their natal group to breed on their own and instead remain among kin in their natal location. As with delayed maturation, delayed dispersal of young means that teenagers, "spinster" aunts, real and honorary uncles will be on hand to help their kin rear young. Flexibility is another criterion for cooperative breeders. Helpers must be ready to shift to breeding mode should the opportunity arise. In marmosets and tamarins—the little South American monkeys that are, besides us, the only full-fledged cooperative breeders among primates—a female has to be ready to be a helper this year and a mother the next. She may have one mate or several. In canids such as wolves or wild dogs, usually only the dominant, or alpha, male and female in a pack reproduce, but younger group members hunt with the mother and return to the den to regurgitate predigested meat into the mouths of her pups. In a fascinating instance of physiological flexibility, a subordinate female may actually undergo hormonal transformations similar to those of a real pregnancy: her belly swells, and she begins to manufacture milk and may help nurse the pups of the alpha pair. Vestiges of cooperative breeding crop up as well in domestic dogs, the distant descendant of wolves. After undergoing a pseudopregnancy, my neighbors' Jack Russell terrier chased away the family's cat and adopted and suckled her kittens. To suckle the young of another species is hardly what Darwinians call an adaptive trait (because it does not lead to survival of the surrogate's own genetic line). But in the environment in which the dog family evolved, a female's tendency to respond when infants signaled their need—combined with her capacity for pseudopregnancy—would have increased the survival chances for large litters born to the dominant female.

According to the late W. D. Hamilton, evolutionary logic predicts that an animal with poor prospects of reproducing on his or her own should be predisposed to assist kin with better prospects so that at least some of their shared genes will be perpetuated. Among wolves, for example, both male and female helpers in the pack are likely to be genetically related to the alpha litter and to have good reasons for not trying to reproduce on their own: in a number of cooperatively breeding species (wild dogs, wolves, hyenas, dingoes, dwarf mongooses, marmosets), the helpers do try, but the dominant female is likely to bite their babies to death. The threat of coercion makes postponing ovulation the better part of valor, the least-bad option for females who must wait to breed until their circumstances improve, either through the death of a higher-ranking female or by finding a mate with an unoccupied territory.

One primate strategy is to line up extra fathers. Among common marmosets and several species of tamarins, females mate with several males, all of which help rear her young. As primatologist Charles T. Snowdon points out, in three of the four genera of Callitrichidae *(Callithrix, Saguinus,* and *Leontopithecus),* the more adult males the group has available to help, the more young survive. Among many of these species, females ovulate just after giving birth, perhaps encouraging males to stick around until after babies are born. (In cotton-top tamarins, males also undergo hormonal changes that prepare them to care for infants at the time of birth.) Among cooperative breeders of certain other species, such as wolves and jackals, pups born in the same litter can be sired by different fathers.

Human mothers, by contrast, don't ovulate again right after birth, nor do they produce offspring with more than one genetic father at a time. Ever inventive, though, humans solve the problem of enlisting help from several adult males by other means. In some cultures, mothers rely on a peculiar belief that anthropologists call partible paternity—the notion that a fetus is built up by contributions of semen from all the men with whom women have had sex in the ten months or so prior to giving birth. Among the Canela, a matrilineal tribe in Brazil studied for many years by William Crocker of the Smithsonian Institution, publicly sanctioned intercourse between women and men other than their husbands—sometimes many men—takes place during villagewide ceremonies. What might lead to marital disaster elsewhere works among the Canela because the men believe in partible paternity. Across a broad swath of South America—from Paraguay up into Brazil, westward to Peru, and northward to Venezuela—mothers rely on this convenient folk wisdom to line up multiple honorary fathers to help them provision both themselves and their children. Over hundreds of generations, this belief has helped children thrive in a part of the world where food sources are unpredictable and where husbands are as likely as not to return from the hunt empty-handed.

The Bari people of Venezuela are among those who believe in shared paternity, and according to anthropologist Stephen Beckerman, Bari children with more than one father do especially well. In Beckerman's study of 822 children, 80 percent of those who had both a "primary" father (the man married to their mother) and a "secondary" father survived to age fifteen, compared with 64 percent survival for those with a primary father alone. Not surprisingly, as soon as a Bari woman suspects she is pregnant, she accepts sexual advances from the more successful fishermen or hunters in her group. Belief that fatherhood can be shared draws more men into the web of possible paternity, which effectively translates into more food and more protection.

But for human mothers, extra mates aren't the only source of effective help. Older children, too, play a significant role in family survival. University of Nebraska anthropologists Patricia Draper and Raymond Hames have just shown that among !Kung hunters and gatherers living in the Kalahari Desert, there is a significant correlation between how many children a parent successfully raises and how many older siblings were on hand to help during that person's own childhood.

Older matrilineal kin may be the most valuable helpers of all. University of Utah anthropologists Kristen Hawkes and James O'Connell and their UCLA colleague Nicholas Blurton Jones, who have demonstrated the important food-gathering role of older women among Hazda hunter-gatherers in Tanzania, delight in explaining that since human life spans may extend for a few decades after menopause, older women become available to care for—and to provide vital food for—children born to younger kin. Hawkes, O'Connell, and Blurton Jones further believe that dating from the earliest days of *Homo erectus,* the survival of weaned children during food shortages may have depended on tubers dug up by older kin.

At various times in human history, people have also relied on a range of customs, as well as on coercion, to line up allomaternal assistance—for example, by using slaves or hiring poor women as wet nurses. But all the helpers in the world are of no use if they're not motivated to protect, carry, or provision babies. For both humans and nonhumans, this motivation arises in three main ways: through the manipulation of information about kinship; through appealing signals coming from the babies themselves; and, at the heart of it all, from the endocrinological and neural processes that induce individuals to respond to infants' signals. Indeed, all primates and many other mammals eventually respond to infants in a nurturing way if exposed long enough to their signals. Trouble is, "long enough" can mean very different things in males and females, with their very different response thresholds.

For decades, animal behaviorists have been aware of the phenomenon known as priming. A mouse or rat encountering a strange pup is likely to respond by either ignoring the pup or eating it. But presented with pup after pup, rodents of either sex eventually become sensitized to the baby and start caring for it. Even a male may gather pups into a nest and lick or huddle over them. Although nurturing is not a routine part of a male's repertoire, when sufficiently primed he behaves as a mother would. Hormonal change is an obvious candidate for explaining this transformation. Consider the case of the cooperatively breeding Florida scrub-jays studied by Stephan Schoech, of the University of Memphis. Prolactin, a protein hormone that initiates the secretion of milk in female mammals, is also present in male mammals and in birds of both sexes. Schoech showed that levels of prolactin go up in a male and female jay as they build their nest and incubate eggs and that these levels reach a peak when they feed their young. Moreover, prolactin levels rise in the jays' nonbreeding helpers and are also at their highest when they assist in feeding nestlings.

As it happens, male, as well as immature and nonbreeding female, primates can respond to infants' signals, although quite different levels of exposure and stimulation are required to get them going. Twenty years ago, when elevated prolactin levels were first reported in common marmoset males (by Alan Dixson, for *Callithrix jacchus),* many scientists refused to believe it. Later, when the finding was confirmed, scientists assumed this effect would be found only in fathers. But based on work by Scott Nunes, Jeffrey Fite, Jeffrey French, Charles Snowdon, Lucille Roberts, and many others—work that deals with a variety of species of marmosets and tamarins—we now know that all sorts of hormonal changes are associated with increased nurturing in males. For example, in the tufted-eared marmosets studied by French and colleagues, testosterone levels in males went down as they engaged in caretaking after the birth of an infant. Testosterone levels tended to be lowest in those with the most paternal experience.

The biggest surprise, however, has been that something similar goes on in males of our own species. Anne Storey and colleagues in Canada have reported that prolactin levels in men who were living with pregnant women went up toward the end of the pregnancy. But the most significant finding was a 30 percent drop in testosterone in men right after the birth. (Some endocrinologically literate wags have proposed that this drop in testosterone levels is due to sleep deprivation, but this would probably not explain the parallel testosterone drop in marmoset males housed with parturient females.) Hormonal changes during pregnancy and lactation are, of course, indisputably more pronounced in mothers than in the men consorting with them, and no one is suggesting that male consorts are equivalent to mothers. But both sexes are surprisingly susceptible to infant signals—explaining why fathers, adoptive parents, wet nurses, and day-care workers can become deeply involved with the infants they care for.

Genetic relatedness alone, in fact, is a surprisingly unreliable predictor of love. What matters are cues from infants and how these cues are processed emotionally. The capacity for becoming emotionally hooked—or primed—also explains how a fully engaged father who is in frequent contact with his infant can become more committed to the infant's well-being than a detached mother will.

But we can't forget the real protagonist of this story: the baby. From birth, newborns are powerfully motivated to stay close, to root—even to creep—in quest of nipples, which they instinctively suck on. These are the first innate behaviors that any of us engage in. But maintaining contact is harder for little humans to do than it is for other primates. One problem is that human mothers are not very hairy, so a human mother not only has to position the baby on her breast but also has to keep him there. She must be motivated to pick up her baby even *before* her milk comes in, bringing with it a host of hormonal transformations.

Within minutes of birth, human babies can cry and vocalize just as other primates do, but human newborns can also read facial expressions and make a few of their own. Even with blurry vision, they engage in eye-to-eye contact with the people around them. Newborn babies, when alert, can see about eighteen inches away. When people put their faces within range, babies may reward this attention by looking back or even imitating facial expressions.

Orang and chimp babies, too, are strongly attached to and interested in their mothers' faces. But unlike humans, other ape mothers and infants do not get absorbed in gazing deeply into each other's eyes.

To the extent that psychiatrists and pediatricians have thought about this difference between us and the other apes, they tend to attribute it to human mental agility and our ability to use language. Interactions between mother and baby, including vocal play and babbling, have been interpreted as proto-conversations: revving up the baby to learn to talk. Yet even babies who lack face-to-face stimulation—babies born blind, say—learn to talk. Furthermore, humans are not the only primates to engage in the continuous rhythmic streams of vocalization known as babbling. Interestingly, marmoset and tamarin babies also babble. It may be that the infants of cooperative breeders are specially equipped to communicate with caretakers. This is not to say that babbling is not an important part of learning to talk, only to question which came first—babbling so as to develop into a talker, or a predisposition to evolve into a talker because among cooperative breeders, babies that babble are better tended and more likely to survive.

If humans evolved as cooperative breeders, the degree of a human mother's commitment to her infant should be linked to how much social support she herself can expect. Mothers in cooperatively breeding primate species can afford to bear and rear such costly offspring as they do only if they have help on hand. Maternal abandonment and abuse are very rarely observed among primates in the wild. In fact, the only primate species in which mothers are anywhere near as likely to abandon infants at birth as mothers in our own species are the other cooperative breeders. A study of cotton-top tamarins at the New England Regional Primate Research Center showed a 12 percent chance of abandonment if mothers had older siblings on hand to help them rear twins, but a 57 percent chance when no help was available. Overburdened mothers abandoned infants within seventy-two hours of birth.

This new way of thinking about our species' history, with its implications for children, has made me concerned about the future. So far, most Western researchers studying infant development have presumed that living in a nuclear family with a fixed division of labor (mom nurturing, dad providing) is the normal human adaptation. Most contemporary research on children's psychosocial development is derived from John Bowlby's theories of attachment and has focused on such variables as how available and responsive the mother is, whether the father is present or absent, and whether the child is in the mother's care or in day care. Sure enough, studies done with this model in mind always show that children with less responsive mothers are at greater risk.

It is the baby, first and foremost, who senses how available and how committed its mother is. But I know of no studies that take into account the possibility that humans evolved as cooperative breeders and that a mother's responsiveness also happens to be a good indicator of her social supports. In terms of developmental outcomes, the most relevant factor might not be how securely or insecurely attached to the mother the baby is—the variable that developmental psychologists are trained to measure—but rather how secure the baby is in relation to *all* the people caring for him or her. Measuring attachment this way might help explain why even children whose relations with their mother suggest they are at extreme risk manage to do fine because of the interventions of a committed father, an older sibling, or a there-when-you-need-her grandmother.

The most comprehensive study ever done on how nonmaternal care affects kids is compatible with both the hypothesis that humans evolved as cooperative breeders and the conventional hypothesis that human babies are adapted to be reared exclusively by mothers. Undertaken by the National Institute of Child Health and Human Development (NICHD) in 1991, the seven-year study included 1,364 children and their families (from diverse ethnic and economic backgrounds) and was conducted in ten different U.S. locations. This extraordinarily ambitious study was launched because statistics showed that 62 percent of U.S. mothers with children under age six were working outside the home and that the majority of them (willingly or unwillingly) were back at work within three to five months of giving birth. Because this was an entirely new social phenomenon, no one really knew what the NICHD's research would reveal.

The study's main finding was that both maternal and hired caretakers' sensitivity to infant needs was a better predictor of a child's subsequent development and behavior (such traits as social "compliance," respect for others, and self-control were measured) than was actual time spent apart from the mother. In other words, the critical variable was not the continuous presence of the mother herself but rather how secure infants felt when cared for by someone else. People who had been convinced that babies need full-time care from mothers to develop normally were stunned by these results, while advocates of day care felt vindicated. But do these and other, similar findings mean that day care is not something we need to worry about anymore?

Not at all. We should keep worrying. The NICHD study showed only that day care was better than mother care if the mother was neglectful or abusive. But excluding such worst-case scenarios, the study showed no detectable ill effects from day care *only* when infants had a secure relationship with parents to begin with (which I take to mean that babies felt wanted) and *only* when the day care was of high quality. And in this study's context, "high quality" meant that the facility had a high ratio of caretakers to babies, that it had the same caretakers all the time, and that the caretakers were sensitive to infants' needs—in other words, that the day care staff acted like committed kin.

Bluntly put, this kind of day care is almost impossible to find. Where it exists at all, it's expensive. Waiting lists are long, even for cheap or inadequate care. The average rate of staff turnover in day care centers is 30 percent per year, primarily because these workers are paid barely the minimum wage (usually less, in fact, than parking-lot attendants). Furthermore, day care tends to be age-graded, so even at centers where staff members stay put, kids move annually to new teachers. This kind of day care is unlikely to foster trusting relationships.

What conclusion can we draw from all this? Instead of arguing over "mother care" versus "other care," we need to make day care better. And this is where I think today's evolution-minded researchers have something to say. Impressed by just how variable child-rearing conditions can be in human societies, several anthropologists and psychologists (including Michael Lamb, Patricia Draper, Henry Harpending, and James Chisholm) have suggested that babies are up to more than just maintaining the relationship with their mothers. These researchers propose that babies actually monitor mothers to gain information about the world they have been born into. Babies ask, in effect, Is this world filled with people who are going to provide for me and help me survive? Can I count on them to care about me? If the answer to those questions is yes, they begin to sense that developing a conscience and a capacity for compassion would be a great idea. If the answer is no, they may then be asking, Can I not afford to count on others? Would I be better off just grabbing what I need, however I can? In this case, empathy, or thinking about others' needs, would be more of a hindrance than a help.

For a developing baby and child, the most practical way to behave might vary drastically, depending on whether the mother has kin who help, whether the father is around, whether foster parents are well-meaning or exploitative. These factors, however unconsciously perceived by the child, affect important developmental decisions. Being extremely self-centered or selfish, being oblivious to others or lacking in conscience—traits that psychologists and child-development theorists may view as pathological—are probably quite adaptive traits for an individual who is short on support from other group members.

If I am right that humans evolved as cooperative breeders, Pleistocene babies whose mothers lacked social support and were less than fully committed to infant care would have been unlikely to survive. But once people started to settle down—10,000 or 20,000 or perhaps 30,000 years ago—the picture changed. Ironically, survival chances for neglected children increased. As people lingered longer in one place, eliminated predators, built walled houses, stored food—not to mention inventing things such as rubber nipples and pasteurized milk—infant survival became decoupled from continuous contact with a caregiver.

Since the end of the Pleistocene, whether in preindustrial or industrialized environments, some children have been surviving levels of social neglect that previously would have meant certain death. Some children get very little attention, even in the most benign of contemporary homes. In the industrialized world, children routinely survive caretaking practices that an Efe or a !Kung mother would find appallingly negligent. In traditional societies, no decent mother leaves her baby alone at any time, and traditional mothers are shocked to learn that Western mothers leave infants unattended in a crib all night.

Without passing judgment, one may point out that only in the recent history of humankind could infants deprived of supportive human contact survive to reproduce themselves. Certainly there are a lot of humanitarian reasons to worry about this situation: one wants each baby, each child, to be lovingly cared for. From my evolutionary perspective, though, even more is at stake.

Even if we manage to survive what most people are worrying about—global warming, emergent diseases, rogue viruses, meteorites crashing into earth—will we still be human thousands of years down the line? By that I mean human in the way we currently define ourselves. The reason our species has managed to survive and proliferate to the extent that 6 billion people currently occupy the planet has to do with how readily we can learn to cooperate when we want to. And our capacity for empathy is one of the things that made us good at doing that.

At a rudimentary level, of course, all sorts of creatures are good at reading intentions and movements and anticipating what other animals are going to do. Predators from gopher snakes to lions have to be able to anticipate where their quarry will dart. Chimps and gorillas can figure out what another individual is likely to know or not know. But compared with that of humans, this capacity to entertain the psychological perspective of other individuals is crude.

The capacity for empathy is uniquely well developed in our species, so much so that many people (including me) believe that along with language and symbolic thought, it is what makes us human. We are capable of compassion, of understanding other people's "fears and motives, their longings and griefs and vanities," as novelist Edmund White puts it. We spend time and energy worrying about people we have never even met, about babies left in dumpsters, about the existence of more than 12 million AIDS orphans in Africa.

Psychologists know that there is a heritable component to emotional capacity and that this affects the development of compassion among individuals. By fourteen months of age, identical twins (who share all genes) are more alike in how they react to an experimenter who pretends to painfully pinch her finger on a clipboard than are fraternal twins (who share only half their genes). But empathy also has a learned component, which has more to do with analytical skills. During the first years of life, within the context of early relationships with mothers and other committed caretakers, each individual learns to look at the world from someone else's perspective.

And this is why I get so worried. Just because humans have evolved to be smart enough to chronicle our species' histories, to speculate about its origins, and to figure out that we have about 30,000 genes in our genome is no reason to assume that evolution has come to a standstill. As gene frequencies change, natural selection acts on the outcome, the expression of those genes. No one doubts, for instance, that fish benefit from being able to see. Yet species reared in total darkness—as are the small, cave-dwelling characin of Mexico—fail to develop their visual capacity. Through evolutionary time, traits that are unexpressed are eventually lost. If populations of these fish are isolated in caves long enough, youngsters descended from those original populations will no longer be able to develop eyesight at all, even if reared in sunlight.

If human compassion develops only under particular rearing conditions, and if an increasing proportion of the species survives to breeding age without developing compassion, it won't make any difference how useful this trait was among our ancestors. It will become like sight in cave-dwelling fish.

No doubt our descendants thousands of years from now (should our species survive) will still be bipedal, symbol-generating apes. Most likely they will be adept at using sophisticated technologies. But will they still be human in the way we, shaped by a long heritage of cooperative breeding, currently define ourselves?

NATALIE ANGIER

Of Altruism, Heroism and Nature's Gifts in the Face of Terror

FROM *THE NEW YORK TIMES*

Perhaps there is no way for science—or any other discipline—to explain why the men who planned and carried out the terrorist attacks of September 11 did what they did. But science may well have an explanation for why so many others risked their lives to try to rescue the victims and prevent more destruction. Writing only one week after that awful day, the always eloquent Natalie Angier finds that evolution has some hopeful things to say about human nature.

For the worldless, formless, expectant citizens of tomorrow, here are some postcards of all that matters today:

Minutes after terrorists slam jet planes into the towers of the World Trade Center, streams of harrowed humanity crowd the emergency stairwells, heading in two directions. While terrified employees scramble down, toward exit doors and survival, hundreds of New York firefighters, each laden with 70 to 100 pounds of lifesaving gear, charge upward, never to be seen again.

As the last of four hijacked planes advances toward an unknown but surely populated destination, passengers huddle together and plot resistance against their captors, an act that may explain why the plane fails to reach its target, crashing instead into an empty field outside Pittsburgh.

Hearing of the tragedy whose dimensions cannot be charted or absorbed, tens of thousands of people across the nation storm their local hospitals and blood banks, begging for the chance to give blood, something of themselves to the hearts of the wounded—and the heart of us all—beating against the void.

Altruism and heroism. If not for these twin radiant badges of our humanity, there would be no us, and we know it. And so, when their vile opposite threatened to choke us into submission last Tuesday, we rallied them in quantities so great we surprised even ourselves.

Nothing and nobody can fully explain the source of the emotional genius that has been everywhere on display. Politicians have cast it as evidence of the indomitable spirit of a rock-solid America; pastors have given credit to a more celestial source. And while biologists in no way claim to have discovered the key to human nobility, they do have their own spin on the subject. The altruistic impulse, they say, is a nondenominational gift, the birthright and defining characteristic of the human species.

As they see it, the roots of altruistic behavior far predate *Homo sapiens,* and that is why it seems to flow forth so readily once tapped. Recent studies that model group dynamics suggest that a spirit of cooperation will arise in nature under a wide variety of circumstances.

"There's a general trend in evolutionary biology toward recognizing that very often the best way to compete is to cooperate," said Dr. Barbara Smuts, a professor of anthropology at the University of Michigan, who has published papers on the evolution of altruism. "And that, to me, is a source of some solace and comfort."

Moreover, most biologists concur that the human capacity for language and memory allows altruistic behavior—the desire to give, and to sacrifice for the sake of others—to flourish in measure far beyond the cooperative spirit seen in other species.

With language, they say, people can learn of individuals they have never met and feel compassion for their suffering, and honor and even emulate their heroic deeds. They can also warn one another of any selfish cheaters or malign tricksters lurking in their midst.

"In a large crowd, we know who the good guys are, and we can talk about, and ostracize, the bad ones," said Dr. Craig Packer, a professor of ecology and evolution at the University of Minnesota. "People are very concerned about their reputation, and that, too, can inspire us to be good."

Oh, better than good.

"There's a grandness in the human species that is so striking, and so profoundly different from what we see in other animals," he added. "We are an amalgamation of families working together. This is what civilization is derived from."

At the same time, said biologists, the very conditions that encourage heroics and selflessness can be the source of profound barbarism as well. "Moral behavior is often a within-group phenomenon," said Dr. David Sloan Wilson, a professor of biology at the State University of New York at Binghamton. "Altruism is practiced within your group, and often turned off toward members of other groups."

The desire to understand the nature of altruism has occupied evolutionary thinkers since Charles Darwin, who was fascinated by the apparent existence of altruism among social insects. In ant and bee colonies, sterile female workers labor ceaselessly for their queen, and will even die for her when the nest is threatened. How could such seeming selflessness evolve, when it is exactly those individuals that are behaving altruistically that fail to breed and thereby pass their selfless genes along?

By a similar token, human soldiers who go to war often are at the beginning of their reproductive potential, and many are killed before getting the chance to have children. Why don't the stay-at-homes simply outbreed the do-gooders and thus bury the altruistic impulse along with the casualties of combat?

The question of altruism was at least partly solved when the British evolutionary theorist William Hamilton formulated the idea of inclusive fitness: the notion that individuals can enhance their reproductive success not merely by having young of their own, but by caring for their genetic relatives as well. Among social bees and ants, it turns out, the sister workers are more closely related to one another than parents normally are to their offspring; thus it behooves the workers to care more about current and potential sisters than to fret over their sterile selves.

The concept of inclusive fitness explains many brave acts observed in nature. Dr. Richard Wrangham, a primatologist at Harvard, cites the example of the red colobus monkey. When they are being hunted by chimpanzees, the male monkeys are "amazingly brave," Dr. Wrangham said. "As the biggest and strongest members of their group, they undoubtedly could escape quicker than the others." Instead, the males jump to the front, confronting the chimpanzee hunters while the mothers and offspring jump to safety. Often, the much bigger chimpanzees pull the colobus soldiers off by their tails and slam them to their deaths.

Their courageousness can be explained by the fact that colobus monkeys live in multimale, multifemale groups in which the males are almost always related. So in protecting the young monkeys, the adult males are defending their kin.

Yet, as biologists are learning, there is more to cooperation and generosity than an investment in one's nepotistic patch of DNA. Lately, they have accrued evidence that something like group selection encourages the evolution of traits beneficial to a group, even when members of the group are not related.

In computer simulation studies, Dr. Smuts and her colleagues modeled two types of group-living agents that would behave like herbivores: one that would selfishly consume all the food in a given patch before moving on, and another that would consume resources modestly rather than greedily, thus allowing local plant food to regenerate.

Researchers had assumed that cooperators could collaborate with genetically unrelated cooperators only if they had the cognitive capacity to know goodness when they saw it.

But the data suggested otherwise. "These models showed that under a wide range of simulated environmental conditions you could get selection for prudent, cooperative behavior," Dr. Smuts said, even in the absence of cognition or kinship. "If you happened by chance to get good guys together, they remained together because they created a mutually beneficial environment."

This sort of win-win principle, she said, could explain all sorts of symbiotic arrangements, even among different species—like the tendency of baboons and impalas to associate together because they use each other's warning calls.

Add to this basic mechanistic selection for cooperation the human capacity to recognize and reward behaviors that strengthen the group—the tribe, the state, the church, the platoon—and selflessness thrives and multiplies. So, too, does the need for group identity. Classic so-called minimal group experiments have shown that when people are gathered together and assigned membership in arbitrary groups, called, say, the Greens and the Reds, before long the members begin expressing amity for their fellow Greens or Reds and animosity toward those of the wrong "color."

"Ancestral life frequently consisted of intergroup conflict," Dr. Wilson of SUNY said. "It's part of our mental heritage."

Yet he does not see conflict as inevitable. "It's been shown pretty well that where people place the boundary between us and them is extremely flexible and strategic," he said. "It's possible to widen the moral circle, and I'm optimistic enough to believe it can be done on a worldwide scale."

Ultimately, though, scientists acknowledge that the evolutionary framework for self-sacrificing acts is overlaid by individual choice. And it is there, when individual firefighters or office workers or airplane passengers choose the altruistic path, that science gives way to wonder.

Dr. James J. Moore, a professor of anthropology at the University of California at San Diego, said he had studied many species, including many different primates. "We're the nicest species I know," he said. "To see those guys risking their lives, climbing over rubble on the chance of finding one person alive, well, you wouldn't find baboons doing that." The horrors of last week notwithstanding, he said, "the overall picture to come out about human nature is wonderful."

"For every 50 people making bomb threats now to mosques," he said, "there are 500,000 people around the world behaving just the way we hoped they would, with empathy and expressions of grief. We are amazingly civilized."

True, death-defying acts of heroism may be the province of the few. For the rest of us, simple humanity will do.

JULIAN DIBBELL

Pirate Utopia

FROM *FEED*

Steganography is the ancient art of hiding messages. In the computer age, steganography has gone digital, with messages hidden in the same zeroes and ones that are used to construct otherwise unassuming images or text. For the tech writer Julian Dibbell, however, steganography is not a shadowy craft but a symbol of the now vanished utopianism that surrounded the early, heady days of the Internet.

In early 2001, *USA Today* broke the shocking news that Osama bin Laden's terrorist organization had infiltrated the world's supply of Web porn, hiding messages for its global operatives deep within the digits of pictures posted on Godless Western triple-X sites. For historically minded readers, the article afforded a moment of wonder at the depths of the national-security establishment's Cold War nostalgia and the media's willingness to indulge it. There, once again, was the old familiar intimacy of the alleged subversion, the thrilling suggestion that the enemy might lurk among us everywhere, sneaking into our bedrooms and our cubicles under cover of cultural trash. "You very well could have a photograph and image with the time and information of an attack sitting on your computer, and you would never know it," one cyberwar expert told *USA Today*'s reporter.

I confess, though, that I got a bit nostalgic myself when I read the story.

Not for the Cold War—I was born too late to enjoy it in the fullness of its Eisenhowerian heyday—but for its Bush-era aftermath. Specifically, I found myself looking back with melancholy fondness upon the summer of 1992, a moment perhaps not equal to the summer of '67 in its hold on the memories of a generation but one which for me, at least, holds much the same sense of freedom and promise in the bubble of its recollection. It was a moment, after all, when radical political thought was just beginning to adjust to the reality of '89, just rising to the challenge of imagining the possibilities that that reality implied. It was a moment, as well, when the Internet, long a distant, reverie-inspiring rumor known firsthand only to military contractors and computer-science majors, was just starting to enter the lives of the rest of us. But most importantly, perhaps, and certainly not at all coincidentally, it was the moment when I first learned it was possible to do with digital communications what Osama bin Laden is now reported to have done.

THE TECHNICAL NAME for it is steganography, from the Greek for "covered writing." It is the art of keeping communications undetected, and it is not to be confused with the related discipline of cryptography. Cryptography assumes that messages will be intercepted and uses codes and ciphers to make sure they can't be understood if they are. But steganography aims for a deeper sort of cover: it assumes that if the message is so much as found to exist, the game is over.

Steganographic techniques are as old, at least, as Herodotus, who documented their use among the Greeks of the fifth century B.C. In Book Seven of *The Histories,* he writes that when Demaratus, a Spartan living in Persia, got wind of the emperor Xerxes' plan to invade Greece, he contrived to tip his compatriots off by sending them a stegotext: he took a pair of folding wooden message tablets, scraped the wax writing surface off them, wrote his message on the wood, then covered his message back over with wax. Persian counter-intelligence never suspected a thing. Nor did the Persians have a clue when Histiaeus of Miletus sent a similarly subversive letter home tattooed onto the scalp of a trusted slave. The messenger arrived safely at his destination and said no more than what he'd been instructed to say: "Shave my head and look thereon."

In contrast with cryptography, a field long given over to high math and puzzle-making abstraction, steganography was always more or less a materials science, its history florid with the range of substances and gadgetry used at one time or another to conceal communications. Simon Singh's *The Code Book* relates that in the first century A.D., Pliny the Elder explained how the milk of the thithymallus plant dried to transparency when applied to paper but darkened to brown when subsequently heated, thus recording one of the earliest recipes for invisible ink. The ancient Chinese wrote notes on small pieces of silk that they then wadded into little balls and coated in wax, to be swallowed by a messenger and retrieved, I guess, at the messenger's gastrointestinal convenience. The sixteenth-century Italian scientist Giovanni Porta proposed a steganographic scheme involving hard-boiled eggs: Write on the shell with a vinegar-and-alum solution and your message passes through to the surface of the egg white, where it can't be read until the shell is peeled away.

In the 1860s, the technology of microfilm was perfected, ushering in a golden age of stealth that reached its manic peak in World War II, when German spies began making heavy use of the microdot. A page of text shrunk down to a one-millimeter speck of film, the microdot could be pasted almost indetectably into any humdrum business letter, hiding out in the shallow well of a typewritten period or comma. With swarms of them passing through the mails, the U.S. government went sort of nuts and started seeing hidden messages everywhere. By the end of the war, censors had either prohibited or tampered with flower deliveries, commercial-radio song requests, broadcast weather reports, postal chess games, children's drawings on their way to Grandma, knitting instructions, and anything else that might too easily encode Axis intelligence. At one point, an entire shipment of watches was held up so officials could spin the dials and wipe out any messages hidden in the positions of the hands.

In the half century since the war, however, the sweeping digitization of communications tech has caused steganography at last to veer away from the material world, joining cryptography in the realms of math and abstraction. As ever, of course, information still reaches the mind in the form of concrete sensory stimuli—as light and sound—but increasingly it is the universal code of binary numbers that shapes that information, and it is in the numbers that hackers and spooks have looked for places to hide still more information.

They have found those places. Digital stego takes a number of forms, but most of them are variations on the most popular technique (the one most likely to be used by bin Laden and company, in fact, if they are really using any at all). It's called "least significant bit" steganography, and to understand how it works you have to think a little about how digital media work. Consider the dots of light that compose an image on a computer screen. Or the slivers of sound that blend together to form a song as it streams from a CD player. Each dot, each sliver, is recorded as a small number of bits—ones and zeroes—maybe sixteen of them, maybe twenty-four. Most of the bits specify crucial information about the color or tone of the sensory blip they represent, but a few stand for nuances the average eye or ear won't even pick up. These latter, the so-called least significant bits (LSBs), are effectively indistinguishable from noise—from the random hiss and blur that shows up in any information channel. And since properly encrypted data is also indistinguishable from noise, it turns out that an untouched digital copy of a song or photograph is very hard to tell from a copy whose LSBs have been overwritten with a well-enciphered message. The collected LSBs of a Radiohead CD, for instance, might encode a completely undetectable blueprint of the Stealth bomber. This sentence might fit imperceptibly inside a small JPEG image embedded in a Web site somewhere.

Indeed, *this* sentence actually does fit inside a small JPEG image on a Web site somewhere—an image on my own Web site, to be precise, which you can see by pointing your browser to <http://www.juliandibbell.com/stego.jpg>. Go on, take a look: The previous sentence is in there right now, braided into the bits with the aid of a free stego program I downloaded, a program called Jsteg.

Or maybe it isn't.

And how would you ever know?

I FIRST LEARNED what digital steganography was, as I said, in 1992. I learned it from a man named Timothy C. May, a forty-year-old microchip physicist who had retired from Intel a wealthy man several years earlier. May was soon to land on the cover of *Wired* magazine's début issue as one of the founding members of the Cypherpunks, a mostly online collective of mostly hardcore technolibertarians united to promote the spread of digital encryption tools in what was then just beginning to be called cyberspace. Cryptography was itself about to hit a sort of big time. In mid-1993, the freshly elected Clinton Administration would make its mark on technological history by announcing a new digital encryption standard, Clipper, equipped with an FBI-accessible back door aimed at heading off precisely what the Cypherpunks aimed to achieve: the complete invulnerability of private electronic communications to government surveillance.

Grippingly chronicled in Steven Levy's new history of digital-age encryption, *Crypto,* the resulting political struggle between Clipper's backers at the FBI, CIA, and NSA, on the one hand, and a coalition of hackers, civil-liberties advocates, and software industrialists, on the other, became the Internet's first great privacy crisis—and arguably never ended. Today's Cypherpunk diehards, for instance, mostly dismissed the recent news about bin Laden's porn habit as yet another attempt by the Three-Letter Agencies to soften up the populace for restrictions on crypto, and they may well have been right. Though the TLAs long ago lost the Clipper battle (unbreakable crypto has become the infrastructural backbone of e-commerce), their panic-mongering pronouncements on terrorists' use of crypto suggests they may not have given up hopes of winning the war.

But back in the summer of '92, all that was future history. Cryptography was still just an obscurely fascinating field I had read about in an old paperback I'd picked up secondhand (David Kahn's crypto-history bible, *The Codebreakers),* and Tim May was just a guy whose obscurely fascinating remarks on the subject I had come across on my local bulletin board's Usenet feed. Out of professional curiosity, I got him on the phone one day and didn't get off for another forty-five minutes, during which time I did very little of the talking. "When Tim May thought about crypto," writes Levy in his chapter on the Cypherpunks, "it was almost like dropping acid"—and when he *talked* about crypto it was almost like you'd drunk from the same spiked punch bowl. He conjured visions of a world in which entire virtual communities disappeared into the dark freedom of impenetrable privacy. A world in which all markets were black, untaxable, and in which the tyranny of the nation-state therefore withered inexorably away. A world always just beneath the surface of this one but at the same time light-years distant, safe behind a wall of math so thick even the NSA's most powerful computers could never crunch through it.

"You can get further away in cyberspace than you could in going to Alpha Centauri," May told me. "Some of these things sound like just a bunch of fucking numbers, but what they really are is they're things which in computability space take more energy to get to than to drive a car to Andromeda."

In a certain light, of course, May's prophecies were just an extreme form of sci-fi geekdom and really not quite my cup of Kool-Aid. As it turned out (and as I should not have been surprised to learn), May was an energetic adept of Extropianism, a scientistic California semi-cult devoted to Ayn Rand, immortality through cryogenics, and the Gnostic dream of uploading human consciousness into computers—all of which was a bit much to swallow for a club-hopping young New Yorker still tipsy on the soft-Marxist politics and anti-positivist literary theory he'd imbibed in college.

And yet at its core May's "crypto anarchist" vision (his phrase) resonated deeply with some of the latest wrinkles in soft Marxism and literary anti-positivism coming out of the theory mills. In particular, it seemed almost to have taken direct inspiration from Hakim Bey's lively anarcho-Baudrillardian classic *The Temporary Autonomous Zone,* a then recently published tract celebrating not the final utopias yearned for in traditional radicalism (and finally junk-piled by the events of '89) but the brief liberatory grace of failed uprisings, transient communes, excellent parties, and other carnivalesque moments smuggled out from under the controlling gaze of the state. Itself inspired by the shadow history of "pirate utopias"—tropical island havens of democratic lawlessness to which eighteenth-century buccaneers repaired between bouts of bloody economic parasitism—Hakim Bey's notion of the temporary autonomous zone, or TAZ, embraced with guarded enthusiasm the possibility of virtual outlaw colonies taking quiet shape amid the burgeoning connections of the world's computer networks. "Islands in the Net," Bey called them, hip enough to borrow the phrase from arch-cyberpunk Bruce Sterling's latest novel.

As it happens, *The Temporary Autonomous Zone* has recently been republished by the MIT Press in *Crypto Anarchy, Cyberstates, and Pirate Utopias,* an anthology of essays on "emerging political structures in cyberspace" that, as its title indicates, includes a couple of Tim May's visionary rants as well. But it didn't take that juxtaposition for me to see the connection. The way I heard it on the phone that day nine years ago, May's project was really just a more pragmatic version of Bey's—an attempt to frame the prospects for online autonomous zones in the only discursive terms Net culture had ever really respected: rough consensus and running code. May laid a whole shopping list of cool hacks and peer-to-peer conspiracies on me—anonymous remailers, untraceable e-cash, zero-knowledge markets in corporate secrets, pirated software, and murder contracts—and each one grabbed my attention like he was telling me the precise date and channel the revolution would be televised on.

But, curiously, none of them captured my imagination like a certain very cool but essentially minor hack he mentioned toward the end of our conversation, almost in passing. The hack was LSB stego, and only now, having seen in the pages of *USA Today* how the very thought of it affects grown national-security experts, can I quite articulate what so thrilled me about it back then. It was the idea that any piece of information I came across on the Net might secretly hold within it yet another piece, which for that matter might contain another one in its turn, and so on and on. It was the way this idea seemed itself to contain all the headiest meanings swirling through that historical moment—the way it metaphorized both Tim May's and Hakim Bey's schemes for hiding micro-utopias beneath the surface of the social. The way it literalized how pregnant with possible futures the post–Cold War world had become and the Net was then becoming. The way it even somehow conceptually resembled the iconic cybercultural image of the day—the fractal Mandelbrot set, with its levels within levels of intricately chaotic structure, swirling psychedelically on a million dorm-room computer screens like so many digital-age lava lamps, blowing minds the same way Tim May had just blown mine.

It was stego. And it was so—I can't think how else to put it—so very 1992.

THESE DAYS, steganography is not very 1992 at all—and, needless to say, neither are these days.

The moment when the Net could serve as an empty screen to project dreams of radical autonomy onto has long since passed. Already in 1995, old-school Net evangelist John Perry Barlow was drawing snickers from the post-soft-Marxist set for his "Declaration of the Independence of Cyberspace," a classic bit of mailing-list bluster that proclaimed to the governments of the industrial world ("you weary giants of flesh and steel") that their laws and their intellectual-property regimes held no sovereignty over "Cyberspace, the new home of Mind." Nowadays, the snickerers would probably find it a balm to be able to entertain the fantasy for just a moment or two, but at this point that is surely beyond the powers of even Barlow's expansive imagination.

And as for those intellectual-property regimes—well, let's just say that if ever there was a genuine pirate utopia online, it was Napster, and that if ever there was an online equivalent to the appearance of His Majesty's gunships in the waters off the last genuine pirate utopia before that—Blackbeard's beachfront shantytown at Nassau in the Bahamas—it was the Ninth Circuit Court of Appeals judgment that effectively handed Napster's ass to the record industry. It's probably too soon to predict with any precision the future of Napster's fifty-eight-million-member "community," but anyone planning on giving it a shot could do worse than look for precedent in the Nassau colony's overnight collapse at the first appearance of the Law. "Blackbeard and 'Calico Jack' Rackham and his crew of pirate women moved on to wilder shores and nastier fates," writes Hakim Bey in one of the gloomier moments of his *TAZ*, "while others meekly accepted the Pardon and reformed."

In the midst of all these less-than-thrilling changes, steganography, sadly and ironically enough, remains a mirror of its times. Rumors of terrorist applications notwithstanding, the majority of interest in steganographic techniques these days comes not from criminal and/or libertarian hackers looking for virtual hidey-holes but from corporate researchers looking for a way to put digital locks on intellectual property. It turns out stego's pretty good for that: by weaving encrypted copyright information and serial numbers into the binary code of photos, songs, and movies, rights owners can sear a sort of virtual brand into their property. These digital watermarks, as they're called, have the usefully paired qualities of being (a) difficult to erase and (b) easy to trace. Some watermark schemes haunting the research journals propose sending mark-hunting Web spiders out to troll for content pirates and ID them for prosecution. One imagines fleets of radio-sensitive vans cruising urban business districts, scanning for stego-marked electromagnetic emissions that would give away the presence of pirated software in nearby offices.

Whether any of these plans will take hold in the real world remains to be seen—and may be beside the point. As in '92, so in '01 it's not so much the quality of the tech as the quality of the dreams behind the tech that give the tenor of the times. And if pirate-chasing stego systems is where the dreams lie these days, then the times they are depressing.

And yet I can't help hoping. The anarcho-visionary energy of the Cypherpunk movement may have dissipated—siphoned off into the mundane importance of making the world safe for online credit-card transactions—but here and there imaginative hackers are still drawn to stego. They code their cool little apps: the aforementioned Jsteg, which hides data in JPEGs, and MP3Stego, which does the same with MP3s. There is Snow, which embeds information in the white space of text documents. And the hilarious Spam Mimic, which translates brief messages into the semi-coherent raving of junk-email-speak. They're so much fun, these programs, that I suppose it's possible that fun is really all they're about. And yet, all the same, their presence makes me suspect there's still an urge out there to drop off the radar, to find that dark freedom Tim May used to rant about.

Heaven knows, they inspire my own small dreams, chief among them the dream that somewhere, someone has embedded the text of Hakim Bey's tribute to pirate utopias in an album's worth of Metallica MP3s and thrown it up on Napster, thereby both flipping the bird to the RIAA and at the same time, by ineradicably marking his own connection to an act of piracy, daring them to come after him. I'd do it myself if I were hacker enough. Or, more to the point, I'd do it if I didn't prefer the fantasy that someone besides me has already been moved by the spirit of '92 to make it happen. It's not that I'm embarrassed by my own political nostalgia. After all, if Hakim Bey has the right idea, then nostalgia is no longer the radical sin it used to be: Utopias exist in the past as well as the future now, and yearning for the ones gone by is pretty much the same as hoping for the next one to come along.

I'd just like to think I'm not the only one doing the hoping, is all.

Carolyn Meinel

Code Red for the Web

FROM *SCIENTIFIC AMERICAN*

The dot-com bubble may have burst, but that hasn't stopped the Internet from growing increasingly pervasive. The more we rely on it, the more vulnerable we are to cyberattacks—viruses, worms, and a host of other demons from the hackers' armamentarium. One such attack occurred in the summer of 2001, when the Code Red worm infiltrated hundreds of thousands of servers. Carolyn Meinel, a "white-hat" hacker, who writes about computer security, suggests that Code Red may be a disquieting preview of coming infractions.

"Imagine a cold that kills. It spreads rapidly and indiscriminately through droplets in the air, and you think you're absolutely healthy until you begin to sneeze. Your only protection is complete, impossible isolation."

Jane Jorgensen, principal scientist at Information Extraction & Transport in Arlington, Virginia, which researches Internet epidemiology for the Defense Advanced Research Projects Agency, isn't describing the latest flu outbreak but an affliction that affects the Web. One such computer disease emerged in July and August of 2001, and it has computer security researchers more worried about the integrity of the Internet than ever before. The consternation was caused by Code Red, a Web worm, an electronic ailment akin to computerized snakebite. Code Red infects Microsoft Internet Information

Servers (IIS). Whereas home computers typically use other systems, many of the most popular Web sites run on IIS. In two lightning-fast strikes, Code Red managed to infiltrate hundreds of thousands of IIS servers in only a few hours, slowing the Internet's operations. Although Code Red's effects have waned, patching the security holes in the estimated six million Microsoft IIS Web servers worldwide and repairing the damage inflicted by the worm have cost billions of dollars.

What really disturbs system administrators and other experts, however, is the idea that Code Red may be a harbinger of more virulent Internet plagues. In the past, Web defacements were perpetrated by people breaking into sites individually—the cyberwarfare equivalent of dropping propaganda leaflets on targets. But computer researchers dread the arrival of better-designed automated attack worms that could degrade or even demolish the World Wide Web.

Further, some researchers worry that Code Red was merely a test of the type of computer programs that any government could use to crash the Internet in times of war. The online skirmishes in the spring of 2001 over the U.S. spy plane incident with China emphasize the dangers. Full-scale cyberwarfare could cause untold damage to the industrialized world. These secret assaults could even enlist your PC as a pawn, making it a "zombie" that participates in the next round of computerized carnage.

Save for the scales on which these computer assaults are waged, individual hacking and governmental cyberwarfare are essentially two sides of the same electronically disruptive coin. Unfortunately, it's hard to tell the difference between them until it's too late.

Often popularly lumped in with viruses, Code Red and some similar pests such as Melissa and SirCam are more accurately called worms in the hacker lexicon. Mimicking the actions of its biological namesake, a software virus must incorporate itself into another program to run and replicate. A computer worm differs in that it is a self-replicating, self-contained program. Worms frequently are far more infectious than viruses. The Code Red worm is especially dangerous because it conducted what are called distributed denial of service (DDoS) attacks, which overwhelm Internet computers with a deluge of junk communications.

During its peak in July 2001, Code Red menaced the Web by consuming its bandwidth, or data-transmission capacity. "In cyberwarfare, bandwidth is a weapon," says Greggory Peck, a senior security engineer for FC Business Systems in Springfield, Virginia, which works to defend U.S. government clients against computer crime. In a DDoS attack, a control computer commands many zombies to throw garbage traffic at a victim in an attempt to use up all available bandwidth. This kind of assault first made the news in 2000 when DDoS attacks laid low Yahoo, eBay and other dot-coms.

These earlier DDoS incidents mustered just hundreds to, at most, thousands of zombies. That's because attackers had to break into each prospective zombie by hand. Code Red, being a worm, spreads automatically—and exponentially. This feature provides it with hundreds of times more zombies and hence hundreds of times more power to saturate all available Internet bandwidth rapidly.

The initial outbreak of Code Red contagion was not much more than a case of the sniffles. In the five days after it appeared on July 12, it reached only about 20,000 out of the estimated half a million susceptible IIS computer servers. It wasn't until five days afterward that Ryan Permeh and Marc Maiffret of eEye Digital Security in Aliso Viejo, California, a supplier of security software for Microsoft servers, discovered the worm and alerted the world to its existence.

On July 19 the worm reemerged in a more venomous form. "More than 359,000 servers were infected with the Code Red worm in less than 14 hours," says David Moore, senior technical manager at the Cooperative Association for Internet Data Analysis in La Jolla, California, a government- and industry-supported organization that surveys and maps the Net's server population. The traffic jam generated by so many computers attempting to co-opt other machines began to overload the capacity of the Internet. By midafternoon, the Internet Storm Center at incidents.org—the computer security industry's watchdog for Internet health—was reporting "orange alert" status. This is one step below its most dire condition, red alert, which signals a breakdown.

Then, at midnight, all Code Red zombies quit searching for new victims. Instead they all focused on flooding one of the servers that hosts the White House Web site with junk connections, threatening its shutdown. "The White House essentially turned off one of its two DNS servers, saying that any requests to whitehouse.gov should be rerouted to the other server," says Jimmy Kuo, a Network Associates McAfee fellow who assisted the White House in finding a solution. Basically, the system administrators dumped all communications addressed to the compromised server. As it turned out, Code Red couldn't cope with the altered Internet protocol address and waged war on the inactive site. "The public didn't notice anything, because any requests went to the other server," Kuo says.

By the close of July 20, all existing Code Red zombies went into a preprogrammed eternal sleep. As the worms lodge only in each computer's RAM memory, which is purged when the machine shuts down, all it took was a reboot to eradicate their remnants. Case closed.

Or was it? A few days later analysts at eEye revealed that if someone were to release a new copy of Code Red at any time between the first through the 19th day of any month (the trigger dates coded in by the original hacker), the infection would take off again.

Over the next 10 days computer security volunteers worked to notify Microsoft IIS users of the vulnerability of their servers. On July 29 the White House held a press conference to implore people to protect their IIS servers against Code Red's attacks. "The mass traffic associated with this worm's propagation could degrade the functioning of the Internet," warned Ronald L. Dick, director of the FBI's National Infrastructure Protection Center. By the next day Code Red was all over the news.

The second coming of Code Red was, as expected, weaker than the first. On August 1, it infected approximately 175,000 servers—nearly all those susceptible and about half the total of the previous episode. A slower infection rate and fewer vulnerable servers held Internet disruptions to a minimum. After a while, the second attack subsided.

But that was not the end. Yet another worm was unleashed on August 4 using the same break-in method as Code Red. The new worm, dubbed Code Red II, installed a backdoor allowing a master hacker to direct the activities of victim computers at will. The worm degraded intranets with "arp storms" (floods of Ethernet packets) and hunted for new victims. In short order, Code Red II disabled parts of the Web-based e-mail provider Hotmail, several cable and digital subscriber-line (DSL) Internet providers and part of the Associated Press news distribution system. As time passed, Code Red II managed to infect many corporate and college intranets. Halfway through August, Code Red II disabled some Hong Kong government internal servers. The most common victim computers were personal Web servers run by Windows 2000 Professional. This rash of disruptions prompted incidents.org to again declare an orange alert. Experts estimate that 500,000 internal servers were compromised.

In mid-August, Computer Economics, a security research company, said that Code Red had cost $2 billion in damage. By the time it is fully purged from the Internet, the computer attack will probably rank among the most expensive in history. Nearly $9 billion was spent to fight 2000's LoveLetter virus, and 1999's Melissa worm assault cost $1 billion to repair.

Of course, Code Red isn't the only worm out there. Some of them are aimed at home computers. A worm called W32/Leaves, for example, permits a remote attacker to control infected PCs in a coordinated fashion, enabling synchronized waves of attacks. (Although Code Red II allows this possibility as well, it lacks the coding that enables remote control.) The Computer Emergency Response Team, a federally funded watchdog organization at Carnegie Mellon University, has received reports of more than 23,000 W32/Leaves zombies. The current total is unknown, but as W32/Leaves continues to propagate, the infected population will probably grow significantly. In July 2001, Britain's Scotland Yard charged an unidentified 24-year-old man with creating W32/Leaves.

"Almost any computer, operating system or software you may buy contains weaknesses that the manufacturer knows lets hackers break in," says Larry Leibrock, a leading researcher in computer forensics and associate dean for technology of the business school of the University of Texas at Austin. Future "federal regulation could require that vendors take the initiative to contact customers and help them upgrade their products to fix security flaws," he continues. "Today, however, it is up to each consumer to hunt down and fix the many ways hackers and cyberwarriors exploit to abuse their computers."

World Cyberwars

Beyond the threat posed by malicious hacker programs is the danger of Internet attacks conducted in a concerted fashion by top computer talent spurred to act by international events. The cyber-battles that broke out over the collision of a Chinese fighter plane that collided with a U.S. Navy EP-3E spy plane in April 2001, give a hint of how such a conflict might play out.

According to accounts in the press, the hacker exchanges began when negotiations for the release of American hostages stalled. On April 9 and 10, attackers defaced two Chinese Web sites with slurs, insults and even threats of nuclear war. During the following week, American hackers hit dozens more Chinese sites. Those supporting China responded by disfiguring one obscure U.S. Navy Web site.

China, however, held a weapon in reserve. In late March the National Infrastructure Protection Center had warned of a new worm on the loose: the 1iOn Worm. Lion, the hacker who founded the hacker group HUC (Honkers Union of China), has taken credit for writing it. Unlike the initial Code Red's preprogrammed zombies, 1iOn's zombies accept new commands from a central computer. Also, 1iOn infects Linux computers, which means it can masquerade as any computer on the Net. This property makes it hard to track down infected servers.

Meanwhile pro-U.S. hack attacks escalated. The official Chinese publication, *People's Daily,* reported that "by the end of April over 600 Chinese Web sites had come under fire." In contrast, Chinese hackers had hit only three U.S. sites during the same period.

In the next few days the Chinese hacker group's HUC, Redcrack, China Net Force, China Tianyu and Redhackers assaulted a dozen American Web sites with slogans such as "Attack anti-Chinese arrogance!" On the first of May several DDoS strikes were initiated. Over the next week Chinese hackers took credit for wrecking about 1,000 additional American Web sites.

On May 7 China acknowledged its responsibility for the DDoS attacks and called for peace in a *People's Daily* news story. It ran: "The Chinese hackers were also urged to call off all irrational actions and turn their enthusiasm into strength to build up the country and safeguard world peace."

U.S. law-enforcement agencies, the White House and U.S. hacker organizations never objected to the American side of this cyberconflict, although the FBI's infrastructure center had warned of "the potential for increased hacker activity directed at U.S. systems."

How to Wage Covert Cyberwarfare

In view of the spy-plane episode, some commentators have wondered whether the U.S. federal government encouraged American hackers to become agents of cyberwar. After all, the U.S. has worked with private groups to wage covert warfare before, as in the Iran/Contra scandal. And links between the two communities have been reported. It's difficult, however, to say exactly how strong the connection between hackers and the government might be. Clearly, the murky world of hacking doesn't often lend itself to certainty. And because it is the policy of the U.S. National Security Agency and various Defense Department cyberwarfare organizations not to comment on Web security matters, these relationships cannot be confirmed. Still, the indications are at least suggestive.

Consider the history of Fred Villella, now an independent computer consultant. According to numerous press reports and his own statements, Villella took part in counterterrorism activities in the 1970s. In 1996 he hired hackers of the Dis Org Crew to help him conduct training sessions on the hacker threat for federal agencies. This gang also helps to staff the world's largest annual hacker convention, Def Con.

Erik Ginorio (known to the hacker world as Bronc Buster) publicly took credit for defacing a Chinese government Web site on human rights in October 1998. This act is illegal under U.S. law. Not only was Ginorio not prosecuted, he says Villella offered him a job. Villella could not be reached for comment.

In another hacker-government connection, Secure Computing in San Jose, California, became a sponsor of Def Con in 1996. According to its 10-K reports to the U.S. Securities and Exchange Commission, Secure Computing was created at the direction of the National Security Agency, the supersecret code-breaking and surveillance arm of the U.S. government. Two years after that, Secure Computing hired the owner of Def Con, Jeff Moss. Several former Villella instructors also staffed and managed Def Con.

Questionable things happen at Def Cons. At the 1999 Def Con, for example, the Cult of the Dead Cow, a hacker gang headquartered in Lubbock, Texas, put on a mediagenic show to promote its Back Orifice 2000 break-in program. Gang members extolled the benefits of "hacking to change the world," claiming that eight-year-olds could use this program to break into Windows servers.

Meanwhile Pieter Zatko, a Boston-area hacker-entrepreneur and a member of the gang, was onstage promoting a software plug-in for sale that increased the power of Back Orifice 2000. According to the Cult's Web site, Back Orifice 2000 was downloaded 128,776 times in the following weeks. On February 15, 2000, President Bill Clinton honored Zatko for his efforts by inviting him to the White House Meeting on Internet Security. Afterward Zatko remained with a small group to chat with the president.

Every year Def Con holds a "Meet the Feds" panel. At its 2000 meeting, Arthur L. Money, former U.S. assistant secretary of defense for command, control, communications and intelligence, told the crowd, "If you are extremely talented and you are wondering what you'd like to do with the rest of your life—join us and help us educate our people [government personnel]."

In 1997 Moss launched the Black Hat Briefings. In hacker lingo, a black hat is a computer criminal. Theoretically, these meetings are intended to train people in computer security. They bear considerable similarity to Def Con, however, only with a $1,000 price tag per attendee. Their talks often appear to be more tutorials in how to commit crime than defend against it. For example, at one session attendees learned about "Evidence-Eliminator," billed as being able to "defeat the exact same forensic software as used by the U.S. Secret Service, Customs Department and Los Angeles Police Department."

It should be noted that the U.S. government does have a formal means to wage cyberwar. On October 1, 2000, the U.S. Space Command took charge of the Computer Network Attack mission for the Department of Defense. In addition, the U.S. Air Force runs its Information Warfare Center research group, located in San Antonio.

Given these resources, why would the U.S. and China encourage cybermilitias? "It's very simple. If you have an unofficial army, you can disclaim them at any time," says Mark A. Ludwig, author of *The Little Black Book of Computer Viruses* and *The Little Black Book of E-mail Viruses.* "If your military guys are doing it and you are traced back, the egg's on your face."

Wherever it came from, the Code Red assault was just a taste of what a concerted cyberwar could become. "I think we can agree that it was not an attempt at cyberwar. The worm was far too noisy and easily detected to be much more than graffiti/vandalism and a proof-of-concept," says Harlan Carvey, an independent computer security consultant based in Virginia.

Stuart Staniford, president of Silicon Defense in Eureka, California, notes, however, that if the zombie computers "had a long target list and a control mechanism to allow dynamic retargeting, [they] could have DDoSed [servers] used to map addresses to contact information, the ones used to distribute patches, the ones belonging to companies that analyze worms or distribute incident response information. Code Red illustrates that it's not much harder for a worm to get *all* the vulnerable systems than it is to get some of them. It just has to spread fast enough."

Code Red already offers deadly leverage for nefarious operators, according to Marc Maiffret, who bills himself as "chief hacking officer" of eEye: "The way the [Code Red] worm is written, it could allow online vandals to build a list of infected systems and later take control of them."

Get enough zombies attacking enough targets, and the entire Internet could become unusable. Even the normal mechanisms for repairing it—downloads of instructions and programs to fix zombies and the ability to shut off rogue network elements—could become unworkable. In addition, hackers constantly publicize new ways to break into computers that could be used by new worms. A determined attacker could throw one devastating worm after another into the Internet, hitting the system every time it struggled back and eventually overpowering it.

"We know how [crashing the Internet] can be done right," says Richard E.

Smith, a researcher with Secure Computing and author of the book *Authentication.* "What I've found particularly disquieting is how little public fuss there's been [about Code Red]. The general press has spun the story as being an unsuccessful attack on the White House as opposed to being a successful attack on several hundred thousand servers: 'Ha, ha, we dodged the bullet!' A cynic might say this demonstrates how 'intrusion tolerant' IIS is—the sites are all penetrated but aren't disrupted enough to upset the owners or generate much press comment. The rest of us are waiting for the other shoe to drop."

DAVID BERLINSKI

What Brings a World into Being?

FROM *COMMENTARY*

This being the digital age, we hear the word "information" constantly. A stream of digital "bits" is information in the technical sense—just as the stock price those bits are coded to convey is another kind of information. Scientists are now even talking about information as an organizing principle of the universe: The physicist's quantum and the biologist's cell are, at heart, bearers of information. Is the new theory of information a sufficient way to explain how our universe was created? The mathematician, philosopher, and popular author David Berlinski is skeptical.

Since their inception in the 17th century, the modern sciences have been given over to a majestic vision: there is nothing in nature but atoms and the void. This is hardly a new thought, of course; in the ancient world, it received its most memorable expression in Lucretius' *On the Nature of Things.* But it has been given contemporary resonance in theories—like general relativity and quantum mechanics—of terrifying (and inexplicable) power. If brought to a successful conclusion, the trajectory of this search would yield a single theory that would subsume all other theories and, in its scope and purity, would be our only necessary intellectual edifice.

In science, as in politics, the imperial destiny drives hard. If the effort to subordinate all aspects of experience to a single set of *laws* has often proved inconclusive, the scientific enterprise has also been involved in the search for universal *ideas.* One such idea is information.

Like energy, indeed, information has become ubiquitous as a commodity and, like energy, inescapable as an idea. The thesis that the human mind is nothing more than an information-processing device is now widely regarded as a fact. "Viewed at the most abstract level," the science writer George Johnson remarked recently in *The New York Times,* "both brains and computers operate *the same way* by translating phenomena—sounds, images, and so forth—into a code that can be stored and manipulated" (emphasis added). More generally, the evolutionary biologist Richard Dawkins has argued that life is itself fundamentally a river of information, an idea that has in large part also motivated the successful effort to decipher the human genome. Information is even said to encompass the elementary particles. "All the quarks and electrons in the cosmic wilds," Johnson writes, "are exchanging information each time they interact."

These assertions convey a current of intellectual optimism that it would be foolish to dismiss. Surely an idea capable of engaging so many distinct experiences must be immensely attractive. But it seems only yesterday that other compelling ideas urged their claims: chaos and nonlinear dynamics, catastrophe theory, game theory, evolutionary entropy, and various notions of complexity and self-organization.

The history of science resembles a collection of ghosts remembering that once they too were gods. With respect to information, a note of caution may well be in order if only because a note of caution is always in order.

II

IF INFORMATION CASTS a cold white light on the workings of the mind in general, it should certainly shed a little on the workings of language in particular.

The words and sentences of Herman Melville's *Moby-Dick,* to take a suggestive example, have the power to bring a world into being. The beginning of the process is in plain sight. There are words on the printed page, and they make up a discrete, one-dimensional, linear progression. Discrete—there are no words between words (as there are fractions between fractions); one-dimensional—each word might well be specified by a single number; linear—as far as words go, it is one thing after another. The end of the process is in sight as well: a richly organized, continuous, three- (or four-) dimensional universe. Although that universe is imaginary, it is recognizably contiguous to our own.

Bringing a world into being is an act of creation. But bringing a world into being is also an activity that suggests, from the point of view of the sciences, that immemorial progression in which causes evoke various effects: connections achieved between material objects, or between the grand mathematical abstractions necessary to explain their behavior.

And therein lies a problem. Words are, indeed, material objects, or linked as abstractions to material objects. And as material objects, they have an inherent power to influence other material objects. But no informal account of what words do as material objects seems quite sufficient to explain what they do in provoking certain experiences and so in creating certain worlds.

In the case of *Moby-Dick,* the chemical composition of words on the printed page, their refractive index, their weight, their mass, and ultimately their nature as a swarm of elementary particles—all this surely plays *some* role in getting the reader sympathetically to see Captain Ahab and imaginatively suffer his fate. The relevant causal pathways pass from the printed ink to our eyes, a river of light then serving to staple the shape of various words to our tingling retinal nerves; thereafter our nervous system obligingly passes on those shapes in the form of various complicated electrical signals. This is completely a physical process, one that begins with physical causes and ends with physical effects.

And yet the experience of reading begins where those physical effects end. It is, after all, an experience, and the world that it reveals is imaginary. If purely physical causes are capable of creating imaginary worlds, it is not by means of any modality known to the physical sciences.

Just how *do* one set of discrete objects, subject to the constraints of a single dimension, give rise to a universe organized in completely different ways and according to completely different principles?

IT IS HERE that information makes its entrance. The human brain, the linguist Steven Pinker has argued in *How the Mind Works,* is a physical object existing among other physical objects. Ordinary causes in the world at large evoke their ordinary effects within the brain's complicated fold and creases. But the brain is, *also,* an information-processing device, an instrument designed by evolution for higher things.

It is the brain's capacity to process information that, writes Pinker, allows human beings to "see, think, feel, choose, and act." Reading is a special case of seeing, one in which information radiates from the printed page and thereafter transforms itself variously into various worlds.

So much for what information does—clearly, almost everything of interest. But what is it, and how does it manage to do what it does? Pinker's definition, although informal, is brisk and to the point. Information, he writes, "is a correlation between two things that is produced by a lawful process." Circles in a tree stump carry information about the tree's age; lines in the human face carry information about the injuries of time.

Words on the page also contain or express information, and as carriers of information they convey the stuff from one place to another, piggy-backed, as it were, on a stream of physical causes and their effects.

Why not? The digital computer is a device that brilliantly compels a variety of discrete artifacts to scuttle along various causal pathways, ultimately exploiting pulsed signals in order to get one thing to act upon another. But in addition to their physical properties, the symbols flawlessly manipulated by a digital computer are capable of carrying and so conveying information, transforming one information-rich stream, such as a data base of proper names, into another information-rich stream, such as those same names arranged in alphabetical order.

The human mind does as much, Pinker argues; indeed, what it does, it does in the same way. Just as the computer transforms one information stream into another, the human mind transforms one source of information—words on the printed page—into another—a world in which whalers pursue whales and the fog lowers itself ominously over the spreading sea.

Thus Pinker; thus almost everyone.

The theory that gives the concept of information almost all of its content was created by the late mathematician Claude Shannon in 1948 and 1949. In it, the rich variety of human intercourse dwindles and disappears, replaced by an idealized system in which an information *source* sends signals to an information *sink* by means of a communication channel (such as a telephone line).

Communication, Shannon realized, gains traction on the real world by means of the firing pistons of tension and release. From far away, where the system has its source, messages are selected and then sent, one after the other—perhaps by means of binary digits. In the simplest possible set-up, symbols are limited to a single digit: *1*, say. A binary digit may occupy one of two states *(on* or *off).* We who are tensed at the system's sink are uncertain whether *1* will erupt into phosphorescent life or the screen will remain blank. Let us assume that each outcome is equally likely. The signal is sent—and then received. Uncertainty collapses into blessed relief, the binary digit *1* emerging in a swarm of pixels. The exercise has conveyed one unit, or bit, of information. And with the definition of a unit in place, information has been added to the list of properties that are interesting because they are *measurable.*

The development of Shannon's theory proceeds toward certain deep theorems about coding channels, noise, and error-reduction. But the details pertinent to this discussion proceed in another direction altogether, where they promptly encounter a roadblock.

"Frequently," Shannon observes, "messages have *meaning:* that is, they refer to or are correlated according to some system with certain physical or conceptual entities." Indeed. Witness *Moby-Dick*, which is *about* a large white whale. For Shannon, however, these "semantic aspects of communication" take place in some other room, not the one where his theory holds court; they are, he writes with a touch of asperity, "irrelevant to the engineering problem." The significance of communication lies only with the fact that "the actual message is one selected from a set of possible messages"—*this* signal, and not some other.

Shannon's strictures are crucial. They have, however, frequently proved difficult to grasp. Thus, in explaining Shannon's theory, Richard Dawkins writes that "the sentence, 'It rained in Oxford every day this week,' carries relatively little information, because the receiver is not surprised by it. On the other hand, 'It rained in the Sahara every day this week' would be a message with high information content, well worth paying extra to send."

But this is to confuse a signal with what it signifies. Whether I am surprised by the sentence "It rained in the Sahara desert every day this week" depends only on my assessment of the source sending the signal. Shannon's theory makes no judgments whatsoever about the subjects treated by various signals and so establishes no connection whatsoever to events in the real world. It is entirely possible that whatever the weather in Oxford *or* the Sahara may be, a given source might send both sentences with equal probability. In that case, they would convey precisely the same information.

The roadblock now comes into view. Under ordinary circumstances, reading serves the end of placing one man's thoughts in contact with another man's mind. On being told that whales are not fish, Melville's readers have learned something *about* whales and so about fish. Their uncertainty, and so their intellectual tension, has its antecedent roots in facts about the world beyond the symbols they habitually encounter. For most English speakers, the Japanese translation of *Moby-Dick*, although conveying precisely the same information as the English version, remains unreadable and thus unavailing as a guide to the universe created by the book in English.

What we who have conceived an interest in reading have required is some idea of how the words and sentences of *Moby-Dick* compel a world into creation. And about this, Shannon's theory says nothing.

For readers, it is the connections that are crucial, for it is those connections themselves—the specific correlations between the words in Melville's novel and the world of large fish and demented whalers—that function as the load-bearing structures. Just how, then, *are* such connections established?

Apparently they just are.

III

If, in reading, every reader embodies a paradox, it is a paradox that in living he exemplifies as well. "Next to the brain," George Johnson remarks, "the most obvious biological information-processor is the genetic machinery of the cell."

The essential narrative is by now familiar. All living creatures divide themselves into their material constituents *and* an animating system of instruction and information. The plan is in effect wherever life is in command: both the reader and the bacterial cell are expressions of an ancestral text, their brief appearance on the stage serving in the grand scheme of things simply to convey its throbbing voice from one generation to another.

Within the compass of the cell itself, there are two molecular classes: the proteins, and the nucleic acids (DNA and RNA). Proteins have a precise three-dimensional shape, and resemble tight, tensed knots. Their essential structure is nonetheless linear; when denatured and then stretched, the complicated jumble of a functional protein gracefully reveals a single filament, a kind of strand, punctuated by various amino acids, one after another.

DNA, on the other hand, is a double-stranded molecule, the two strands turned as a helix. Within the cell, DNA is wound in spools and so has its own complicated three-dimensional shape; but like the proteins, it also has an essentially linear nature. The elementary constituents of DNA are the four nucleotides, abbreviated as A, C, G, and T. The two strands of DNA are fastened to one another by means of struts, almost as if the strands were separate halves of a single ladder, and the struts gain purchase on these strands by virtue of the fact that certain nucleotides are attracted to one another by means of chemical affinities.

The structure of DNA as a double helix endows one molecule with two secrets. In replicating itself, the cell cleaves its double-stranded DNA. Each strand then reconstitutes itself by means of the same chemical affinities that held together the original strands. When replication has been concluded, there are two double-stranded DNA molecules where formerly there was only one, thus allowing life on the cellular level to pass from one generation to the next.

But if DNA is inherently capable of reproducing itself, it is also inherently capable of conveying the linear order of its nucleotides to the cell's amino acids. In these respects, DNA functions as a template or pattern. The mechanism is astonishingly complex, requiring intermediaries and a host of specialized enzymes to act in concert. But whatever the details, the central dogma of molecular biology is straight as an arrow. The order of nucleotides within DNA is read by the cell and then expressed in its proteins.

Read by the cell? Apparently so. The metaphor is inescapable, and so hardly a metaphor. As the DNA is read, proteins form in its wake, charged with carrying on the turbulent affairs of the cell itself. It was an imaginary reader, nose deep in Melville's great novel who suggested the distinction between what words do as material causes and what they achieve as symbols. The same distinction recurs in biology. Like words upon the printed page, DNA functions in any number of causal pathways, the tic of its triplets inducing certain biochemical changes and suppressing others.

And this prompts what lawyers call a leading question. We quite know what DNA is: it is a macromolecule and so a material object. We quite know what it achieves: apparently everything. Are the two sides of this equation in balance?

The cell is, after all, a living system. It partakes of all the mysteries of life. The bacterium *Escherichia coli,* for example, contains roughly 2,000 separate proteins, and every one of them is mad with purpose and busy beyond belief. Eucaryotic cells, which contain a nucleus, are more complicated still. Chemicals cross the cell membrane on a tight schedule, consult with other chemicals, undertake their work, and are then capped in cylinders, degraded and unceremoniously ejected from the cell. Dozens of separate biochemical systems act independently, their coordination finely orchestrated by various signaling systems. Enzymes prompt chemical reactions to commence and, work done, cause them to stop as well. The cell moves forward in time, functional in its nature, continuous in its operations.

Explaining all this by appealing to the causal powers of a single molecule involves a disturbing division of attention, rather as if a cathedral were seen suddenly to rise from the head of a carrot. Nonetheless, many biologists, on seeing the carrot, are persuaded that they *can* discern the steps leading to the cathedral. Their claim is often presented as a fact in the textbooks. The difficulty is just that, while the carrot—DNA, when all is said and done—remains in plain sight, subsequent steps leading to the cathedral would seem either to empty in a computational wilderness or to gutter out in an endless series of inconclusive causal pathways.

FIRST, THE COMPUTATIONAL WILDERNESS. Proteins appear in living systems in a variety of three-dimensional shapes. Their configuration is crucial to their function and so to the role they play in the cell. The beginning of a causal process is once again in plain sight—the linear order expressed by a protein's amino acids. And so, too, is the end—a specific three-dimensional shape. It is the mechanism in the middle that is baffling.

Within the cell, most proteins fold themselves into their proper configuration within seconds. Folding commences as the protein itself is being formed, the head of an amino-acid chain apparently knowing its own tail. Some proteins fold entirely on their own; others require molecular chaperones to block certain intermediate configurations and encourage others. Just how a protein manages to organize itself in space, using only the sequence of its own amino acids, remains a mystery, perhaps the deepest in computational biology.

Mathematicians and computer scientists have endeavored to develop powerful algorithms in order to predict the three-dimensional configuration of a given protein. The most successful of these algorithms gobble the computer's time and waste prodigally its power. To little effect. Protein folding remains a mystery.

Just recently, IBM announced the formation of a new division, intended to supply computational assistance to the biological community. A supercomputer named Blue Gene is under development. Operating at processing speeds 100 times faster than existing supercomputers, the monster will be dedicated largely to the problem of protein folding.

The size of the project is a nice measure of the depth of our ignorance. The slime mold has been slithering since time immemorial, its proteins folding themselves for just as long. No one believes that the slime mold accomplishes this by means of supercomputing firepower. The cell is not obviously an algorithm, and a simulation, needless to say, is not obviously an explanation. Whatever else the cell may be doing, it is not using Monte Carlo methods or consulting genetic algorithms in order to fold its proteins into their proper shape. The requisite steps are chemical. No other causal modality is available to the cell.

If these chemical steps were understood, simulations would be easy to execute. The scope of the research efforts devoted to simulation suggest that the opposite is the case: simulations are difficult to achieve, and the requisite chemical steps are poorly understood.

IF COMPUTATIONS are for the moment intractable, every analysis of the relevant causal pathways is for the moment inconclusive.

As they are unfolding, proteins trigger an "unfolding protein response," one that alerts an "intracellular signaling system" of things to come. It is this system that in turn "senses" when unfolded proteins accumulate. The signal sent, the signaling system responds by activating the transcription of still other genes that provide assistance to the protein struggling to find its correct three-dimensional shape. Each step in the causal analysis suggests another to come.

But no matter the causal pathways initiated by DNA, some *overall* feature of living systems seems stubbornly to lie beyond their reach. Signaling systems must themselves be regulated, their activities timed. If unfolding proteins require chaperones, these must make their appearance in the proper place; their formation requires energy, and so, too, do their degradation and ejection from the cell. Like the organism of which it is a part, the cell has striking global properties. It is *alive.*

Our own experience with complex dynamical systems, such as armies in action (or integrated microchips), suggests that in this regard command and coordination are crucial. The cell requires what one biologist has called a "supreme controlling and coordinating power." But if there is such a supreme system, biologists have not found it. The analysis of living systems is, to be sure, a science still in its infancy. My point, however, is otherwise, and it is general.

Considered strictly as a material object, DNA falls under the descriptive powers of biochemistry, its causal pathways bounded by chemical principles. Chemical actions are combinatorial in nature, and *local* in their effect. Chemicals affect chemicals within the cell by means of various weak affinities. There is no action at a distance. The various chemical affinities are essentially arrangements in which molecules exchange their parts irenically or like seaweed fronds drift close and then hold fast.

But command, control, and coordination, if achieved by the cell, would represent a phenomenon incompatible with its chemical activities. A "supreme controlling and coordinating power" would require a device receiving signals from every part of the cell and sending its own universally understood signals in turn. It would require, as well, a universal clock, one that keeps time globally, and a universal memory, one that operates throughout the cell. There is no trace of these items within the cell.

Absent these items, it follows that the cell quite plainly has the ability to organize itself *from itself*, its constituents bringing order out of chaos on their own, like a very intricate ballet achieved without a choreographer. And what holds for the cell must hold as well for the creatures of which cells are a part. One biologist has chosen to explain a mystery by describing it as a fact. "Organisms," he writes, "from daisies to humans, are naturally endowed with a remarkable property, an ability to *make themselves*."

Naturally endowed?

Just recently, the biologist Evelyn Fox Keller has tentatively endorsed this view. The system of control and coordination that animates the cell, she observes in *The Century of the Gene*, "consists of, and lives in, the interactive complex made up of genomic structures and the vast network of cellular machinery in which those structures are embedded." This may well be so. It is also unprecedented in our experience.

We have no insight into such systems. No mathematical theory predicts their existence or explains their properties. How, then, *do* a variety of purely local chemical reactions manage to achieve an overall and global mode of functioning?

INFORMATION NOW MAKES its second appearance as an analytic tool. DNA is a molecule—that much is certain. But it is also, molecular biologists often affirm, a library, a blueprint, a code, a program, or an algorithm, and as such it is quivering with information that it is just dying to be put to good use. As a molecule, DNA does what molecules do; but in its secondary incarnation as *something else,* DNA achieves command of the cell and controls its development.

A dialogue first encountered on the level of matter (DNA as a molecule and nothing more) now reappears on the level of metaphor (DNA as an information source). Once again we know what DNA is like, and we know what it does: apparently everything. And the question recurs: are the sides of this equation in balance?

Unfortunately, we do not know and cannot tell.

Richard Dawkins illustrates what is at issue by means of a thought experiment. "We have an intuitive sense," he writes,

> that a lobster, say, is more complex (more "advanced," some might even say more "highly evolved") than another animal, perhaps a millipede. Can we *measure* something in order to confirm or deny our intuition? Without literally turning it into bits, we can make an approximate estimation of the *information contents* [emphasis added] of the two bodies as follows. Imagine writing a book describing the lobster. Now write another book describing the millipede down to the same level of detail. Divide the word-count in one book by the word-count in the other, and you have an approximate estimate of the relative information content of lobster and millipede.

These statements have the happy effect of enforcing an impression of quantitative discipline on what until now have been a series of disorderly concepts. Things are being measured, and that is always a good sign. The comparison of one book to another makes sense, of course. Books are made up of words the way computer programs are made up of binary digits, and words and binary digits may both be counted.

It is the connection outward, from these books or programs to the creatures they describe, that remains problematic. What level of detail is required? In the case of a lobster, a very short book comprising the words "Yo, a lobster" is clearly not what Dawkins has in mind. But adding detail to a description—and thus length—is an exercise without end; descriptions by their very nature form an infinitely descending series.

Information is entirely a static concept, and we know of no laws of nature that would tie it to other quantitative properties. Still, if we cannot answer the question precisely, then perhaps it might be answered partially by saying that we have reached the right level of descriptive detail when the information in the book—that is, the lobster's DNA—is roughly of the same order of magnitude as the information latent in everything that a lobster is and does. This would at least tell us that the job at hand—constructing a lobster—is doable insofar as information plays a role in getting *anything* done.

Some biologists, including John Maynard Smith, have indeed argued that the information latent in a lobster's DNA *must* be commensurate with the information latent in the lobster itself. How otherwise could the lobster get on with the business at hand? But this easy response assumes precisely what is at issue, namely that it is *by means of information* that the lobster gets going in the first place. Skeptics such as ourselves require a direct measurement, a comparison between the information resident in the lobster's DNA and the information resident in the lobster itself. Nothing less will do.

DNA is a linear string. So far, so good. And strings are well-defined objects. There is thus no problem in principle in measuring their information. It is there for the asking and reckoned in bits. But what on earth is one to count in the case of the lobster? A lobster is not discrete; it is not made of linear symbols; and it occupies three or four dimensions and not one. Two measurements are thus needed, but only one is obviously forthcoming.

The unhappy fact is that we have in general no noncircular way of specifying the information in any three- (or four-) dimensional object except by an appeal to the information by which it is generated. But this appeal makes literal sense only when strings or items like strings are in concourse. The request for a direct comparison between what the lobster has to go on—its DNA—and what it is—a living lobster—ends with only one measurement in place, the other left dangling like a mountain climber's rope.

We are thus returned to our original question: how do symbols—words, strings, DNA—bring a world into being?

Apparently they just do.

IV

ONE MIGHT HOPE that in one discipline, at least, the situation might be different. Within the austere confines of mathematical physics, where a few pregnant symbols command the flux of space and time, information as an idea might come into its own at last.

The laws of physics have a peculiar role to play in the economy of the sciences, one that goes beyond anything observed in psychology or biology. They lie at the bottom of the grand scheme, comprising principles that are not only fundamental but irreducible. They must provide an explanation for the behavior of matter in all of its modes, and so they must explain the *emergence* as well as the organization of material objects. If not, then plainly they would not explain the behavior of matter in all of its modes, and, in particular, they would not explain its *existence.*

This requirement has initiated a curious contemporary exercise. Current cosmology suggests that the universe began with a big bang, erupting from nothing whatsoever 15 billion years ago. Plainly, the creation of something from nothing cannot be explained in terms of the behavior of material objects. This circumstance has prompted some physicists to assign a causative role to the laws of physics *themselves.*

The inference, indeed, is inescapable. For what else is there? "It is hard to resist the impression," writes the physicist Paul Davies, "of something—some influence capable of transcending space-time and the confinements of relativistic causality—possessing an overview of the entire cosmos at the instant of its creation, and manipulating all the causally disconnected parts to go bang with almost exactly the same vigor at the same time."

More than one philosopher has drawn a correlative conclusion: that, in this regard, the fundamental laws of physics enjoy attributes traditionally assigned to a deity. They are, in the words of Mary Hesse, "universal and eternal, comprehensive without exception (omnipotent), independent of knowledge (absolute), and encompassing all possible knowledge (omniscient)."

If *this* is so, the fundamental laws of physics cannot themselves be construed in material terms. They lie beyond the system of causal influences that they explain. And in this sense, the information resident in those causal laws is richer—it is more abundant—than the information resident in the universe itself. Having composed one book describing the universe to the last detail, a physicist, on subtracting that book from the fundamental laws of physics, would rest with a positive remainder, the additional information being whatever is needed to bring the universe into existence.

WE ARE NOW at the very limits of the plausible. Contemporary cosmology is a subject as speculative as scholastic theology, and physicists who find themselves irresistibly drawn to the very largest of its intellectual issues are ruefully aware that they have disengaged themselves from any evidential tether, however loose. Nevertheless, these flights of fancy serve a very useful purpose. In the image of the laws of nature zestfully wrestling a universe into existence, one sees a peculiarly naked form of information—naked because it has been severed from every possibility of a material connection. Stripped of its connection to a world that does not yet exist, the information latent in the laws of physics is nonetheless capable of *doing* something, by bringing the universe into being.

A novel brings a world into creation; a complicated molecule an organism. But these are the low taverns of thought. It is only when information is assigned the power to bring something into existence from *nothing whatsoever* that its essentially magical nature is revealed. And contemplating magic on this scale prompts a final question. Just how did the information latent in the fundamental laws of physics unfold itself to become a world?

Apparently it just did.

TIM FOLGER

Quantum Shmantum

FROM *DISCOVER*

One of the paradoxes of quantum physics is that elementary particles can seemingly exist in more than one state at once. One interpretation of why this is so is known as the "many-worlds" theory: Each of those quantum states represents an entire universe. The maverick physicist David Deutsch adamantly takes this theory to its inevitable conclusion: that at every moment, we all exist in an infinite number of other universes. The journalist Tim Folger catches up to Deutsch—or one version of him, at least.

At three o'clock on a warm summer afternoon, I arrive as scheduled at David Deutsch's home in Oxford, England. Deutsch, one of the world's leading theoretical physicists, a distinguished fellow of the British Computer Society and champion of what must certainly be the strangest scientific worldview ever created, is something of a recluse. He likes to sleep late and warned me not to come too early. Although I'm on time, my knocks on his door go unanswered. The house is dark and quiet. The doorbell doesn't seem to be working. After about 10 minutes a light goes on in an upstairs window, followed by the sound of running water. I knock harder, which at last triggers activity on the other side of the door. I hear feet pounding down stairs; the door opens, and Deutsch asks me to come in.

Piles of precariously stacked books line the route to his office, rising from the floor like stalagmites. A large poster of a brooding Albert Einstein hangs on one wall. Deutsch sits, sipping orange juice. He is slender, with birdlike attentiveness, and for someone who hardly ever leaves his home, surprisingly friendly and open. He looks much younger than 48. If his arguments, which have won over more than a few of his colleagues, turn out to be correct, our meeting is also occurring countless times in innumerable parallel universes, all in perfect accord with the uncanny laws of quantum theory.

Few physicists deny the validity of these laws, although they might not agree with Deutsch's interpretation of them. The laws insist that the fundamental constituents of reality, such as protons, electrons, and other subatomic particles, are not hard and indivisible. They behave like both waves and particles. They can appear out of nothing—a pure void—and disappear again. Physicists have even managed to teleport atoms, to move them from one place to another without passing through any intervening space. On the quantum scale, objects seem blurred and indistinct, as if created by a besotted god. A single particle occupies not just one position but exists here, there, and many places in between. "That quantum theory is outlandish, everyone agrees," says Deutsch. It seems completely in conflict with the world of big physics according to Newton and Einstein.

To grapple with the contradictions, most physicists have chosen an easy way out: They restrict the validity of quantum theory to the subatomic world. But Deutsch argues that the theory's laws must hold at every level of reality. Because everything in the world, including ourselves, is made of these particles, and because quantum theory has proved infallible in every conceivable experiment, the same weird quantum rules must apply to us. We, too, must exist in many states at once, even if we don't realize it. There must be many versions of late-rising David Deutsches, Earth, and the entire universe. All possible events, all conceivable variations on our lives, must exist, says Deutsch. We live not in a single universe, he says, but in a vast and rich "multiverse."

He knows the idea takes some getting used to, especially when one pauses to consider what it means on an everyday level. For starters, it solves once and for all the ancient question of whether we have free will. "The bottom line is that the universe is open," Deutsch says. "In the relevant sense of the word, we have free will."

We also have every possible option we've ever encountered acted out somewhere in some universe by at least one of our other selves. Unlike the traveler facing a fork in the road in Robert Frost's poem "The Road Not

Taken," who is "sorry that I could not travel both/And be one traveler," we take all the roads in our lives. This has some unsettling consequences and could explain why Deutsch is reluctant to venture from his house.

Driving a car, for example, becomes extremely hazardous, because it's almost certain that somewhere in some other universe the driver will accidentally hit and kill a child. So should we never drive? Deutsch thinks it's impossible to control the fate of our other selves in the multiverse. But if we're cautious, other copies of us may decide to be cautious. "There's also the argument that because the child's death will happen in some universes, you ought to take more care when doing even slightly risky things," he says.

Coming from a physicist of lesser stature, such startling views might be dismissed. But Deutsch possesses impeccable credentials. While still in his early thirties he created the theoretical framework for an entirely new discipline called quantum computation. Spurred by those ideas, researchers around the globe are attempting to construct a fundamentally different type of computer that is powerful almost beyond imagining.

Deutsch himself is more interested in convincing physicists that quantum theory has to be taken into consideration in the everyday world than he is in seeing a quantum computer built. Physicists may argue about what the theory means, but fortunately for the rest of us they have no qualms about working with it. By some estimates, 30 percent of the United States' gross national product is said to derive from technologies based on quantum theory. Without the insights provided by quantum mechanics, there would be no cell phones, no CD players, no portable computers. Quantum mechanics is not a branch of physics; it *is* physics.

And yet more than a century after it was first proposed by German scientist Max Planck, physicists who work with the theory every day don't really know quite what to make of it. They fill blackboards with quantum calculations and acknowledge that it is probably the most powerful, accurate, and predictive scientific theory ever developed. But as Deutsch wrote in an article for the *British Journal for the Philosophy of Science:* "Despite the unrivalled empirical success of quantum theory, the very suggestion that it may be literally true as a description of nature is still greeted with cynicism, incomprehension, and even anger."

To understand why the theory presents a conceptual challenge for physicists, consider the following experiment, based on an optical test first performed in 1801 by Thomas Young:

In the experiment, particles of light—photons—stream through a single vertical slit cut into a screen and fall onto a piece of photographic film placed some distance behind the screen. The image that develops on the film isn't surprising—simply a bright, uniform band. But if a second slit is cut into the screen, parallel to the first, the image on the film changes in an unexpected way: In place of a uniformly bright patch, the photons now form a pattern of alternating bright and dark parallel lines on the film. Dark lines appear in areas that were bright when just one slit was open. Somehow, cutting a second slit for the light to shine through prevents the photons from hitting areas on the screen they easily reached when only one slit was open.

Physicists usually explain the pattern by saying that light has a dual nature; it behaves like a wave, although it consists of individual photons. When light waves emerge from the two slits, overlapping wave crests meet at the film to create the bright lines; crests and troughs cancel out to produce the dark lines.

But there's a problem with this explanation: The same pattern of light and dark lines gradually builds up even when photons pass one at a time through the slits, as if each photon had somehow spread out like a wave and gone through both slits simultaneously. That clearly isn't the case, because the distance between the two slits can be hundreds, thousands, or in principle, any number of times greater than the size of a single photon. And if that isn't confusing enough, consider this: If detectors are placed at each slit, they register a photon traveling through only one of the slits, never through both at the same time.

Yet the photons behave as if they had traveled through both slits at once. The same baffling result holds not just for photons but also for particles of matter, such as electrons. Each seems able to exist in many different places at once—but only when no one is looking. As soon as a physicist tries to observe a particle—by placing a detector at each of the two slits, for example—the particle somehow settles down into a single position, as if it knew it was being detected.

Most physicists, when pressed, will usually say that the lesson quantum mechanics has for us is that our concepts of how a particle should behave simply don't match reality. But Deutsch believes that the implications of the theory are clear: If in every case a particle—be it a photon, an electron, or any other denizen of the quantum world—appears to occupy more than one position at a time, then it clearly *does* occupy many positions at once. And thus so do we, and so does everything else in the universe.

But is that an awfully big conclusion to draw from a simple pattern of light and shadow? Deutsch responds by pointing out that a similarly huge assumption—that the universe is expanding—is based on subtle light and shadow observations. Yet hardly any physicist anywhere disputes it.

Under normal circumstances we never encounter the multiple realities of quantum mechanics. We certainly aren't aware of what our other selves are doing. Only in carefully controlled conditions, as in the two-slit experiment, do we get a hint of the existence of what Deutsch calls the multiverse. That experiment offers a rare example of two overlapping realities, in which photons in one universe interfere with those in another. In our universe, we see a photon passing through one slit that seems to interact with another, invisible photon traveling through the second slit. In another universe, the photon that we see is invisible to the physicist in that world, while the one that we can't see is the photon the otherworldly physicist detects. Peculiar? Deutsch believes there is no alternative way of looking at quantum mechanics. "When it comes to a conflict about what a theory of physics says and what we are expecting, then physics has to win."

Deutsch is not the originator of the multiverse concept. That credit goes to Hugh Everett, whose 1957 Princeton doctoral thesis first presented what has come to be called the "many worlds" interpretation of quantum mechanics.

In creating the many worlds view, Everett was trying to solve the problem of why we see only one of the multiple states in which a particle can exist. Some years before Everett's work, physicists had crafted an ad hoc explanation that to this day remains the standard way of coping with quantum phenomena. In the conventional view, the very act of our observation causes all the possible states of a particle to "collapse" abruptly into a single value, which specifies the position, say, or energy of the particle. To understand how this works, imagine that the particle is an e-mail message. When the message is sent, there are multiple possible outcomes: The e-mail could reach its intended destination; any number of people could get it by mistake; or the sender might receive a notice that the message could not be delivered. But when one outcome is observed, all other possibilities with regard to the e-mail delivery collapse into one reality.

To some physicists the notion of collapse is an unsightly addition to quantum mechanics, tacked on to smooth over the uncomfortable fact that the theory mandates multiple states for every particle in existence. And the collapse model creates its own problems: Because it says our observations affect the outcomes of experiments, it assigns a central role to consciousness. "It's an unpleasant thing to bring people into the basic laws of physics," says Steven Weinberg, a Nobel laureate at the University of Texas.

Everett labored to move beyond those laws, arguing that nothing like a quantum collapse ever occurs and that human consciousness does not determine the outcome of experimental results. He said the collapse only seems to happen from our limited perspective. Everett believed that all quantum states are equally real and that if we see only one result of an experiment, other versions of us must see all the remaining possibilities.

Bryce DeWitt, the physicist who coined the term "many worlds" to describe this perplexing idea, remembers his first reaction to Everett's paper. "I was shocked but delighted," he says. By contrast, other physicists greeted Everett's theory with resounding indifference. "The article appeared, and that was the end of it," says DeWitt. "Just total silence."

The cool reception apparently didn't faze Everett. Although he left physics to work on classified projects for the United States government, he remained convinced until his death in 1982 that he was right about quantum mechanics. And if the many worlds theory is true, Deutsch, for one, believes that other copies of Everett might remain alive somewhere in the multiverse.

In 1976, a few years before he vanished from this corner of the multiverse, Everett, at DeWitt's invitation, visited the University of Texas, where Deutsch was then a graduate student. Like DeWitt before him, Deutsch became a convert. "I don't think there are any interpretations of quantum theory other than many worlds," he says. "The others deny reality."

Deutsch argues that physicists who use quantum mechanics in a utilitarian way—and that means most physicists working in the field today—suffer from a loss of nerve. They simply can't accept the strangeness of quantum reality. This is probably the first time in history, he says, that physicists have refused to believe what their reigning theory says about the world. For Deutsch, this is like Galileo refusing to believe that Earth orbits the sun and using the heliocentric model of the solar system only as a convenient way to predict the positions of stars and planets in the sky. Like modern physicists, who speak of photons as being both wave and particle, here and there at once, Galileo could have argued that Earth is both moving and stationary at the same time and ridiculed impertinent graduate students for questioning what that could possibly mean.

"This dilemma of whether you should accept that the world really is the way a theory says it is or whether you should just think of the theory as a manner of speaking, has occurred with every fundamentally new scientific theory right back to Copernicus," says Deutsch.

"I'm not quite sure why physicists should be more ready to believe in planets in distant galaxies than to believe in Everett's other worlds," says DeWitt. "Of course the number of parallel universes is really huge. I like to say that some physicists are comfortable with little huge numbers but not with big huge numbers."

Indeed, most other physicists view the many worlds route as a road best not taken. Steven Weinberg, paraphrasing Winston Churchill's quip about democracy, says: "It's a miserable idea except for all the other ideas." So which road does Weinberg choose? "I don't know," he says. "I think I come out with the pragmatic people who say, 'Oh, the hell with it. I'm too busy.' "

Christopher Fuchs, a research physicist at Lucent Technologies' Bell Labs, believes that quantum mechanics doesn't tell us so much about the world itself as it does about our interaction with the world. "It represents our interface with reality," he says. "I don't think it goes further than that." Fuchs believes that odd properties of quantum mechanics, such as the apparent ability of particles to exist in many places at once, merely reflect our ignorance of the world and are not true features of reality. "When a quantum state collapses, it's not because anything is happening physically, it's simply because this little piece of the world called a person has come across some knowledge, and he updates his knowledge," he says. "So the quantum state that's being changed is just the person's knowledge of the world, it's not something existent in the world in and of itself."

In Fuch's view, quantum mechanics describes a reality that shrinks away from us when we probe it too closely. "There's a certain ticklishness to the world," he says. "It is this extreme sensitivity of quantum systems that keeps us from ever knowing more about them than can be captured with the formal structure of quantum mechanics."

To Deutsch, such arguments are just complex rationalizations for avoiding the most straightforward implications of quantum theory. "It's a tenable point of view to say I don't know what the world is like," he says. "The obvious question then is what is in fact happening in reality? If quantum theory is true, it puts heavy constraints on what the world can be like."

The most serious consequence of refusing to consider the many worlds view is that physicists will never advance to a new, deeper understanding of nature. Deutsch adds: "What one can hope for in the long run is that a new theory will be facilitated by understanding this present theory. Once you understand the existing theory, you have a handle on what you can change in it. Whereas if you don't understand it, if it's just a set of equations, then it's astronomically unlikely that you will happen upon a better theory."

In the meantime, Deutsch is optimistic that the refined application of quantum mechanics principles will produce a tool that could bolster his argument for the existence of parallel universes. Many physicists around the world, including a team at Oxford with whom Deutsch works, are trying to build a quantum computer that would manipulate atoms or photons and exploit the particles' abilities to exist simultaneously in more than one state. Those quantum properties would tremendously increase the speed and capacity of the computer, allowing it to complete tasks beyond the reach of existing machines. In fact, says Deutsch, a quantum computer could in theory perform a calculation requiring more steps than there are atoms in the entire universe.

To do that, the computer would have to be manipulating and storing all that information somewhere. Computation is, after all, a physical process; it uses real resources, matter and energy. But if those resources exceed the amount available in our universe, then the computer would have to be drawing on the resources of other universes. So Deutsch feels that if such a computer is built, the case for many worlds will be compelling.

IT'S ALMOST SEVEN O'CLOCK in the evening. Deutsch has been answering questions for nearly four hours and has not yet had breakfast. He invites me to his conservatory, and we clamber over book stalagmites to a glassed-in porch facing his backyard for steak and orange juice. Deutsch muses about why people have trouble accepting strange new ideas. "I must say I don't understand the whole psychology of why people like some scientific theories and not others," he says.

He pauses briefly while he lights the "barbecue in a box" on which he will grill his breakfast. While the meat sizzles, he answers one last question: What if quantum theory is wrong?

"I'm sure that quantum theory will be proved false one day, because it seems inconceivable that we've stumbled across the final theory of physics. But I would bet my bottom dollar that the new theory will either retain the parallel universe feature of quantum physics, or it will contain something even more weird."

OLIVER MORTON

Shadow Science

FROM *WIRED*

In the past few years, astronomers have discovered dozens of planets orbiting distant stars. Most of these planets, however, are massive, and none of them has been directly observed: Their existence is proven primarily by discerning the minute gravitational effects they have on the stars themselves. Earth-like planets are simply too small to be detected by this method. Science writer Oliver Morton has found one astronomer who has come up with a way to not only detect such planets but allow us to see them as well.

Lenox Laser is a highly focused specialty engineering company, the self-proclaimed leader in small-hole technology. If you want small holes punched through something, Lenox, based in Glen Arm, Maryland, is the place to go. The company's Web site is awash with air slits, conflats (don't ask), calibrated gaskets, orifices both standard and customized, luers (see conflats), filters, and all that is holey. If the people at Lenox don't already make what you want, they'll be happy to do a custom job—they're quite proud of having drilled a hole through FDR's eye on a dime as a publicity stunt. But they've rarely heard anything as strange as a 1999 request from NASA's Ames Research Center for a stainless steel plate, pierced by a precisely described but seemingly random array of 1,600 holes, some as small as three-thousandths of a millimeter, or 3 microns, in diameter. A man named Bill Borucki wanted Lenox to make him a starry sky. A little one, to practice with.

For Borucki, the Lenox Laser contract was just one in an endless stream of tasks necessary to get Kepler, the project he champions, off the CAD/CAM desk and into orbit. So far, all these efforts have been in vain. Again and again, Borucki has proposed his Kepler mission to NASA's higher-ups. Again and again, he has been turned down. And each time he's been turned down, Borucki has gone away, improved his proposal, and come right back for another try. He does this because he is convinced that his spacecraft is capable of the first great astronomical discovery of the 21st century: *Kepler* represents the easiest way to find Earth-like planets in Earth-like orbits around other stars.

At first sight, you'd expect this to be an easy deal to close. The first planets around stars other than our own were discovered in the mid-'90s; since then, the study of such "extrasolar" planets has become one of the hottest fields in astronomy. Finding Earth-like extrasolar planets—which would be much smaller than the ones already discovered, and farther from the stars they orbit—has become one of NASA's clearest long-term goals. Dan Goldin, NASA's administrator, has waxed lyrical on the subject. "Shouldn't we set our sights on finding out whether there is a nurturing environment beyond our solar system?" he asked an audience at the American Geophysical Union's spring meeting in 1994. "Perhaps, just perhaps, the next generation's legacy will be an image of a planet 30 light-years from Earth."

You could argue that NASA owes the world such an image to dream on, having effectively quashed earlier dreams of nurturing environments closer to home. Missions have now visited the four rocky inner planets, the asteroids that lie beyond them, the gas giants even farther out, and the ice giants, Uranus and Neptune, farther out still. Only anomalous little Pluto, on the fringes, remains to be seen. And spacecraft have revealed all these places to be profoundly *un*nurturing. While life elsewhere in this solar system—deep beneath the Martian surface, perhaps, or in the ice-locked ocean wrapped around Jupiter's moon, Europa—is not ruled out, it is going to be hard to get to. If you have to drill for it, don't expect results until the 2010s at the earliest, with the '20s or '30s more likely.

If finding life nearby is hard, so too is looking for Earth-like planets around other stars. Producing actual images, as Goldin suggested, is far beyond current capabilities. Researchers using the two established methods of planet hunting never even see the planets involved; they simply infer their presence from planets' effects on their parent stars. And these methods work only for large planets; to actually detect light from a small planet like Earth will require space-based telescopes far more sophisticated than the current generation. A set of four space telescopes, all larger than the Hubble, might do the trick; so might a single telescope with a mirror 11 times bigger than the Hubble's and considerably smoother. A Terrestrial Planet Finder (TPF), using some such technology, is the centerpiece of NASA's long-term astronomy plans—but don't expect to see it fly in less than 10 years, and count on a bill well over a billion dollars.

Compared with all this, Borucki's *Kepler* is almost too good to be true. Its technology is mostly developed. Its design is simple. It could fly within five years of getting the go-ahead. It is budgeted at just $299 million, which puts it in the same price range as the cheapest successful missions to other planets. And Borucki and his colleagues claim that by using *Kepler* to look at huge numbers of stars for years at a time and carefully sifting through the terabytes of data thus produced, they can deliver evidence of Earth-like planets—not just a few, but a few hundred.

Still, Borucki has had to fight two parallel battles. He's had to come up with a design for a spacecraft with unique capabilities, *and* he's had to convince his colleagues and superiors the design can work. "People in the community said it couldn't be done," says Borucki, "and that was very effective at stopping us for a long time." But success on the first front has slowly led to success on the second. He has convinced his bosses at the Ames Research Center. He has convinced the engineers at the spacecraft maker Ball Aerospace, who have become his partners. He has sold the idea to more and more fellow scientists. One of them is Alan Boss, an expert on extrasolar planets at the Carnegie Institution of Washington, who has now joined Kepler's scientific team. "We want to make sure that the objects Terrestrial Planet Finder is going to be looking for—namely, terrestrial planets—really are out there," says Boss. "We need a sort of sanity check. So something like Kepler is really important. The time has come for it to go forward."

Most crucially, Borucki has made inroads at NASA headquarters. At the end of 2000, after three previous proposals had been turned away, Borucki's quiet, reasonable approach and his team's attention to detail finally got Kepler onto a shortlist for NASA's next planetary exploration. That's a great achievement. But the list has three missions on it, and only one will get funded: The odds are still less than even.

So why has Borucki kept trying? I asked him once as we chatted in his office in the Ames Space Science Building.

His eyes found the floor as I posed the question; he sighed. Then he quickly looked up, and it was as though the man in front of me had come truly into focus for the first time. His voice was no louder than before, but his eyes were locked on mine, and he spoke with a sincerity of the sort that in most lives is reserved for declarations of love. "I know for a fact it will work. I have known that for over a decade. It's a matter of will."

Bill Borucki is not an imposing man; you'd overlook him in a crowd. He speaks quietly, and when he's excited it shows mostly in pace rather than volume. He's balding, and the hair on each side of his head often sticks up like that of Dilbert's boss. Even with so little in the way of raw material, though, Borucki still manages to keep this hair in a distracted-scientist sort of disarray. During one conversation, I got the unnerving impression that one side of his head had just been licked by a cow.

Borucki came to NASA's Ames Research Center in Mountain View, California, just north of San Jose, in the early '60s; it was his first job after getting his MS at the University of Wisconsin-Madison. He started off working on the design for the Apollo heat shield, which involved using the center's spectacular gas guns to accelerate samples of heat-shield material up to orbital speeds, then catching them to see the effects. Later, he worked on the heat shields for spacecraft returning from Mars—a much harder problem, he says, because they would come in at much higher speeds: "It turned out that if we had a heat shield, and then we made the structure of the spacecraft out of the same material, and then we made the couches the astronauts were sitting on out of heat shield, and then we made the astronauts out of heat shield, we could get an astronaut's nose about two-thirds of the way through the atmosphere before it burned up." Heat shields, he assures me, have improved since then.

Ames is an interdisciplinary sort of place, so after heat shields Borucki went on to study atmospheres, both those around other planets and the one above Earth. He studied the effects of supersonic aircraft on the ozone layer. And at much the same time, in the late '70s, he became interested in the work of a colleague, David Black.

Black had started his own career looking at the remnants of the solar system's formation found in meteorites. But he soon decided that studying a single planetary system was not good science; to understand planetary systems, how they form, what they are made of, and so on, you needed a bigger sample. So at Ames, Black ran workshops on looking for planetary systems elsewhere—planetary systems being born in disks of gas and dust, and mature planetary systems already neatly arranged.

Seeing planets around other stars, however, is inherently difficult: Planets are dim, stars are bright, and seeing dim things next to bright ones is hard. Various techniques for overcoming this problem had been studied by ambitious astronomers before Black. The two most promising involved watching the star, not the planet. Stars with big planets are like hammer throwers who never release the hammer: The large hammer thrower moves in a small circle to balance the small hammer moving in a large circle. Very precise measurements of a star's velocity (which can be calculated from the spectrum of its light) might show this tiny, planet-induced circling moving the star toward us and then away; very precise measurements of the star's position in the sky might reveal the same effect making it swing from side to side. It is the first method—radial velocity—that has led to the discovery of 50 or so Jupiter-sized planets from ground-based observatories during the past five years. Unfortunately, there's a limit to the velocities this method can detect. It will never be able to see the minuscule motions caused by Earth-sized worlds swinging round a star.

The second method—precision astrometry, which reveals a star's movements from side to side—requires taking measurements of a star's position with an accuracy that today's Earth-based systems can't achieve. The expensive and extremely complicated Space Interferometry Mission (SIM) that NASA has on its books for launch in 2006 should be precise enough in its astrometry to pick up the effects of planets a bit smaller than those seen with the radial velocity method, but it won't be able to see the effects of planets as small as Earth unless they move in very particular orbits around very nearby stars.

The main alternative Black saw to these indirect approaches was direct imaging: Build a telescope big enough to gather a detectable amount of light from the planet, and at the same time find a way to block out the light from the star. This is the option that, over time, has developed into the planned TPF. But there was another alternative, what you might call a semidirect method: transit photometry. Black talked about it with Borucki, and Borucki became interested. By the mid-'80s Borucki was writing papers on it. By the late '80s he was designing spacecraft to use the technique and trying it out on Earth.

A transit is the passage of a planet's shadow across Earth. To astronomers in the 17th and 18th centuries, transits of Venus and Mercury across the face of the Sun were vital to measurements of the solar system's size. (It was to observe a transit of Venus that Captain Cook made the first of his voyages to the South Pacific.) The transits Borucki is interested in, though, occur when a planet orbiting another star moves between that star and Earth, casting its shadow across an earthly (or space-based) telescope. Such a transit will dim the star's light for about 10 hours. Photometry measures this dimming, thereby detecting the presence of a planet. Stars get a touch brighter and a touch dimmer all the time, but unlike most such variations, a transit will repeat itself almost exactly the next time the planet comes round the star, in a rhythm set by the planet's "year." These rhythmical dimmings are what *Kepler's* photometry is designed to pick out.

There are three problems with this method. The first is a matter of time. Because transits last only 10 hours every year or so, to pick them up you have to watch a star continuously, hour by hour. (For radial velocity and precision astrometry searches, you can take data on a much more relaxed schedule, because the movements you watch are continuous.)

The second problem is a matter of geometry. Imagine looking at the Sun from a random viewpoint a hundred light-years away. From most viewpoints you will never see a transit of Earth, because you, Earth, and the Sun will never lie in a straight line. Only if your viewpoint is in the same plane as Earth's orbit will Earth's shadow ever fall across your eye. The chances of that are a bit less than 200 to 1. The odds of our seeing an Earth-like planet's transits of any other star are similarly slim.

The third problem is one of photometric sensitivity. Even if you're in the right place to see it, a planetary transit will cut a star's brightness by only a smidgen; looking at the Sun, you would find that Earth's passage across its face dimmed it by only about 1 part in 12,000. That's a small effect: Light passing through clear window-glass is dimmed 300 times that much.

Kepler's design is a response to these three problems. The answer to the time question is simple but radical. Rather than turning from star to star, as other telescopes do, *Kepler* will stare unblinkingly at a single patch of sky for four years, gathering photometric data on the stars in this area almost continuously. This would be impossible for an earthbound telescope, because Earth's rotation means that stars rise and set. But for a space-based telescope it's easy. The Hubble needs complicated, fallible gyroscopes and all sorts of guidance systems to point itself at all the different galaxies its users want to observe. Once *Kepler* is pointed at the right bit of sky, it can simply be left alone; falling freely around the Sun, it will point in the same direction for thousands of years. The *Kepler* design thus ends up with only a handful of moving parts: a couple of very small momentum wheels to keep the solar cells pointing at the Sun, and a mechanism that keeps its antenna pointed at Earth.

The geometry problem is solved with numbers. The telescope at *Kepler's* heart is specifically designed for a wide field of view. Though its mirror is less than half the size of the Hubble's, it is configured to see much more of the sky. "The Hubble's field of view is the size of a grain of sand between your fingers at arm's length," says Borucki. "Ours is the size of your open hand at arm's length." Pointed in the right direction—the team has chosen an area in the constellation Cygnus—*Kepler* can watch stars by the hundreds of thousands. The telescope will project their images onto a set of charge-coupled devices, the technology used in digital cameras. Astronomical CCDS—semiconductor detectors that astronomers use to turn the photons of starlight into electrons that can be processed digitally—are particularly large and sensitive examples of the craft.

In the first year of operation, *Kepler*'s CCDs will monitor about 170,000 of the stars in its field of view; after that, they will concentrate on 100,000 particularly Sun-like stars that have shown themselves to shine with a pleasingly steady light. With that many targets, 1-in-200 odds of spotting a planet are pretty acceptable: The Kepler team expects to see from 50 to 650 Earth-like planets, enough to provide statistics about which sorts of stars produce earths and which don't. Sometimes it will see more than one planet in a system: If it were looking at *our* Sun from far away and seeing transits of Earth, for example, it might see transits of Venus as well. Mercury and Mars would be missed because they are small; the outer planets would be missed because they take more than five years to go around the Sun, so a five-year mission wouldn't see the regular pattern of their transits.

A telescope that, if studying our own solar system, would pick up at best two planets out of nine is clearly not perfect. But as exploration of our own solar system has shown, some planets are more interesting than others. What makes Earth interesting—and what allows us to be here taking an interest in it—is that it sits in the Sun's habitable zone, a ring where the warmth the star gives off is great enough to allow water on the surface of a planet with an atmosphere, but not strong enough to boil it. By looking around stars like ours for planets with orbits of about a year, *Kepler* would seek out planets in the stars' habitable zones. It would search for the most interesting ones of all: planets where there might be creatures like us.

In 1992, Borucki presented *Kepler*'s predecessor design, called Fresip (Frequency of Earth-Sized Inner Planets), for possible inclusion in NASA's plans. In 1993, he took the Fresip design to a meeting of NASA's Toward Other Planetary Systems Science Working Group. By something less than a coincidence, the meeting was chaired by David Black, who had first brought transits to Borucki's attention; Black has been pursuing avenues to extrasolar planets ever since his time at Ames. But the connection didn't help. The working group already had proposals for various space telescopes capable of precision astrometry, including the one that would evolve into SIM. Fresip, a new idea, got nowhere.

Borucki's next attempt to get to space was through NASA's Discovery program for small planetary spacecraft. Discovery spacecraft are chosen through a wide-ranging competition. A scientist—the principal investigator, or PI—puts together a team, which includes one of NASA's research centers and an industrial partner, and proposes a mission. The proposals are judged on scientific merit and technical readiness; the most promising are put on a shortlist and given some money to conduct feasibility studies. Six months later, one or two winners are chosen. Since the program started in the early 1990s, there have been three rounds of competition; the fourth is currently under way. Each round sees about 30 applications. Some PIs have resubmitted once, or even twice, before either being picked or giving up. Just two missions have thrown their hats into the ring all four times: Kepler and Vesat, a Venus orbiter.

Every time Kepler entered the contest, its science was judged first-rate, but its technology was not. NASA's evaluation panels for the Discovery program keep their deliberations confidential, but they do debrief the unsuccessful applicants on what counted against them. "Every time," according to Borucki, "NASA had a new and more ingenious reason for not accepting Kepler." In 1994 the hurdle was cost: Discovery missions have a cost cap that stands at $300 million. "The first time," says Borucki, "they said, 'This is a space telescope about 40 percent the size of the Hubble, which cost $2 billion, so this is going to cost 40 percent of that; it will break the budget cap.' After that, we got three different organizations to cost the project, so NASA believed our figure," which was safely under the cap.

"The second time," Borucki continues, "the objection was that the CCDs couldn't do the job. So before applying to the Discovery program again, we went out and made measurements with the CCDs we wanted to use, published the measurements, and showed the CCDs would work. The *third* time, in 1998, they said, 'Sure, the CCDs can do that and the costs are reasonable, but how can you show it'd all work together?' "

This time, though, the NASA judges were intrigued enough to do something unprecedented—they didn't put Kepler on the shortlist, but they gave it some money anyway. In fact, they gave it considerably more than they gave the shortlisted missions for their feasibility studies. They wanted the Ames team to build a working mock-up of the whole system and demonstrate that it could function before returning in two years. The management at Ames added a matching half-million dollars, and in the basement of the Space Science Building, Borucki and a handful of colleagues started to build their own private universe and a system with which to observe it.

The result is meant to be the technological equivalent of a lawyer's closing argument. It's a case designed to take you through the *Kepler* system from beginning to end, in a way that removes all reasonable doubts. The argument stands about 10 feet high. It is isolated from the lab environment by shock absorbers on the floor and by a set of frameworks holding polystyrene foam and aluminum plating: Each plate's temperature is controlled separately to mimic the situation on the spacecraft. Each heating element is cooled by a fan, held in place by an independent frame to prevent its whirring from jostling the precisely arranged test rig. At the same time, though, little motors inside the system impart just the amount of jiggle that would be expected from vibrations on the craft—one of various ways of simulating sources of "noise" that might corrupt the data. Borucki takes noise very seriously. "It's ubiquitous," he says. "You think 'Death and taxes'? Add noise."

At the bottom of the rig is a lamp that illuminates the inside of a 20-inch white scattering sphere: not just white, says Borucki, but "perfectly white. You couldn't see it if you looked at it." A hole in the first sphere opens into a larger scattering sphere, an arrangement that spreads out the light from the lamp as evenly as possible. Mounted on top of the larger sphere is the star plate. This is the component that came from Lenox Laser: a piece of stainless steel with tiny holes in it. The light streams through these holes, mimicking the stars of varying brightness that *Kepler* will be observing.

The holes come in many sizes. There are holes that let through the same number of photons a second as *Kepler*'s camera will see from its target stars. There are smaller holes that let through less light, just as there will be dimmer stars in *Kepler*'s field of view to confuse things—more noise. Some of the holes are doubled, because some of the stars will be double stars. Optical fibers attached to some of the holes in the plate produce a few brighter lights, which correspond to the stars in *Kepler*'s field of view that are bright enough to see with the naked eye, and bright enough to pose problems to its delicate camera—loud noise. "We're trying to make this as bad as we can," says Borucki, "to show that no matter what goes wrong, it will work fine." All in all, the craftspeople at Lenox drilled 1,600 stars into the plate, according to Borucki's exacting specifications.

Forty-two of these "stars" are capable of producing "transits." Changing the intensity of the light coming through the holes in the plate was in some ways the hardest part of the demonstration. Controlling brightness at a level of better than 1 part in 10,000 is a next-to-impossible task (the state-of-the-art lamp used to illuminate the test bed is only controllable to about 10 times that amount). The solution turned out to be simple but finicky. Underneath some of the holes is a very fine wire through which a current can be passed. When the current flows through the wire it warms the metal, which expands by a few millionths of a millimeter to block out just enough light to simulate the passage of a planet in front of a star.

Looking down on all this is a prototype of the *Kepler* camera built around a CCD chip—about 21/2 centimeters across and 5 centimeters long—on which there is an array of 4 million separate CCD pixels. Each of these pixels is a tiny well in which falling photons are converted into electrons.

Some of the 4 million pixels on the CCD will be hit by light from the pinhole stars; these photons will knock electrons free. As more photons hit a given pixel, more free electrons accumulate. These electrons need to be constantly siphoned off: A pixel that fills up with electrons is of no use, because once full—in the test rig's CCD there's room for 600,000 electrons per pixel—it can no longer measure changes in brightness. The camera has no shutter. ("We don't want a shutter," says Borucki. "If we don't have a shutter, it can't get stuck.") The data is read off the chip at a megapixel per second, and it comes off somewhat blurred: The carefully orchestrated jitter of the test rig means that the images of the stars can move back and forth between pixels. A little more noise is added to this data to mimic the effects of cosmic rays penetrating the detector and sending spurious signals, and then the data is fed into the test rig's computer. Comparisons between "frames" are used to subtract out the blur—so the output represents the amount of light seen by each pixel every three seconds.

Borucki says this constantly repeated sampling is where the precision needed to spot transits comes from: "You just measure things hundreds of thousands of times in the same way." Traces of transits are sought through constant comparisons between the brightness of each individual star and the brightness of the large number of similar stars in the field at that time. A transit will show up as the dimming of one star in relation to the aggregated light of its peers. It's much simpler to see these discrepancies than to calibrate the dimmings according to an absolute scale of brightness. *Kepler* is a Zen spacecraft, stripped of distractions and living in the moment.

This metal-and-plastic argument seems to have done its work. In January,

NASA announced that Kepler had finally reached the Discovery program's shorter-than-usual shortlist. And its case had been bolstered outside the lab, too, when astronomers observed transits of a very large planet orbiting close to the star HD209458. Solar researcher Tim Brown of the High Altitude Observatory in Boulder, Colorado, followed up these findings with the Hubble, which showed the effect even more dramatically. When his data was presented at the International Astronomical Union's summer 2000 meeting in Manchester, England, it caused a genuine stir: "Really, really beautiful data," says the Carnegie Institution's Alan Boss. The beautiful data was a simple graph with all its points clustered close to a single line—a line level to the left and right but with a deep, clear, flat-bottomed, symmetrical dip in the middle. The shadow of another world.

But even if the technology looks good and the thirst for earths is growing ever stronger, the deal may not be closed. The other two missions on the Discovery shortlist—a spacecraft that would orbit Jupiter's poles to look for clues to its interior structure, and a spacecraft that would study two large and very different asteroids, Ceres and Vesta—wouldn't be there if they didn't have a shot at the prize. The questions they address may not be as fundamental as Kepler's, but is profundity enough to win over the Discovery program when it makes its final selection in late 2001?

Left to himself, Bill Borucki would probably not be looking for planets. He'd probably be looking for lightning. He loves lightning. When he moved out West from Wisconsin in the 1960s, he and his wife spent their first vacation in Arizona, watching lightning. The lightning literature, which deals with the most extreme temperatures known on Earth, helped him in his heat-shield research. He has simulated lightning in the lab. He has looked for lightning on Venus (a project that first got him involved in the details of doing photometry from a spacecraft). And he will look for lightning on Saturn's moon, Titan, when the *Huygens* spacecraft lands there in 2004. At his home he has a small collection of fulgurites, tubes of glass formed when lightning strikes desert sands.

But Borucki has not been left to himself. For 17 years he has been in the company of an idea that simply will not let him go. From the moment he met it, he's known that it was a good idea; it has offered a way to answer a really important question, and it has needed a champion to help it realize its potential. Talking with Borucki, I find myself expecting him to betray a touch of resentment that the idea has dominated his life for so long and that he has nothing to show for it. If he ever does have anything to show for it—if Kepler makes it off the shortlist and into flight—the results will be available only after an additional four years of development and four years after that of taking data—each day starting with the worry that something might go wrong. It'll be a decade he might have spent far more enjoyably, and quite likely the last decade of his productive career.

There seems to be no trace of such resentment, though. The nearest thing to it is the slightly mischievous glee with which Borucki speculates that he might build the spacecraft perfectly and still not find any Earth-like planets—for the simple reason that there aren't any there. "Imagine it," he beams. "An empty galaxy!"

And while the idea has been teasing Borucki along, failing to embitter him, it has also drawn in other champions. A team based at Stanford has tried unsuccessfully to get NASA to back missions that would use similar technology to measure the pulsations of stars and their planetary transits. Attempts to observe extrasolar planets crossing the face of stars—projects which would be satisfied with a single planet—are under development in France, Denmark, and Canada. Kepler is more ambitious than they are, however, because it is Borucki's intention not just to use the idea, but to do it justice. Kepler represents the best first-generation transit-observing telescope you could make. "If you want to find Earth-like planets, that's our niche. We don't compete with people doing other things. But we do *that* superbly."

And yet, no one knows whether *Kepler* itself will ever drift observantly in and out of cones of shadow light-years long—the Discovery judges have yet to decide.

Say they decide for Kepler, and sometime in 2005 a countdown reaches zero and *Kepler* heads into the sky. Ignore for a moment the roar of the Delta 2925–10's 10 engines. Discount the $299 million carefully spent, which could have been spent as prudently on studying the heart of a giant planet or puzzling out the origins of asteroids. Forget those questions about humanity's place in the universe, or the cosmic significance of life. Think instead of an unremarkable-looking man in an unremarkable office, and of defeat after defeat. And remember the words, measured and clear: "I know for a fact it will work. I have known that for over a decade. It's a matter of will."

STEVEN WEINBERG

Can Science Explain Everything? Anything?

FROM *THE NEW YORK REVIEW OF BOOKS*

Some philosophers have drawn a distinction between the concept of "explanation" and "description." Science, they would claim, can describe elements of the natural world but not explain them. For many scientists, this is a distinction without a difference. The Nobel Prize–winning physicist Steven Weinberg, an eloquent advocate for science's place in the realm of ideas, takes this distinction at face value—in order to demonstrate that science does indeed explain something.

One evening a few years ago I was with some other faculty members at the University of Texas, telling a group of undergraduates about work in our respective disciplines. I outlined the great progress we physicists had made in explaining what was known experimentally about elementary particles and fields—how when I was a student I had to learn a large variety of miscellaneous facts about particles, forces, and symmetries; how in the decade from the mid-1960s to the mid-1970s all these odds and ends were explained in what is now called the Standard Model of elementary particles; how we learned that these miscellaneous facts about particles and forces could be deduced mathematically from a few fairly simple principles; and how a great collective *Aha!* then went out from the community of physicists.

After my remarks, a faculty colleague (a scientist, but not a particle physicist) commented, "Well, of course, you know science does not really explain things—it just describes them." I had heard this remark before, but now it took me aback, because I had thought that we had been doing a pretty good job of explaining the observed properties of elementary particles and forces, not just describing them.[1]

I think that my colleague's remark may have come from a kind of positivistic angst that was widespread among philosophers of science in the period between the world wars. Ludwig Wittgenstein famously remarked that "at the basis of the whole modern view of the world lies the illusion that the so-called laws of nature are the explanations of natural phenomena."

It might be supposed that something is explained when we find its cause, but an influential 1913 paper by Bertrand Russell had argued that "the word 'cause' is so inextricably bound up with misleading associations as to make its complete extrusion from the philosophical vocabulary desirable."[2] This left philosophers like Wittgenstein with only one candidate for a distinction between explanation and description, one that is teleological, defining an explanation as a statement of the purpose of the thing explained.

E. M. Forster's novel *Where Angels Fear to Tread* gives a good example of teleology making the difference between description and explanation. Philip is trying to find out why his friend Caroline helped to bring about a marriage between Philip's sister and a young Italian man of whom Philip's family disapproves. After Caroline reports all the conversations she had with Philip's sister, Philip says, "What you have given me is a description, not an explanation." Everyone knows what Philip means by this—in asking for an explanation, he wants to learn Caroline's purposes. There is no purpose revealed in the laws of nature, and not knowing any other way of distinguishing description and explanation, Wittgenstein and my friend had concluded that these laws could not be explanations. Perhaps some of those who say that science describes but does not explain mean also to compare science unfavorably with theology, which they imagine to explain things by reference to some sort of divine purpose, a task declined by science.

This mode of reasoning seems to me wrong not only substantively, but also procedurally. It is not the job of philosophers or anyone else to dictate meanings of words different from the meanings in general use. Rather than argue that scientists are incorrect when they say, as they commonly do, that they are explaining things when they do their work, philosophers who care about the meaning of explanation in science should try to understand what it is that scientists are doing when they say they are explaining something. If I had to give an a priori definition of explanation in physics I would say, "Explanation in physics is what physicists have done when they say *Aha!*" But a priori definitions (including this one) are not much use.

As far as I can tell, this has become well understood by philosophers of science at least since World War II. There is a large modern literature on the nature of explanation, by philosophers like Peter Achinstein, Carl Hempel, Philip Kitcher, and Wesley Salmon. From what I have read in this literature, I gather that philosophers are now going about this the right way: they are trying to develop an answer to the question "What is it that scientists do when they explain something?" by looking at what scientists are actually doing when they *say* they are explaining something.

Scientists who do pure rather than applied research commonly tell the public and funding agencies that their mission is the explanation of something or other, so the task of clarifying the nature of explanation can be pretty important to them, as well as to philosophers. This task seems to me to be a bit easier in physics (and chemistry) than in other sciences, because philosophers of science have had trouble with the question of what is meant by an explanation of an event (note Wittgenstein's reference to "natural phenomena") while physicists are interested in the explanation of regularities, of physical principles, rather than of individual events.

Biologists, meteorologists, historians, and so on are concerned with the causes of individual events, such as the extinction of the dinosaurs, the blizzard of 1888, the French Revolution, etc., while a physicist only becomes interested in an event, like the fogging of Becquerel's photographic plates that in 1897 were left in the vicinity of a salt of uranium, when the event reveals a regularity of nature, such as the instability of the uranium atom. Philip Kitcher has tried to revive the idea that the way to explain an event is by reference to its cause, but which of the infinite number of things that could affect an event should be regarded as its cause?[3]

Within the limited context of physics, I think one can give an answer of sorts to the problem of distinguishing explanation from mere description, which captures what physicists mean when they say that they have explained some regularity. The answer is that we explain a physical principle when we show that it can be deduced from a more fundamental physical principle. Unfortunately, to paraphrase something that Mary McCarthy once said about a book by Lillian Hellman, every word in this definition has a questionable meaning, including "we" and "a." But here I will focus on the three words that I think present the greatest difficulties: the words "fundamental," "deduced," and "principle."

THE TROUBLESOME WORD "fundamental" can't be left out of this definition, because deduction itself doesn't carry a sense of direction; it often works both ways. The best example I know is provided by the relation between the laws of Newton and the laws of Kepler. Everyone knows that Newton discovered not only a law that says the force of gravity decreases with the inverse square of the distance, but also a law of motion that tells how bodies move under the influence of any sort of force. Somewhat earlier, Kepler had described three laws of planetary motion: planets move on ellipses with the sun at the focus; the line from the sun to any planet sweeps over equal areas in equal times; and the squares of the periods (the times it takes the various planets to go around their orbits) are proportional to the cubes of the major diameters of the planets' orbits.

It is usual to say that Newton's laws explain Kepler's. But historically Newton's law of gravitation was deduced from Kepler's laws of planetary motion. Edmund Halley, Christopher Wren, and Robert Hooke all used Kepler's relation between the squares of the periods and the cubes of the diameters (taking the orbits as circles) to deduce an inverse square law of gravitation, and then Newton extended the argument to elliptical orbits. Today, of course, when you study mechanics you learn to deduce Kepler's laws from Newton's laws, not vice versa. We have a deep sense that Newton's laws are more fundamental than Kepler's laws, and it is in that sense that Newton's laws explain Kepler's laws rather than the other way around. But it's not easy to put a precise meaning to the idea that one physical principle is more fundamental than another.

It is tempting to say that more fundamental means more comprehensive. Perhaps the best-known attempt to capture the meaning that scientists give to explanation was that of Carl Hempel. In his well-known 1948 article written with Paul Oppenheim, he remarked that "the explanation of a general regularity consists in subsuming it under another more comprehensive regularity, under a more general law."[4] But this doesn't remove the difficulty. One might say for instance that Newton's laws govern not only the motions of planets but also the tides on Earth, the falling of fruits from trees, and so on, while Kepler's laws deal with the more limited context of planetary motions. But that isn't strictly true. Kepler's laws, to the extent that classical mechanics applies at all, also govern the motion of electrons around the nucleus, where gravity is irrelevant. So there is a sense in which Kepler's laws have a generality that Newton's laws don't have. Yet it would feel absurd to say that Kepler's laws explain Newton's, while everyone (except perhaps a philosophical purist) is comfortable with the statement that Newton's laws explain Kepler's.

This example of Newton's and Kepler's laws is a bit artificial, because there is no real doubt about which is the explanation of the other. In other cases the question of what explains what is more difficult, and more important. Here is an example. When quantum mechanics is applied to Einstein's general theory of relativity one finds that the energy and momentum in a gravitational field come in bundles known as gravitons, particles that have zero mass, like the particle of light, the photon, but have a spin equal to two (that is, twice the spin of the photon). On the other hand, it has been shown that any particle whose mass is zero and whose spin is equal to two will behave just the way that gravitons do in general relativity, and that the exchange of these gravitons will produce just the gravitational effects that are predicted by general relativity. Further, it is a general prediction of string theory that there must exist particles of mass zero and spin two. So is the existence of the graviton explained by the general theory of relativity, or is the general theory of relativity explained by the existence of the graviton? We don't know. On the answer to this question hinges a choice of our vision of the future of physics—will it be based on space-time geometry, as in general relativity, or on some theory like string theory that predicts the existence of gravitons?

THE IDEA OF EXPLANATION as deduction also runs into trouble when we consider physical principles that seem to transcend the principles from which they have been deduced. This is especially true of thermodynamics, the science of heat and temperature and entropy. After the laws of thermodynamics had been formulated in the nineteenth century, Ludwig Boltzmann succeeded in deducing these laws from statistical mechanics, the physics of macroscopic samples of matter that are composed of large numbers of individual molecules. Boltzmann's explanation of thermodynamics in terms of statistical mechanics became widely accepted, even though it was resisted by Max Planck, Ernst Zermelo, and a few other physicists who held on to the older view of the laws of thermodynamics as free-standing physical principles, as fundamental as any others. But then the work of Jacob Bekenstein and Stephen Hawking in the twentieth century showed that thermodynamics also applies to black holes, and not because they are composed of many molecules, but simply because they have a surface from which no particle or light ray can ever emerge. So thermodynamics seems to transcend the statistical mechanics of many-body systems from which it was originally deduced.

Nevertheless, I would argue that there is a sense in which the laws of thermodynamics are not as fundamental as the principles of general relativity or the Standard Model of elementary particles. It is important here to distinguish two different aspects of thermodynamics. On one hand, thermodynamics is a formal system that allows us to deduce interesting consequences from a few simple laws, wherever those laws apply. The laws apply to black holes, they apply to steam boilers, and to many other systems. But they don't apply everywhere. Thermodynamics would have no meaning if applied to a single atom. To find out whether the laws of thermodynamics apply to a particular physical system, you have to ask whether the laws of thermodynamics can be deduced from what you know about that system. Sometimes they can, sometimes they can't. Thermodynamics itself is never the explanation of anything—you always have to ask why thermodynamics applies to whatever system you are studying, and you do this by deducing the laws of thermodynamics from whatever more fundamental principles happen to be relevant to that system.

In this respect, I don't see much difference between thermodynamics and Euclidean geometry. After all, Euclidean geometry applies in an astonishing variety of contexts. If three people agree that each one will measure the angle between the lines of sight to the other two, and then they get together and add up those angles, the sum will be 180 degrees. And you will get the same 180-degree result for the sum of the angles of a triangle made of steel bars or of pencil lines on a piece of paper. So it may seem that geometry is more fundamental than optics or mechanics. But Euclidean geometry is a formal system of inference based on postulates that may or may not apply in a given situation. As we learned from Einstein's general theory of relativity, the Euclidean system does not apply in gravitational fields, though it is a very good approximation in the relatively weak gravitational field of the earth in which it was developed by Euclid. When we use Euclidean geometry to explain anything in nature we are tacitly relying on general relativity to explain why Euclidean geometry applies in the case at hand.

IN TALKING ABOUT DEDUCTION, we run into another problem: Who is it that is doing the deducing? We often say that something is explained by something else without our actually being able to deduce it. For example, after the development of quantum mechanics in the mid-1920s, when it became possible to calculate for the first time in a clear and understandable way the spectrum of the hydrogen atom and the binding energy of hydrogen, many physicists immediately concluded that all of chemistry is explained by quantum mechanics and the principle of electrostatic attraction between electrons and atomic nuclei. Physicists like Paul Dirac proclaimed that now all of chemistry had become understood. But they had not yet succeeded in deducing the chemical properties of any molecules except the simplest hydrogen molecule. Physicists were sure that all these chemical properties were consequences of the laws of quantum mechanics as applied to nuclei and electrons.

Experience has borne this out; we now can in fact deduce the properties of fairly complicated molecules—not molecules as complicated as proteins or DNA, but still some fairly impressive organic molecules—by doing complicated computer calculations using quantum mechanics and the principle of electrostatic attraction. Almost any physicist would say that chemistry is explained by quantum mechanics and the simple properties of electrons and atomic nuclei. But chemical phenomena will never be entirely explained in this way, and so chemistry persists as a separate discipline. Chemists do not call themselves physicists; they have different journals and different skills from physicists. It's difficult to deal with complicated molecules by the methods of quantum mechanics, but still we know that physics explains why chemicals are the way they are. The explanation is not in our books, it's not in our scientific articles, it's in nature; it is that the laws of physics require chemicals to behave the way they do.

Similar remarks apply to other areas of physical science. As part of the Standard Model, we have a well-verified theory of the strong nuclear force—the force that binds together both the particles in the nucleus and the particles that make up those particles—known as quantum chromodynamics, which we believe explains why the proton mass is what it is. The proton mass is produced by the strong forces that the quarks inside the proton exert on one another. It is not that we can actually calculate the proton mass; I'm not even sure we have a good algorithm for doing the calculation, but there is no sense of mystery about the mass of the proton. We feel we know why it is what it is, not in the sense that we have calculated it or even can calculate it, but in the sense that quantum chromodynamics can calculate it—the value of the proton mass is entailed by quantum chromodynamics, even though we don't know how to do the calculation.

It can be very important to recognize that something has been explained, even in this limited sense, because it can give us a strategic sense of what problems to work on. If you want to work on calculating the proton mass, go ahead, more power to you. It would be a lovely show of calculational ability, but it would not advance our understanding of the laws of nature, because we already understand the strong nuclear force well enough to know that no new laws of nature will be needed in this calculation.

ANOTHER PROBLEM with explanation as deduction: in some cases we can deduce something without explaining it. That may sound really peculiar, but consider the following little story. When physicists started to take the big bang cosmology seriously one of the things they did was to calculate the production of light elements in the first few minutes of the expanding universe. The way this was done was to write down all the equations that govern the rates at which various nuclear reactions took place. The rate of change of the quantity (or "abundance," as physicists say) of any one nuclear species is equal to a sum of terms, each term proportional to the abundances of other nuclear species. In this way you develop a large set of linked differential equations, and then you put them on a computer that produces a numerical solution.

When these equations were solved in the mid-1960s by James Peebles and then by Robert Wagoner, William Fowler, and Fred Hoyle, it was found that after the first few minutes one quarter of the mass of the universe was left in the form of helium, and almost all the rest was hydrogen, with other elements present only in tiny quantities. These calculations also revealed certain regularities. For instance, if you put something in the theory to speed up the expansion, as for instance by adding additional species of neutrinos, you would find that more helium would be produced. This is somewhat counterintuitive—you might think speeding up the expansion of the universe would leave less time for the nuclear reactions that produce helium, but in fact the calculations showed that it increased the amount of helium produced.

The explanation is not difficult, though it can't easily be seen in the computer printout. While the universe was expanding and cooling in the first few minutes, nuclear reactions were occurring that built up complex nuclei from the primordial protons and neutrons, but because the density of matter was relatively low these reactions could occur only sequentially, first by combining some protons and neutrons to make the nucleus of heavy hydrogen, the deuteron, and then by combining deuterons with protons or neutrons or other deuterons to make heavier nuclei like helium. However, deuterons are very fragile; they're relatively weakly bound, so essentially no deuterons were produced until the temperature had dropped to about a billion degrees, at the end of the first three minutes. During all this time neutrons were changing into protons, just as free neutrons do in our laboratories today.

When the temperature dropped to a billion degrees, and it became cold enough for deuterons to hold together, then all of the neutrons that were still left were rapidly gobbled up into deuterons, and the deuterons then into helium, a particularly stable nucleus. It takes two neutrons as well as two protons to make a helium nucleus, so the number of helium nuclei produced at that time was just half the number of remaining neutrons. Therefore the crucial thing that determines the amount of helium produced in the early universe is how many of the neutrons decayed before the temperature dropped to a billion degrees. The faster the expansion went, the earlier the temperature dropped to a billion degrees, so the less time the neutrons had to decay, so the more of them were left, and so the more helium was produced. That's the explanation of what was found in the computer calculations; but the explanation was not to be found in the computer-generated graphs showing the abundance in relation to the speed of expansion.

Further, although I have said that physicists are only interested in explaining general principles, it is not so clear what is a principle and what is a mere accident. Sometimes what we think is a fundamental law of nature is just an accident. Kepler again provides an example. He is known today chiefly for his famous three laws of planetary motion, but when he was a young man he tried also to explain the diameters of the orbits of the planets by a complicated geometric construction involving regular polyhedra. Today we smile at this because we know that the distances of the planets from the sun reflect accidents that occurred as the solar system happened to be formed. We wouldn't try to explain the diameters of the planetary orbits by deducing them from some fundamental law.

In a sense, however, there is a kind of approximate statistical explanation for the distance of the earth from the sun.[5] If you ask why the earth is about a hundred million miles from the sun, as opposed, say, to two hundred million or fifty million miles, or even further, or even closer, one answer would be that if the earth were much closer to the sun then it would be too hot for us and if it were any further from the sun then it would be too cold for us. As it stands, that's a pretty silly explanation, because we know that there was no advance knowledge of human beings in the formation of the solar system. But there is a sense in which that explanation is not so silly, because there are countless planets in the universe, so that even if only a tiny fraction are the right distance from their star and have the right mass and chemical composition and so on to allow life to evolve, it should be no surprise that creatures that inquire into the distance of their planet from its star would find that they live on one of the planets in this tiny fraction.

This kind of explanation is known as anthropic, and as you can see it does not offer a terribly useful insight into the physics of the solar system. But anthropic arguments may become very important when applied to what we usually call the universe. Cosmologists increasingly speculate that just as the earth is just one of many planets, so also our big bang, the great expansion of the universe in which we live, may be just one of many bangs that go off sporadically here and there in a much larger mega-universe. They further speculate that in these many different big bangs some of the supposed constants of nature take different values, and perhaps even some of what we now call the laws of nature take different forms. In this case, the question why the laws of nature that we discover and the constants of nature that we measure are what they are would have a rough teleological explanation—that it is only with this sort of big bang that there would be anyone to ask the question.

I certainly hope that we will not be driven to this sort of reasoning, and that we will discover a unique set of laws of nature that explain why all the constants of nature are what they are. But we have to keep in mind the possibility that what we now call the laws of nature and the constants of nature are accidental features of the big bang in which we happen to find ourselves, though constrained (as is the distance of the earth from the sun) by the requirement that they have to be in a range that allows the appearance of beings that can ask why they are what they are.

CONVERSELY, IT IS ALSO POSSIBLE that a class of phenomena may be regarded as mere accidents when in fact they are manifestations of fundamental physical principles. I think this may be the answer to a historical question that has puzzled me for many years. Why was Aristotle (and many other natural philosophers, notably Descartes) satisfied with a theory of motion that did not provide any way of predicting where a projectile or other falling body would be at any moment during its flight, a prediction of the sort that Newton's laws do provide? According to Aristotle, substances tend to move to their natural positions—the natural position of earth is downward, the natural position of fire is upward, and water and air are naturally somewhere in between, but Aristotle did not try to say how fast a bit of earth drops downward or a spark flies upward. I am not asking why Aristotle had not discovered Newton's laws—obviously someone had to be the first to discover these laws, and the prize happened to go to Newton. What puzzles me is why Aristotle expressed no dissatisfaction that he had not learned how to calculate the positions of projectiles at each moment along their paths. He did not seem to realize that this was a problem that anyone ought to solve.

I suspect that this was because Aristotle implicitly assumed that the rates at which the elements move to their natural places are mere accidents, that they are not subject to rules, that you couldn't say anything general about them (except that heavy objects fall faster than light ones), that the only things about which one could generalize were questions of equilibrium—where objects will come to rest. This may have reflected a widespread disdain for change on the part of the Hellenic philosophers, as shown for instance in the work of Parmenides, which was admired by Aristotle's teacher Plato. Of course Aristotle was wrong about this, but if you imagine yourself in his times, you can see how far from obvious it would have been that motion is governed by precise mathematical rules that might be discovered. As far as I know, this was not understood until Galileo began to measure how long it took balls to roll various distances down an inclined plane. It is one of the great tasks of science to learn what are accidents and what are principles, and about this we cannot always know in advance.

So now that I have deconstructed the words "fundamental," "deduce," and "principle," is anything left of my proposal, that in physics we say that we explain a principle when we deduce it from a more fundamental principle? Yes, I think there is, but only within a historical context, a vision of the future of science. We have been steadily moving toward a satisfying picture of the world. We hope that in the future we will have achieved an understanding of all the regularities that we see in nature, based on a few simple principles, laws of nature, from which all other regularities can be deduced. These laws will be the explanation of whatever principles (such as, for instance, the rules of the Standard Model or of general relativity) can be deduced directly from them, and those directly deduced principles will be the explanations of whatever principles can be deduced from them, and so on. Only when we have this final theory will we know for sure what is a principle and what an accident, what facts about nature are entailed by what principles, and which are the fundamental principles and which are the less fundamental principles that they explain.

I HAVE NOW done the best I can to say whether science can explain anything, so let me take up the question whether science can explain everything. Clearly not. There certainly always will be accidents that no one will explain, not because they could not be explained if we knew all the precise conditions that led up to them, but because we never will know all these conditions. There are questions like why the genetic code is precisely what it is or why a comet happened to hit the earth 65 million years ago in just the place it did rather than somewhere else that will probably remain forever outside our grasp. We cannot explain, for example, why John Wilkes Booth's bullet killed Lincoln while the Puerto Rican nationalists who tried to shoot Truman did not succeed. We might have a partial explanation if we had evidence that one of the gunmen's arms was jostled just as he pulled the trigger, but, as it happens, we don't. All such information is lost in the mists of time; events depend on accidents that we can never recover. We can perhaps try to explain them statistically; for example, you might consider a theory that Southern actors in the mid-nineteenth century tended to be good shots while Puerto Rican nationalists in the mid-twentieth century tended to be bad shots, but when you only have a few singular pieces of information it's very difficult to make even statistical inferences. Physicists try to explain just those things that are not dependent on accidents, but in the real world most of what we try to understand does depend on accidents.

Further, science can never explain any moral principle. There seems to be an unbridgeable gulf between "is" questions and "ought" questions. We can perhaps explain why people think they should do things, or why the human race has evolved to feel that certain things should be done and other things should not, but it remains open to us to transcend these biologically based moral rules. It may be, for example, that our species has evolved in such a way that men and women play different roles—men hunt and fight, while women give birth and care for children—but we can try to work toward a society in which every sort of work is as open to women as it is to men. The moral postulates that tell us whether we should or should not do so cannot be deduced from our scientific knowledge.

There are also limitations on the certainty of our explanations. I don't think we'll ever be certain about any of them. Just as there are deep mathematical theorems that show the impossibility of proving that arithmetic is consistent, it seems likely that we will never be able to prove that the most fundamental laws of nature are mathematically consistent. Well, that doesn't worry me, because even if we knew that the laws of nature are mathematically consistent, we still wouldn't be certain that they are true. You give up worrying about certainty when you make that turn in your career that makes you a physicist rather than a mathematician.

Finally, it seems clear that we will never be able to explain our most fundamental scientific principles. (Maybe this is why some people say that science does not provide explanations, but by this reasoning nothing else does either.) I think that in the end we will come to a set of simple universal laws of nature, laws that we cannot explain. The only kind of explanation I can imagine (if we are not just going to find a deeper set of laws, which would then just push the question farther back) would be to show that mathematical consistency requires these laws. But this is clearly impossible, because we can already imagine sets of laws of nature that, as far as we can tell, are completely consistent mathematically but that do not describe nature as we observe it.

For example, if you take the Standard Model of elementary particles and just throw away everything except the strong nuclear forces and the particles on which they act, the quarks and the gluons, you are left with the theory known as quantum chromodynamics. It seems that quantum chromodynamics is mathematically self-consistent, but it describes an impoverished universe in which there are only nuclear particles—there are no atoms, there are no people. If you give up quantum mechanics and relativity then you can make up a huge variety of other logically consistent laws of nature, like Newton's laws describing a few particles endlessly orbiting each other in accordance with these laws, with nothing else in the universe, and nothing new ever happening. These are logically consistent theories, but they are all impoverished. Perhaps our best hope for a final explanation is to discover a set of final laws of nature and show that this is the only logically consistent rich theory, rich enough for example to allow for the existence of ourselves. This may happen in a century or two, and if it does then I think that physicists will be at the extreme limits of their power of explanation.

NOTES

1. This article is based on a talk given at a symposium on "Science and the Limits of Explanation" at Amherst in the autumn of 2000.

2. "On the Notion of Cause," reprinted in *Mysticism and Logic* (Doubleday, 1957), p. 174.

3. There is an example of the difficulty of explaining events in terms of causes that is much cited by philosophers. Suppose it is discovered that the mayor has paresis. Is this explained by the fact that the mayor had an untreated case of syphilis some years earlier? The trouble with this explanation is that most people with untreated syphilis do not in fact get paresis. If you could trace the sequence of events that led from the syphilis to the paresis, you would discover a great many other things that played an essential role—perhaps a spirochete wiggled one way rather than another way, perhaps the mayor also had some vitamin deficiency—who knows? And yet we feel that in a sense the

mayor's syphilis is the explanation of his paresis. Perhaps this is because the syphilis is the most dramatic of the many causes that led to the effect, and it certainly is the one that would be most relevant politically.

4. Carl Hempel and Paul Oppenheim, "Studies in the Logic of Confirmation," *Philosophy of Science,* Vol. 15, No. 135 (1948), pp. 135–175; reprinted with some changes in *Aspects of Scientific Explanation and Other Essays in the Philosophy of Science* (Free Press, 1965).

5. Professor R. J. Hankinson of the University of Texas has directed my attention to Galen for an early example of this "explanation." Of course, writing 1400 years before Copernicus, Galen was concerned to explain the position of the sun rather than that of the earth. In "On the Usefulness of the Parts of the Body" he compared his explanation of the sun's position to the explanation of the position of the human foot at the end of the leg—both sun and foot are placed by the creator where they would do the most good.

Although these explanations are teleological in a way that has been abandoned by modern science, Galen's analogy was better than he could have realized. Just as the earth is one of a vast number of planets, whose distances from their stars is largely a matter of chance, so the position of the foot is the outcome of a vast number of chance mutations in the evolution of our vertebrate ancestors. An organism produced by a chain of chance mutations that put its feet in its mouth would not survive to pass its genes on to its descendants, just as a planet that by chance condensed too close to or too far from its star would not be the home of philosophers.

Nicholas Wade

The Eco-Optimist

FROM *THE NEW YORK TIMES*

A former Greenpeace member, vegetarian, and self-described leftist, Bjørn Lomborg is not the kind of person you would expect to have written a book debunking some of the most cherished tenets of the environmental movement. Could he be right? New York Times *correspondent Nicholas Wade tells the story of the man—and the controversial theories—behind* The Skeptical Environmentalist.

The news from environmental organizations is almost always bleak. The climate is out of whack. Insidious chemicals taint food and drink. Tropical forests are disappearing. Species are perishing en masse. Industrial poisons pollute air, earth and water. Ecosystems are being stressed to the breaking point by the greedy, wasteful consumption of the Western lifestyle and its would-be imitators.

So it is a surprise to meet someone who calls himself an environmentalist but who asserts that things are getting better, that the rate of human population growth is past its peak, that agriculture is sustainable and pollution is ebbing, that forests are not disappearing, that there is no wholesale destruction of plant and animal species and that even global warming is not as serious as commonly portrayed.

Strange to say, the author of this happy thesis is not a steely-eyed economist at a conservative Washington think tank but a vegetarian, backpack-toting academic who was a member of Greenpeace for four years. He is Dr. Bjørn Lomborg, a 36-year-old political scientist and professor of statistics at the University of Aarhus in Denmark. Dr. Lomborg arrived at this position, much to his own astonishment, through a journey that began in a Los Angeles bookshop in February 1997.

Dr. Lomborg was leafing through an issue of *Wired* magazine and started reading an interview with Dr. Julian L. Simon, a University of Maryland economist who argued in several books that population was unlikely to outrun natural resources.

But Dr. Simon, who died in 1998, is more widely known for his solution to the airline overbooking problem (having airlines pay passengers to take a later flight) and for a 1980 bet with Dr. Paul Ehrlich, president of Stanford University's Center for Conservation Biology. He bet that any five metals chosen by Dr. Ehrlich would be cheaper in 1990; Dr. Ehrlich lost on all five.

Dr. Lomborg felt sure that Dr. Simon's arguments on the plenitude of natural resources were "simple American right-wing propaganda," though presented with enough seriousness that they would be worth rebutting. Back in Aarhus, he started nightly study sessions with his statistics students to debunk Dr. Simon's contentions, using figures drawn from reports of the World Bank, the Food and Agriculture Organization, the United States Environmental Protection Agency, the International Panel on Climate Change and other gatherers of official facts.

"Three months into the project, we were convinced that we were being debunked instead," Dr. Lomborg said. "Not everything he said is right. He has a definite right-wing slant. But most of the important things were actually correct."

Dr. Lomborg has presented his findings in *The Skeptical Environmentalist,* published in September 2001 by Cambridge University Press. The primary targets of the book, a substantial work of analysis with almost 3,000 footnotes, are statements made by environmental organizations like the Worldwatch Institute, the World Wildlife Fund and Greenpeace. He refers to the persistently gloomy fare from these groups as the Litany, a collection of statements that he argues are exaggerations or outright myths.

Dr. Lomborg also chides journalists, saying they uncritically spread the Litany, and he accuses the public of an unfounded readiness to believe the worst.

"The Litany has pervaded the debate so deeply and so long," Dr. Lomborg writes, "that blatantly false claims can be made again and again, without any references, and yet still be believed." This is the fault not of academic environmental research, which is balanced and competent, he says, but rather of "the communication of environmental knowledge, which taps deeply into our doomsday beliefs."

To understand the world as it is, Dr. Lomborg says, it is necessary to look at long-term global trends that tell more of the whole story than short-term trends and are less easy to manipulate.

For example, the Worldwatch Institute, in its 1998 "State of the World" report, said, "The world's forest estate has declined significantly in both area and quality in recent decades." But the longest data series of annual figures available from the United Nations' Food and Agriculture Organization shows that global forest cover has in fact increased, to 30.89 percent in 1994 from 30.04 percent of global land cover in 1950. The Worldwatch report goes on to claim that because of soaring demand for paper, "Canada is losing some 200,000 hectares of forest a year." The cited reference, however, says that "in fact Canada grew 174,000 more hectares of forest each year," Dr. Lomborg writes.

Janet Abramovitz, Worldwatch's forest expert, said the world forest cover had shrunk significantly in the last 20 years. She based that contention on a different, shorter series of Food and Agriculture Organization statistics but declined to cite a percentage. The institute's figure on Canadian forest loss was an error, she said.

In its report for 2000, the Worldwatch Institute cited the dangers it had foreseen in 1984—"record rates of population growth, soaring oil prices, debilitating levels of international debt and extensive damage to forests from the new phenomenon of acid rain"—and lamented that "we are about to enter a new century having solved few of these problems."

But in his book, Dr. Lomborg cites figures from the United States Census Bureau, the International Monetary Fund, the World Bank and the European Environment Agency to show that the rate of world population growth has actually been dropping sharply since 1964; the level of international debt decreased slightly from 1984 to 1999; the price of oil, adjusted for inflation, is half what it was in the early 1980's; and the sulfur emissions that generate acid rain (which has turned out to do little if any damage to forests, though some to lakes) have been cut substantially since 1984.

In an interview, the president of the Worldwatch Institute, Christopher Flavin, agreed that progress had been made in the four problems cited in the institute's 1984 report, but he said that had been mentioned in other institute reports. "If you read through our materials as a whole," Mr. Flavin said, "many of these improvements are acknowledged."

Dr. Lomborg has also been unable to find strong support in the official statistics for the regular predictions of disaster from Dr. Ehrlich. "In the course of the 1970's," Dr. Ehrlich wrote in *The Population Bomb,* published in 1968, "the world will experience starvation of tragic proportions—hundreds of millions of people will starve to death."

Although world population has doubled since 1961, Dr. Lomborg writes, calorie intake has increased by 24 percent as a whole and by 38 percent in developing countries.

Dr. Lomborg also takes issue with some global warming predictions. In assessing how waste gases could warm the world's climate, he says, there are four wild cards that affect the climate change models.

One is the multiplier effect of carbon dioxide—as it heats the atmosphere a little, the air can hold more water, and that heats the atmosphere a lot more. How much more is in question, but Dr. Lomborg cites satellite and weather balloon data that seem to weaken the case for a strong multiplier effect.

The other three wild cards, Dr. Lomborg says, are the role of clouds, the effect of aerosols and the effect of the sunspot cycle on earth's climate.

Dr. Lomborg believes that when it comes to computer models of climate change, the International Panel on Climate Change deals all four wild cards in a way that exaggerates the effect of greenhouse gases. This means, in his view, that the actual warming will be at the cooler end of the panel's predicted range.

He contends that the internationally agreed Kyoto targets for reducing carbon dioxide emissions will impose vast costs for little result. A more effective approach, according to Dr. Lomborg, would be to increase research on alternative sources of energy, like solar and fusion.

But Dr. Kevin Trenberth of the National Center for Atmospheric Research said there were flaws in the satellite data, but that a new analysis, soon to be published, was likely to point toward a strong multiplier effect for carbon dioxide.

Dr. Michael Oppenheimer, an expert on global warming at Environmental Defense, agrees that clouds and aerosols are still weak points in the climate models, but says Dr. Lomborg's contention on the effects of the sunspot cycle is not widely accepted. As to Dr. Lomborg's policy recommendations, Dr. Oppenheimer said that investing in technology alone was "like betting the farm on a policy in which we have less confidence than emissions reduction." In his view, a broad-based technology push would turn into a pork-barrel program and be far less efficient than the technology that would develop in response to a requirement to reduce emissions.

The Skeptical Environmentalist portrays several other elements of the Litany as little more than urban myths. One is the prediction that the world's forests and a large number of species are headed for catastrophe.

Dr. Lomborg believes that forest loss has been less serious than is often described—only 20 percent since the dawn of agriculture, not 67 percent, as stated by the World Wildlife Fund. He also puts the present annual rate of loss at 0.46 percent, as calculated by the Food and Agriculture Organization, rather than at 2 percent or more, the figure cited by many environmentalists.

Given that the forests are not doing that badly, he is skeptical of claims that the world is about to lose half or more of its species. The often quoted figure that 40,000 species are lost every year comes from a 1979 article by Dr. Norman Myers, an ecologist at Oxford University. But this figure, Dr. Lomborg says, was not based on any evidence, just on Dr. Myers's conjecture that one million species might be lost from 1975 to 2000, which works out to be 40,000 species a year.

The International Union for the Conservation of Nature, which maintains the Red Book of endangered species, concluded in 1992 that the extinction figures for mammals and birds were "very small" and that the total extinction rate, assuming 30 million species, was probably 2,300 species a year.

Nonetheless, Dr. Lomborg says, Dr. Myers repeated his estimate in 1999 with a warning that "we are into the opening stages of a human-caused biotic holocaust."

Dr. Myers confirmed in an interview that the figure of 40,000 extinctions a year had come from his estimate. He said that it was an illustration used to make his argument clear and that he gave figures only "when I am speaking to a political leader or policy maker who says that in order to sell his message, he absolutely must have some number."

The International Union for the Conservation of Nature's estimate was too low, Dr. Myers said, because it considered a species extinct only after none of its members had been sighted for 50 years. "All I am trying to do is to demonstrate that we are in the opening phase of a mass extinction," he said.

Though no longer a member of Greenpeace, Dr. Lomborg still counts himself as an environmentalist and portrays his critique as based on the outlook of a leftist. "I'm a left-wing guy," he says, "and a vegetarian because I don't want to kill animals—you can't play the 'he's right-wing so he's wrong' argument."

He believes that the environment must be protected and that regulation is often necessary. But exaggerating problems distorts society's priorities, he says, and makes it hard for society to make the best decisions.

Writing about environmentalists, he says, "The worse they can portray the environment, the easier it is for them to convince us that we need to spend more money on the environment rather than on hospitals, child day care, etc."

Those who abandon long-held faiths are often strident advocates of their new views. But Dr. Lomborg displays little of the convert's zeal. His aim is not to preach free-market solutions for every problem or to deny that threats to the environment exist.

His motive, he says, is simply to document that the facts, in his view, tell a far brighter story than the Litany. Thomas Malthus argued in 1798 that population growth was certain to outrun food supply. As Dr. Lomborg sees it, Malthus's gloomy predictions still hold an iron grip over many minds, and are still wrong.

DARCY FREY

George Divoky's Planet

FROM *THE NEW YORK TIMES MAGAZINE*

The black guillemot is an Arctic seabird of interest primarily to ornithologists. Their patterns of breeding and migration, however, tell a bigger story: the extent of climate change on Earth. Every summer for thirty years, an intrepid scientist named George Divoky has come to the desolate reaches of Cooper Island, off the north coast of Alaska, to study these birds and chart global warming. The equally intrepid writer Darcy Frey, no stranger to covering men under stress, braves frigid temperatures, stir-crazy-inducing isolation, and an occasional polar bear, to track the lengths to which a dedicated scientist will go in the quest for knowledge.

I. In Which George Tries to Build a Fence

This is a story about global warming and a scientist named George Divoky, who studies a colony of Arctic seabirds on a remote barrier island off the northern coast of Alaska. I mention all this at the start because a reader might like to come to the point, and what could be more urgent than the very health and durability of this planet we call Earth? However, before George can pursue his inquiry into worldwide climate change; before he can puzzle out the connections between a bunch of penguinesque birds on a flat, snow-covered, icebound island and the escalating threat of droughts, floods and rising global temperatures, he must first mount a defense—his only defense in this frozen, godforsaken place—against the possibility of being consumed, down to the last toenail, by a polar bear while he sleeps. He must first build a fence.

Cooper Island, June 4, 2 o'clock in the morning. The sky is a cold slab of gray, the air temperature hovers in the upper 20's and the wind—always the wind—howls across hundreds of miles of sea ice with such unremitting force that George has disappeared beneath a hat, two hoods and a thick fleece face mask covering all but his bespectacled eyes. Standing near the three small dome tents that make up his field camp on Cooper, George raises a pair of binoculars and begins to scan for bears. Past the island's north beach, a wind-scarred plain of sea ice stretches uninterrupted to the pole. To the south, the nearest tree stands 200 miles away on the far side of the Brooks Range. Here, some 330 miles north of the Arctic Circle, with the sun making a constant parabolic journey around the sky, George surveys a view that replicates in all directions: the snow-covered island merges with the sea ice at its shores, the dazzling sheets of sea ice stretch to meet a pale gray dome of sky. Surrounded by a vast, undulating whiteness, he appears to be standing in the middle of the Arctic Ocean. He appears to be standing on the tops of cirrus clouds.

"So, . . ." he says, and the rest of his words are carried off by the wind.

"*What?*"

"I said, *so maybe we should put up the polar-bear fence before we get too fatigued!*"

Heading fast toward fatigue, I tell him that's a fine idea indeed, and exactly how many polar bears does he figure might be out there on the ice? George, who spends each summer on Cooper Island, is cheerfully indifferent to its dangers and discomforts and reassuring to those who aren't. Discussing a recent incident in which some Inupiat Eskimos had to shoot a bear that wandered into their nearby whaling camp, I consider it bad news—there are bears in the vicinity!—while George thinks of it as good: yes, but now there's one less bear.

Still scanning the faint horizon line, George insists there's nothing to get worked up about. For the most part, bears stay several miles offshore, where they can gorge on ringed and bearded seals. If a bear *were* to come to this island, he points out, its massive 800-pound frame would stand out against the sky like an approaching blimp. Even if a bear were to wander into camp, he goes on, we are sufficiently armed with a shotgun, cans of compressed pepper spray (mace) and a flarelike device known as a screamer-banger. Nonetheless, George confirms that at least one big bear shows up on Cooper each summer, usually to scavenge the beach for dead, washed-up seals; and furthermore, that an encounter can be so unpleasant that you do have to figure the odds a little differently. "I was out here once and injured the tendon in my knee," he says. "I couldn't walk, I couldn't stay warm. I kept thinking, *If it's true predators key in on the weak and the infirm. . . .*" He shrugs and gestures to the chilling evidence of predation right by our feet—a caribou skull, several seal vertebrae, a scattering of gull's feathers and the sun-bleached skeleton of a clean-picked walrus. "I guess it's safe to assume that most of our fellow Americans are south of us, sound asleep, a lot warmer than we are, and not preparing to put up a polar-bear fence."

Cooper is one of six barrier islands stretching off the coast of Point Barrow, Alaska, where the United States—along with continental North America—comes to a chilly, desolate end. Three miles tip to tip, the island is nothing more than a snow-covered strip of sand and gravel frozen into the Arctic pack ice, its only vertical relief an odd cityscape of rusted 55-gallon drums and destroyed ammunition boxes left here by the United States Navy sometime after the Korean War. In 1972, George came out to Cooper as a young ornithologist and discovered a rare colony of black guillemots—pigeon-size, stiff-legged seabirds—nesting in the abandoned drums and boxes. And for many years he pursued a rather esoteric study of them—mate selection, age of first breeding, "the kind of thing that's of interest to about 20 ornithologists," he says now. Then, almost by accident, he discovered that his birds were picking up on another kind of frequency, and that if he watched and listened with great care, they could tell him about something no less consequential than the climatic fate of the earth.

In coming to the Alaskan Arctic year after year, George is following the logic of many other scientists—that to understand Earth's mysterious and changing climate, you should go directly to its extremes. In the last two decades, scores of researchers have come to the nearby town of Barrow, hoping to learn why the Arctic is warming so significantly and how the changing polar climate may affect the planet as a whole—if the Arctic sea ice were to one day disappear, it would cause drastic changes in the climate of the Northern Hemisphere. But while many scientists gather their data from remote sensing devices—satellites, buoys, robotic airplanes—or come to this frozen, inhospitable region on brief, well-equipped trips before returning to the comforts of the "Lower Forty-Eight," George spends three months of each year sleeping in a small yellow dome tent, warming himself over a two-burner propane stove and crawling around on his hands and knees, up to his binoculars in guillemot scat. Thousands of miles away from the debates on greenhouse-gas emissions, relatively unknown even within the scientific community, George, now 55, has come out to Cooper Island every year for more than 25 years, often with no financial or institutional support. It is not too much to say that he has staked his entire adult life on this barren gravel bar and its avian inhabitants. And now, as he continues to scan the island with his binoculars under bright, 2 A.M. skies, it is not too much to say that his life is staked on whether he can successfully erect that polar-bear fence.

The fence became an essential part of George's repertory for survival after he awoke in his tent one night to the *crunch, crunch, crunch* of approaching footsteps. When he crawled out to investigate, however, he thought he must have been dreaming: he was alone on the island. The next morning, the 16-inch tracks in the snow told him otherwise. If it's possible to see your life flash before your eyes in retrospect, that's what happened to George: a big female bear and her cub had walked within 20 yards. While the cub stayed put, the mother came up to the doorway of his tent, evidently sniffing George's placid, sleeping head before he woke, unzipped his sleeping bag and inadvertently scared both bears away.

"All right, let's see what we've got here," George says, shaking the many parts of the fence out of its canvas bag onto the snow. George's fence is made up of 30 three-foot-high garden stakes that he and I, fueled by the adrenaline of our recent arrival, now try to place in a circle around our tents. Each stake has a small pulley, through which we thread a fine piece of cord. The cord makes a perimeter around the tents and meets up at the home stake, which is connected to a spring-loaded mousetrap, an eight-volt battery and a car alarm with a large plastic horn. The fence operates on the principle that a bear wandering into camp will push the cord, the cord will trigger the mousetrap, and the closing of the mousetrap will complete an electrical circuit that turns on the alarm, thereby waking George and me to the dangers at hand.

That, at any rate, is the principle. For more than an hour, we struggle to drive the garden stakes firmly into the snow, and each time we test the fence by pushing at the cord, a stake pulls loose and the cord slackens, preventing the triggering of the mousetrap. Then, when George turns his attention to the car-alarm horn, a sharp crack echoes in the air, followed by the gentle sound of his voice: "Uh-oh. I think I broke it."

Although George has spent more summers on Cooper Island than he often cares to count, he seems, upon first meeting, an unlikely candidate for the solitary hardships of fieldwork in the high Arctic. Handsome, boyish, with disheveled hair and a face deeply creased by abundant laughter in subfreezing temperatures, George lives nine months of the year in Seattle, and there is about him the unmistakable air of the overcaffeinated urban neurotic. He wakes up talking and, rushing to get the words out, keeps up a rapid, digressive chatter—about George Bush and the Kyoto Protocols, the challenges of romantic commitment and the latest from Philip Roth—almost until the moment, 18 hours later, that he falls directly asleep. It seems a waste of his conversational gifts for him to be on Cooper Island alone. It's also somewhat alarming. When I first met George in Barrow last February, I watched in wonder at his interactions with the mechanical world—forgetting to keep his car engine running when the outside temperature dropped to 30 below; or, when he parked and did remember to plug his engine block into an electric heater, forgetting to unplug before driving off—*snap!*

But for almost three decades, he has hurled himself to the very ends of the earth and met its risks and challenges with tireless enthusiasm. When his new stove broke on the first day of one field season, he kept a series of bonfires going all summer; when he ran out of freshwater, he placed tin cups around his tent to catch the rain; when his radio broke, he tried calling for help with an old signal mirror from World War II. (He also tried spelling HELP on the beach with large pieces of driftwood, but came up one log short, inadvertently announcing to the skies a pressing need for KELP.)

Now, holding the two halves of the car alarm in his hands, he shrugs and says, "To be honest, the fence doesn't keep bears out; it just lets you know that one is about to eat you. Really, it's just a placebo." He looks around at our camp—one tent for cooking and two for sleeping that we've pitched far enough apart for privacy while also keeping, in George's words, "within screaming distance." For our campsite, he chose what seemed the safest, most sheltered location near the island's south shore—out of harm's way, and back from the thick, upended slabs of sea ice that have rammed up against the island to the north. But as far as I can see, we are nothing if not *in* harm's way: like a Bedouin camp in the desert, our three yellow tents are the only signs of life in this white-on-white landscape—the only signs of food, come to think of it, to an animal so wily that it stalks prey by sliding on its belly behind a moving block of ice and is said to raise its white paw to cover its black nose for camouflage in the snow. "Any bear that shows up on this island is probably very hungry or very deranged," George says fatalistically, "and there's not much you can do to keep Charles Manson out of the suburbs."

II: In Which George Rides on Boats, Planes and Snowmobiles

IF THIS STORY IS, truly, about a flock of seabirds in the midst of worldwide climate change, then a reader may be moved to ask: Where are the birds? What's with all the snow and ice? And why does George seem less concerned about the tangible threat of polar bears today than he does with a few intangible degrees of global warming tomorrow? To begin to answer those questions, it may help to review the scientific argument about human-driven climate change: that our endless consumption of fossil fuels is pumping vast amounts of carbon dioxide and other heat-trapping gases into the atmosphere, causing global temperatures to rise. It may also help to brush up on some geography regarding the very top of the world.

Whether the glaciers of Greenland will continue to melt and the southern oceans rise up to flood Bangladesh, whether Cape Cod will erode to a sand spit and the American prairie dry out like the Mojave, whether thunderstorms will one day reach Antarctica and sparrows the North Pole—whether all the disasters predicted by climatologists and their computer models eventually come to pass, one piece of the puzzle is already in place: the earth's climate will change first—and change most substantially—in the Arctic, that enigmatic expanse of snow and ice, of ancient peoples and unspeakably hostile temperatures that spans the top of the world.

Despite its crucial role in managing the earth's climate, however, scientists know surprisingly little about the Arctic compared with the world's other oceans. Covered by ice, nearly impassable by ship, the Arctic is still earth's least-explored frontier. On a planet that has been thoroughly mapped, ship captains in the Arctic still make up their charts as they go along; two years ago, a Navy submarine carrying a crew of scientists passed over a drowned, mile-high mountain no one knew was there. Now that Russian science has gone bankrupt, Canadian research is suffering extreme budget cuts and most U.S. money for polar research gets funneled to the McMurdo research station in Antarctica, some of the best work to fill in the blank spots of the Arctic is being done by a small but hardy group of scientists associated with the Barrow Arctic Science Consortium, which uses this thriving, mostly Inupiat whaling community as its base.

Like most bush communities in northern Alaska, Barrow is accessible only by air, and on June 1, George and I met up in the Fairbanks airport for the 90-minute flight. We had clear skies as we left the wooded foothills around

Fairbanks and climbed steadily toward the Brooks Range. Watching the last of the boreal forest finger its way into protected valleys, we flew over the snow-and-glacier country of the Brooks—ridge after ridge of granite peaks and frozen valleys so remote that they still exist largely without name. Cresting the mountains and passing the imaginary Arctic Circle, we saw the peaks turn to foothills, the foothills level off to the North Slope's frozen coastal plain. In north-central Alaska, the coastal plain stretches for more than 150 miles, and our plane followed the course of wild, braided rivers over a snow-covered, wind-swept landscape desolate as the moon. On this day, the Arctic pack ice was still jammed up against shore, and from the plane it was impossible to tell exactly where land ended and sea began. But eventually a hodgepodge of dusty streets and low-slung buildings came into view, and we touched down in Barrow, perhaps the only town in the United States that would be lost in a sea of whiteness on the 1st of June.

Cooper Island is just 25 miles beyond Barrow, with the ice of the Arctic Ocean against its north shore and the ice of Elson Lagoon to the south, but the fact that those 25 miles include—at various times throughout the summer—solid ice, junked-up ice, choppy 33-degree water, driving snow, sleet, dense fog and 40-mile-an-hour gusts makes travel to and from the island about as predictable as the behavior of Cooper's occasional bears.

To start his field season, George usually snowmobiles to Cooper over the frozen lagoon, as we planned to do two days after our arrival in Barrow. By mid-June, however, the lagoon ice starts to break up and Cooper begins to resemble an island with water lapping at its shores. For the remainder of the summer George is dependent on some kind of air transport for travel and resupply. The phrase "air transport," with its connotations of scheduled departures and uniformed pilots, is misleading. George has built what he proudly refers to as a landing strip on Cooper, intended for small propeller planes equipped with tundra tires. In point of fact, it's a clearing in the gravel with two automobile tires to indicate the start and two crossed, wooden boards marking the end. And unless the plane is a "tail dragger," the soft gravel will make it land nose first into the ground. "Every time I walk back to camp, I drag my boots over my landing strip," George says. "Do that for 20 years, and you have a nice place to land a plane."

The first time George flew by small plane to Alaska, in 1972, he was traveling with three other biologists; within four years all had died in separate plane accidents. Of the 25 bush pilots who have flown him to and from Cooper over the years, five are no longer alive. Nonetheless, George remains sanguine about the perils of flying, often with some odd pilots at the controls. He has flown with the foolish (a pilot who took off with a chipped propeller after trying to fix it with a nail file), and he has flown with the idiotic (a pilot who refused to turn back when propane started leaking into the cabin). And once he planned to fly back to Barrow with a pilot who couldn't find the island beneath a 50-foot cloud bank and zoomed back and forth while George, on the ground, used his radio to guide him. ("You're to the north. . . . Now you're too far south. . . . *O.K., now you're right above me!"*) The only way George refuses to travel is by boat, ever since he was coming back from Cooper on a windy day and his 12-foot Boston whaler took in volumes of spray over the bow. While George bailed furiously, he and the captain watched the outboard motor slip right off the stern and sink to the bottom of the lagoon. Eventually they were towed to shore by Barrow Search and Rescue. "I'm ready to have X number of minutes of terror if a plane goes down," George says, "but just the thought of going in that water—it's 33 degrees. It's a toxic substance. It might as well be an acid bath."

My appreciation for George's fortitude only grew as I tried, while still in Barrow, to arrange a return flight from Cooper in two weeks' time. I called the local air-charter service that used to perform beach landings with a Cessna 185, but a pilot there told me they had suspended off-runway landings "after we rolled that plane up into an itty-bitty ball." When I reached another pilot with his own light, fixed-wing aircraft, he asked in a challenging tone where on the island I expected him to land. I explained about the landing strip, the tires, the X marking the spot. "Sure, sure," he replied and turned down the job as well. With so many recent small-plane crashes on the North Slope keeping pilots quivering on the ground, I eventually found myself with one last option: a search-and-rescue pilot named Gary Quarles, who owned a small vintage helicopter with a rebuilt engine, though he seemed as reluctant as all the rest to take on the potential complications of the job.

At this point, George stepped in on my behalf, and I saw for myself how he has managed to conduct a 27-year, poorly financed study in remote Alaska largely by sheer force of personality alone. Quarles has the pilot's requisite drop-dead sobriety, but George invited him over for coffee and a chat. He told Quarles about his birds. ("They're sort of like pigeons, but their black feathers have a subtle green iridescence—it's very impressive.") And he described all the methods he has devised over the years to survive on Cooper. ("It's like we're inventing a whole new culture out there.") And with a nifty bit of anthropomorphism, he explained what brings the birds back to Cooper Island each year. ("Frankly, I know what they're going through. I once drove from East Lansing, Michigan, to Sunapee, New Hampshire, to visit my girlfriend, and I actually blew out the engine of my car thinking, I've got to get there! *I've got to get there!*") By the time Quarles reached the bottom of his coffee cup, he was laughing and on board, promising not only to pick me up, but also to run resupply for George throughout the summer.

"Well, I guess it's optimistic to think Gary will actually come through for us," George said after Quarles left. "But then evolution selected humans for optimism, didn't it? If Gary can't do the job," he went on, "I bet the folks at Search and Rescue will come get you. To be honest, this is how it always goes. Every year I fly to Barrow never knowing how I'm going to get out to the island and back. But I've learned it's just a matter of putting yourself in the next situation, putting yourself in a position where a favorable outcome is not only likely, but absolutely necessary." He stopped and smiled. "See, unless you're stuck out on Cooper on June 15, no one's going to think of coming out to rescue you on the 15th."

On the morning of our departure, after a fitful night of sleep, I met up with George at the Barrow Arctic Science Consortium for a final check of equipment, then to load three long wooden sleds with our gear: food, tents, stoves, pots, pans, water jugs, shotgun, radio, Global Positioning System, sleeping bags and mounds of fleece and down. Dave Ramey, who runs operations for BASC and moonlights as George's Jewish mother, alternately worried and scolded him for leaving everything to the last minute, for not upgrading his field camp after all these years. "Really, George, why don't you just build a cabin out there?" he said. "It's crazy to be crawling in and out of tents all the time. And you really ought to use an Arctic oven—that would warm you right up."

Ramey and two of his colleagues had agreed to take us out, so we hitched the three loaded sleds to the backs of three snowmobiles and, after pushing hard from behind, ran and jumped on back. Within minutes of leaving Barrow, we hit thick fog and white-out conditions, and as we plowed across the ice-covered lagoon, we often lost sight of each other, hearing only the drone of engines in the mist before a wraithlike, hooded figure and a headlamp came back into view, as if through a desert sandstorm. For the next hour and a half, we roared across the smooth lagoon ice, tunneling through dense sheets of fog, and riding shotgun on the back of a sled, gazing out upon the absolute, featureless whiteness, I was hard pressed to say whether we were moving at 5 or 50 miles an hour; or whether the black, elongated bodies of ringed seals resting by their breathing holes were a stone's throw or a full mile away; or whether we weren't, after all, floating through a cloud bank at 30,000 feet. One summer, George came out to Cooper Island and, lacking landmarks of any kind, mistakenly set up his camp on the snow-covered lagoon—and spent five days waiting for land to appear beneath him before he realized his *terra* was no more *firma* than a slowly melting sheet of ice. He beat a quick path to shore. On this day we had Dave Ramey and a GPS to guide us, and after crossing 25 miles of lagoon ice, we reached a slight rise in elevation, no higher than the back of a breaching whale, and some bare strips of gravel, which confirmed we had reached the shores of Cooper Island.

After helping to unload the sleds, Ramey and his colleagues started up their snowmobiles for the return trip to Barrow. And after waving them goodbye, George and I watched the red taillights disappear into the white fog, then stood and listened to the whine of engines diminish, then die completely in the wind. Alone with George, I looked around at the snow-covered island with its scattering of bones and feathers, the fog hemming in our sightlines to within 100 feet. Unable to tell where land ended and sea began, peering through the mist for bears and up at the sky for birds, I groped for the comforts of the familiar, or at least the analogous. Go to the Alps and you'll recognize the Himalayas. Spend time in Malibu and you'll know what's up on Montauk. But this bleak, bewildering place, in which wind rippling a pool of melt water shows up as a sign of life—what did Cooper Island echo if you'd not yet been to Mars? George saw me shivering and said, "Eat and drink everything in sight, and you'll start to thermo-regulate." So I cinched my hood a little tighter, adjusted my face mask and, beginning to comprehend what it meant to be a creature of Earth's temperate zones, shoveled trail mix into my mouth. But in truth I'd had the shivers ever since Barrow, when Dave Ramey took me aside to deliver some forceful words. "We'll get you out to Cooper Island today," he said. "But the rest is up to you. I don't know what George may have promised you," he went on, his tone suggesting a dozen messy situations George had dragged him into. "But unless you've got reliable air transport lined up, you should expect to be out there till the ice melts out. I'm serious. Don't expect Search and Rescue, or me, or anyone else to come get you." He looked at me hard. "If you go out to Cooper Island today, you're on your own."

III: In Which George First Encounters the Mystery of the Little Black Bird

COOPER ISLAND, June 6, 12:30 in the morning. The sky is full of slanting rain and freezing fog, the air temperature can't seem to reach 30 and the wind continues to blow out of the northeast with such numbing regularity that we've begun using it for support; if it stopped in an instant, we'd both pitch face-forward into the snow. Standing at the edge of the frozen Arctic Ocean, George lifts a high-powered telephoto scope up to his right eye and begins to scan the skies, wondering when his birds will leave their winter habitat on the sea ice and make their annual trip to Cooper Island to breed. George's scope is black and cylindrical and strapped with silver duct tape to the butt of an old wooden rifle. To brace the rifle stock against his shoulder and peer through the chunky, mortarlike scope, he must squint and lean back, bending slightly at the knees. Beneath his hat, his hoods and his thick fleece face mask, he looks like a nearsighted bugler, blowing reveille into a gale.

With no sign of the birds just yet and weather conditions such as they are, we appear to have two choices: walk or freeze. We choose to walk, while George begins to tell the story of how a heap of Navy trash—some of it stamped with the words PLEASE DO NOT DESTROY, THESE BOXES ARE REUSABLE—came to be used by several hundred black guillemots and one lone scientist whose narrow, ornithological study eventually led him onto the trail of worldwide climate change.

Except for the fact that the dead walrus by our campsite had not yet decomposed, Cooper looked much the same when George first came upon it in the summer of 1972. Back then, five years after oil was discovered at Prudhoe Bay, environmentalists were on alert for the possibility of tanker spills, and George, 26 and already an ardent ornithologist, was hired by the Smithsonian Institution to go up and down the northern Alaskan coast to identify any vulnerable seabird habitats. Traveling the coastline on a Coast Guard icebreaker, he was dropped off one day on Cooper. Remarkably, the sun was out, catching the surface of all the Navy debris, and as he walked along, marveling at this strangely picturesque collection of scattered wood and rusted metal, two black guillemots, startled by the crunch of gravel, suddenly flew out from beneath an ammo box. Since guillemots don't normally breed in this part of the Arctic—they are cavity nesters whose natural habitats, rocky cliffs and headlands do not exist for more than 500 miles—it was, in George's words, "definitely a hit." He looked around and found eight more pairs breeding in the boxes, at which point, he says, "I almost wet my pants." Before he left that day, he flipped over a piece of wood, creating a new nest site, and when he returned to Cooper three weeks later, he looked under the planks and found eggs.

George couldn't get enough money to return to Cooper for three more years, but in 1975, when funds for another assessment project became available, he asked to be sent back. When he arrived, he found 18 pairs of guillemots breeding in the boxes and hoping to grow the colony, George began creating—and naming—new nest sites in earnest: Freshman Housing, the Condos, Married Student Housing. . . . "Not only were there these odd places," he says, "but you could *create* odd places and have them breed there." To a degree that was unusual even in the field of ornithology, George became obsessed, and over the next 10 years, he came out to Cooper each summer—creating nest sites, banding the birds and, because guillemots are a long-lived species displaying fidelity to both their mates and their nest sites, coming to know them as individuals. One bird, nicknamed WOGy because of its white-orange-gray leg band, has returned to her same nest site on Cooper for more than 20 years. "It was like a suburban street in the 1950's," George says. "I knew who lived in every house." By 1978, there were 70 guillemots on Cooper Island, and by 1981 the population was up to 220. By 1990, almost 600 birds were scrambling around the drums and ammo boxes, looking for places to breed.

Back then, George was thinking no more about global warming than any well-informed person with an interest in the natural world. Sharing the island with the birds and the bears, he had other things on his mind, and if he did pause to consider the Arctic climate, he assumed that it was static. Because he always got out to Cooper after the snow had melted and the birds had already occupied their nest sites, he also assumed that his seabirds timed their arrival on Cooper Island to the summer solstice, their reproductive schedule cued by the lengthening days.

In 1984, however, he happened to come out to Cooper earlier than usual, when the nest boxes were still covered by snow, and seeing that, he began to wonder if the birds' breeding habits were prompted not by the lengthening photo period, as he'd originally thought, but by access to their nest cavities—like clockwork, the first egg in the colony always showed up exactly 14 days after the birds occupied their nests.

Scientific paradigms don't shift overnight, however, and more than 10 years of grinding, repetitive fieldwork would pass before George came to understand exactly what that meant. In 1995, in response to Vice President Al

Gore's task force on climate change, a call went out for data sets: did anyone have information that would shed light on regional climate change? George acquired National Weather Service data on when the snow melted at Barrow and plotted the dates on a graph. Then he looked at his own data on when the first egg showed up on Cooper Island and plotted those dates as well. The correlation leapt off the page: from 1975 to 1995, snow was melting in northern Alaska, on average, five days earlier each decade. Over those same 20 years, the date his guillemots laid their eggs was occurring, on average, five days earlier each decade. In fact, since guillemots require at least an 80-day snow-free summer in which to copulate, ovulate, hatch and fledge their chicks—and there were rarely 80 snow-free days in northern Alaska until the 1960's—they wouldn't even be this far north were it not for warmer temperatures. Expanding their range, playing with the edge of a changing climate, his guillemots, he realized, were tracking the region's snow melt on an annual basis. And an earlier date of snow melt was, in effect, an indication that the seasons were in flux; that in a mere 20 years, the brief Arctic summer was now arriving 10 days earlier; and perhaps most important, that climate change was having a biological effect, leaving a fingerprint on a species living in a seemingly remote, pristine environment thousands of miles away from the industrial hand of man.

And over the past few years, George has come to believe that the warmer Arctic climate is changing not only his birds' breeding habits but also the species in a far more profound way. Traditionally, the birds come back to Cooper Island over three successive nights, trying to gauge the degree of snow melt from the air, and there are dangers at every turn. If George's birds arrive on the island when it is still covered by snow, they can't take refuge in their dark, protected nest cavities and, with black feathers set off against the white snow, they run a high risk of predation; one summer a snowy owl showed up during the birds' mass arrival and swiped a guillemot out of the flock, eating it offshore and nicely demonstrating why getting to the island too early poses such a danger. If, on the other hand, the guillemots arrive too late, they risk losing their longtime mates, which is what happened one summer when one of George's favorite birds, Black-White-Black, returned to Cooper two days late only to find its mate, White-Red-White, otherwise engaged. Major battles ensued as Black-White-Black tried to fight his way back into his longtime nest. "It was something to watch," George recalls. "There was blood on top of the nest site and guillemot footprints in the blood!" Caught between the perils of early and late arrival, knowing that their survival and reproductive success depends on perfect timing, the guillemots of Cooper Island have had to develop a highly sophisticated method for gauging snow melt. Those that can do it have persisted; those that can't have not. In effect, the warming of the North American Arctic has already precipitated natural selection for birds that can assess climate change in a highly sensitive way.

Just what a hormonally crazed, falcon-fearing guillemot actually looks like became visible to me on our very first night on the island, when George suddenly looked up at the sky at 2 A.M. and said with considerable excitement, "I think I hear guillemots!" I heard or saw not a thing. Again, George looked and listened: "Yup, they're definitely up there somewhere." And a moment later, I had my first sighting: 12 black birds in a tight flock, flying skittishly over the island at 500 feet, their rapid wing beats making a faint, wobbly Doppler sound like the hum of a dozen tiny windmills. Keeping in tight, antipredator formation, determined not to descend into owl or falcon range, they swooped over the snow-covered island, assessing the degree of snow melt on their nest sites and, evidently displeased with what they saw, disappeared as quickly as they'd arrived, dispersing like dust motes into the flat gray sky.

On the following night, with low fog engulfing the island, just one bird showed up, checked out the dismal conditions and flew back out to sea. But tonight, as we walk back to camp after touring the island in the steady rain, eight of them are back, flying lower and less nervously than they were two nights before. Directly above their nest sites, they set their wings and come dive-bombing toward the ground, but pull out of the dive at a hundred feet, caught between their hormonal urge to breed and their terror of landing so conspicuously on the white snow. Then, out of nowhere comes a second flock, merging with the first, flying with agitation, a few individual birds dropping from the flock with a stronger urge to land, but joining the safety of the group once again. Suddenly, their numbers have grown to 20, flying in tight concentric circles at 50 feet—around and around, dipping, rising, surveying the ground still half-covered in snow. I take my eyes away to write some notes, and when I look up again there are 40, groups merging and breaking apart, darting like schools of minnows, keeling from black wing to white underwing, visible to invisible against the slate gray sky. For half an hour, we watch the complex choreography. And at 12:52 A.M., one bird breaks from the group and heads toward ground. "This guy's going to land!" George cries out. "This guy's going to land!" Feet splayed, wings pulled back, the first guillemot of the season arrives on Cooper with a great fluttering fanfare of wings. George looks through his scope and, recognizing the bird, lets out a hoot. "That's a chick that fledged here—a Cooper product, back in the 'hood!"

After three nights alone on the island, George is delighted that his birds are back. But he's also mindful of the grim chronology: when he began his study almost 30 years ago, snow persisted on Cooper, preventing his birds from gaining access to their nests until the last week of June. Over those same 30 years, despite natural year-to-year variability, the birds have arrived, on average, five days earlier each decade. "The great thing about guillemots is that they're birds, they're nonpolitical, they have no choice but to react," he says. "Every weather station in the Arctic should have a bunch of guillemots nearby so that if skeptics doubt the weather data, you can point to the date the first egg gets laid in the colony." Tonight, it is June 6. The birds are back three days earlier than they were the year before.

IV: In Which George Impersonates a Guillemot

UNLIKE ANTARCTICA, a continent surrounded by ocean, the Arctic is mostly ocean ringed by land—the frozen, inhospitable fringes of Alaska, Canada, Greenland, Iceland, Scandinavia and Russia. And therein lies the simple reason for its crucial climatic role. For as long as human memory can recall, the majority of the Arctic Ocean has been covered, year-round, with a nine-foot-thick mosaic of sea ice as vast as the continental United States. Constantly moving, buckling, melting and refreezing, this blindingly white pack ice is remarkably efficient in reflecting solar radiation back into space before the sun's rays can overheat the region. The Arctic Ocean also serves as a kind of heat vent for the entire planet, taking the solar radiation that gets absorbed by the tropics and the temperate zones and, once it has moved poleward, releasing it to the atmosphere. But every year, as the 24-hour polar night shifts to 24-hour summer sunlight, more than half the pack ice melts, and when that white ice changes to dark, open water, the exposed ocean, instead of reflecting the sunlight, absorbs it and begins to warm the overlying air. And if, as a result of ever-increasing fossil-fuel emissions, the Arctic climate became too warm, it would create a "positive feedback loop": as the ice receded, the ocean would absorb more heat, potentially melting more ice until a cycle of heating and melting eliminated the permanent pack ice. Some computer models show that if atmospheric carbon dioxide were to double, the planet would heat up enough to melt the Arctic's summer sea ice by 2050. And if that forecast were to come true, extreme changes in the temperature and salinity of the Arctic and Nordic Seas would follow. In fact, the Arctic sea ice plays a crucial role in the circulation of ocean water for the entire planet; according to one theory, if the pack ice were to melt away completely, the fresh, frigid water cascading out of the Arctic and into the North Atlantic would stop the transport of warm water from the tropics to the high latitudes, shutting down the Gulf Stream and changing climate patterns throughout the Northern Hemisphere.

With those facts, figures and drop-dead predictions in the forefront of their minds, the scientists associated with the Barrow Arctic Science Consortium work with a special sense of mission. Walt Oechel, from San Diego State University, heads a team that flies a one-man airplane over much of Alaska's North Slope in order to measure how much carbon dioxide is produced and absorbed by the thousands of square miles of tundra—he was the first to discover that the tundra, once thought to be an absorbing sink for atmospheric carbon dioxide, had, at some point in the 1980's, become a source, in effect pumping vast new amounts of the gas into the atmosphere. Bernie Zak, from the Sandia National Laboratories in Albuquerque, N.M., runs the Barrow site of a Department of Energy project that, along with sites in the Great Plains and the tropical western Pacific, seeks to measure the role that clouds play in Earth's heat-exchange processes. Dan Endres, living year-round in Barrow, runs the Climate Monitoring and Diagnostics Laboratory station here, the government agency that, along with stations in Hawaii, American Samoa and Antarctica, produced perhaps the most famous and persuasive piece of global-warming data: the graph showing 20th-century global temperatures running—and jumping—in tandem with worldwide emissions of carbon dioxide. And 25 miles away, there's George in his hat and gloves, huddled over his Coleman stove, with a week's growth of beard, breath pluming from his nostrils and lips starting to crack from the cold.

Within a week of our arrival, Cooper Island has lost much of its snow cover, the shoreline is beginning to emerge from its nine-month encasement in the sea ice and the mass of birds is back—some 235 black guillemots set to breed in the dark cavities of the rusted 55-gallon drums and destroyed ammunition boxes littered across the flank of Cooper Island. To lower the risk of predation, and to save the brightest daylight hours for fishing out on the sea ice, the birds show up on Cooper only at night, usually sometime after 12, and roost on the north side of the island's main pond, calling to each other with a high-pitched, melancholy whistle. Then, once they've achieved the safety of numbers, they disperse to their nest sites and commence their breeding activities: courting and head-bobbing, strutting and exploring their nest cavities and, of course, copulating wildly. By day, without the guillemots in attendance, the island looks like what it is—a gravel beach with trash on it. At night, however, with the birds teeming at their nest sites, the place is transformed by a hundred scenes of carnal bliss: Cooper Island, 90210.

Now that the birds are back, George picks up his pace, walking all night in the wind and rain, taking detailed notes of which birds have returned, which nest sites are occupied, who is mating with whom—all written into one of his yellow field journals with a pen taped to a footlong tent stake, the better to manipulate it while wearing two pairs of gloves. Because the birds are gone by day and here by night, George reverses his sleeping habits so that he can observe the birds from midnight until noon, then rest in the afternoon. And so that I can observe George, I do, too. But whereas the 24-hour light, the day-for-night sleep schedule and the ceaseless wind and cold all leave me disoriented—a victim of what scientists here call Arctic brain fuzz, in which higher brain functions seem to shut down as the body works to stay warm—George slips easily out of the diurnal rhythms of civilization and into the surreal, Cooper-driven universe, waking up cheerful and energetic after six hours of afternoon sleep. Since the sun never sets—and won't for more than a month—long, undifferentiated stretches of time pass on Cooper, marked only by golden, low-angle light as the sun approaches the ocean at midnight and then, some five hours later, by a gradual brightening of the sky, followed by the snow buntings' chimelike music—a dawn chorus in a place with no dawn.

We do what we can to domesticate and structure the endlessly unspooling days. Waking at 10 P.M. to a breakfast of oatmeal or pancakes, we tune the shortwave radio to NPR's *All Things Considered.* Then, after working through the night, we sit down to a dinner of Dinty Moore beef stew at 8 A.M., accompanied by *Morning Edition.* At 12, not quite certain whether it's noon or midnight, we call it a day, wish each other good night and head off to our wind-blasted tents. There, we crawl inside two sleeping bags apiece and, truly warm for the first time in 18 hours, fall instantly to sleep. Many animals that live in or migrate to the Arctic each summer have special adaptive features. Polar bears have their eight inches of blubber, ankle-high willows possess scores of extra leaves to soak up the constant light, certain birds shut down their adrenal glands for the season so that their stress response does not become the death of them. Meanwhile, George, living like a large, flightless guillemot in his own low-lying cavity, seems to manage just fine without those adaptations.

After three days alone on the island, we are joined by George's field assistant, Tamara Enz. Extravagantly competent, no less hardy than George, Tamara immediately sets out to fix all of his half-baked projects—remounting the radio antenna to improve our communication with Barrow, shoveling snow into trash bags to avoid a late-summer drought of drinking water. Refusing to work all summer crawling in and out of a three-foot-high cook tent, she also builds a structure called a weatherport—essentially a piece of canvas stretched over an arched metal frame. The weatherport affords us a place to retreat from the wind and permits the luxury of cooking in a standing position. But because it's larger than the old cook tent, we can't seem to heat it with the propane stove; when the outside temperature is 28 degrees, the weatherport's temperature rarely reaches above 32—it's like entering a walk-in meat freezer with the uncomfortable sensation that we are its meat. Sitting beneath the flickering light of the propane lantern while the weatherport creaks in the wind like a ship on high seas, we huddle around the radio, listening eagerly for the weather report, and the report is always the same: highs in the upper 20's, chance of rain, flurries and fog, wind out of the northeast at 15 to 20 miles per hour. Tamara, who at 35 has spent most of her working life in field camps from Maine to Alaska, tries to put the Cooper experience in perspective for me. "Here there's no camp cook, there's no place to go and you're on call all the time," she says. "You spend all day in the rain, the wind and the cold. And to warm up, you walk around in the rain, the wind and the cold." George smiles at the description and says: "I've been doing this for so long, I've lost the ability to assess what's uncomfortable. I mean, it's 32 degrees in here, and I'm working in gloves, but basically I feel good. Sure, my feet feel a little funny, and I'm losing sensation in my lower lip, and for dinner I'm drinking hot Jell-O and eating Wheat Chex melted in chicken bouillon"—he toasts the air with his cup—"but at 32 degrees, it tastes like mother's milk!"

Over the many years George has studied his guillemots, he has developed what he refers to as the Cooper culture—the practice of surviving on the fewest resources possible. For the most part, poverty has been the mother of invention. Though George now gets financing from the National Oceanic and Atmospheric Administration and has income from his off-season work as a seabird consultant for the council investigating the Valdez oil spill, for many years he had to cobble together his field-season budget from a half-dozen sources or, when his funds dried up completely, raise the money himself. During much of the 1980's, he had no funds, no assistant, no radio, and he lived on a diet of oatmeal and rice. To minimize his resupply costs, he'd cache his unused cans of food and fuel in the permafrost, then draw himself a map so that he could dig them up the following year. And he has taught himself time-honored survival techniques used by the Inupiats, like supplying himself with drinking water by melting multiyear sea ice, or waiting for freshwater to float to the top of the lagoon during breakup.

But even now, when he has the money to prosecute his research with less day-to-day hardship, and at a time in life when many of his friends are complaining of their aches and pains, George chooses the ascetic path. "People are always saying to me, 'Why don't you build a shack?' But I need to have a personal relationship with the birds. I need to be in their environment, to experience what the guillemots do, to know what it feels like to leave your cavity in the wind and the rain." Among Arctic scientists, many of whom have their data relayed to them by computer, George is an anomaly, and when he attends conferences on polar science, he sometimes gets miffed. "I once heard someone give a paper on trace metals in Arctic water," he says, "and it was clear the guy had never even been to the Arctic. I thought, *I've passed more Arctic water through my bladder than you'll see in a lifetime!*"

George looks out the weatherport door into a swirl of fog and freezing rain. "Do you think it's strange that I've left a series of beautiful women in April and May—one of whom wanted me to spend the summer at *her mother's Long Island summer house*—to come up here?" He shakes his head. "Basically it all comes down to the yellow field books at the end of the year. I mean, I actually *broke up* with someone here in 1980, but I look at the field books and think: *1980, now that was a good year!*"

In the field of ornithology, you can find other examples of scientists who have kept long-term studies going, year after year. In Great Britain, starting in the 1930's, the naturalist Ronald Lockley studied shearwaters and puffins for 20 years. In New Brunswick, Charles Huntington, a professor from Bowdoin College, has kept a study of Leach's storm petrels going since the 1950's. But in the nascent field of climatology, rare is the scientist with anything longer than a 5- or 10-year data set. That the guillemots come back year after year is, on the whole, less surprising than the fact that George does, too. In the 27 years that have elapsed since he first began his study, the dead walrus by our campsite has lost all its skin, fat and muscle; the young boy who once brought out supplies to Cooper grew up to become the mayor of the North Slope Borough; and Cooper Island itself has eroded a quarter of a mile to the northwest—George's old campsite and airstrip on the east end are now completely underwater. When George wanders the beach, strange, ancient-seeming objects catch his eye, and he picks them up, marveling at how they could have reached Cooper Island—before he remembers bringing them out himself more than two decades before. All of which has put him in a unique position to track changes in the Arctic climate and to make sense of large, seemingly random events that take years to figure out, like the one that began 10 years ago when the population of his guillemot colony gradually and mysteriously began to drop.

Throughout the 1980's, almost 650 birds were coming to Cooper Island each summer, and with 85 percent over-winter survival, competition for the 200 or so nest sites was fierce—one bird came back 11 years in a row before it could breed. "It was like rent control in New York," George says. "They were all waiting for someone to die." Then in 1995, he passed by two nest sites and saw the same male going back and forth, pairing with two females. "I didn't believe my eyes," he recalls. "I'd never seen a female without a male, I'd never seen bigamy, and as all males know, you can't keep one female happy. . . ." Looking around his colony that year, he saw 10 more females who owned a nest site, but couldn't attract a mate. Something, apparently, was decreasing adult survival.

A second clue came four years later, when George noticed a lot of sibling aggression among chicks—a sign that food was scarce. And at several nest sites, he saw something else he'd never seen before: orphaned chicks, not yet able to fledge, starving and walking away from their nests toward shore in search of foster parents and food. "It was pretty disturbing. I picked up one chick, and it was more stress than he could take—he died right in my hands." Something was decreasing adult survival, and something, apparently, was killing off chicks.

If George's colony had consisted of any bird besides a black guillemot, his investigation of this gradual population drop might never have led him to look for answers in the Arctic environment. Arctic terns, for example, winter in Antarctica and fly 25,000 miles each year—through the tropics, through the temperate zones—before arriving in the Arctic to breed; anything between the earth's two poles could cause them to die off. Similarly, glaucous gulls, though not particularly migratory, feed on human sources of food like dumps and carrion, and therefore pick up anthropogenic contaminants that could confound interpretation. George's guillemots, on the other hand, spend their whole annual cycle in the Arctic—wintering from September to May in the pack ice of the Chukchi and Beaufort Seas, then coming to Cooper Island to breed. As George puts it, "They're not temperate-zone birds just slumming it in the Arctic." The guillemots feed at the ice edge all year long, where prey is most plentiful; they feed only on other Arctic organisms like cod and zooplankton; and they have a long, 80-day breeding cycle that they must wedge into the brief Arctic summer, which starts with the snow melt in June and goes right up to the first snowfall of the year in September. They are creatures, in other words, wholly dependent on snow and ice habitats sustained by them, restricted by them and adapted to them—they are captive to the Arctic environment and thus the first to reflect a change.

And George, no less captive to the snow and ice of Cooper Island, began to see a correlation: when the polar pack ice remained up against the shores of Cooper Island, as it did in the 1970's and 1980's, his guillemots—able to feed easily from the nearby ice edge with its great density of prey—had fabulous breeding success. When warmer summer temperatures caused the pack ice to pull offshore and retreat northward out of sight, as it did through the 1990's, his birds were unable to reach the ice edge and began to die off. By 1999, when a series of papers came out describing a major retreat and thinning of the Arctic pack ice due not only to gradually warmer temperatures, but also to a decadelong upper atmospheric shift called the Arctic Oscillation, George was in a position to put the pieces together. His colony was not merely tracking the advancement of snow melt and the earlier arrival of summer; it was also articulating a change in the very makeup of the Arctic itself—the shrinking of the polar ice cap—with all its potentially drastic worldwide consequences.

George, of course, is not the only scientist tracking the physical and biological affects of a warmer climate—permafrost melting, coastlines eroding, moose expanding northward, walruses losing ice habitat on which to pup and hunt. But when it comes to relating such observations to the larger, slow-moving story of climate change, those data sets are useful only in direct proportion to their longevity and depth. And whereas some scientists find the long-distance work of climatology tedious and lacking in the kinds of signal events that grab people's attention, George is undeterred, married as he is to the year-to-year process.

It's interesting to consider: if George had begun his study five years ago, he would have missed the advancement in snow melt, the advancement of summer. If he'd begun his work as far back as 1990, he could never have connected the colony's population drop with the retreat of the pack ice. Having amassed a continuous, eyewitness data set for 27 years, however, he was watching the climate change not only in year-to-year increments but also in shifts from decade to decade, which has enabled him to see through the static of natural climatic cycles like the Arctic Oscillation, which may warm the region one decade, then cool it the next.

"I sometimes forget that there's no other island in the Arctic where someone has gone back for 27 years," he says, cinching his hood and drawing up his face mask, preparing to leave the weatherport. "Now I feel totally obligated to keep the study going. People say, 'Couldn't you just take a year off?' But if I skip a year, then it's lost. I only have one chance." He opens the weatherport door, and the wind nearly snaps it off its hinges. "I don't want to sound like the Old Man of the Arctic, but I can remember the year the ice did this, the year the snow did that. I have the data set, I've got the numbers. It may be that I came back to Cooper all these years simply because I have minor attention deficit disorder, or all this summer sunlight has given me an addiction to high serotonin levels. But it meant something to me in a way that wasn't abstract. I was there. I saw it. And I saw it because I was there, living in my tent."

V: In Which George Sees into the Past

The farther north you travel in this hemisphere, the more you hear conversations about the climate getting conducted not in the future, but in the present tense. In Whitehorse, the capital of Canada's Yukon Territory, officials now hold an annual exposition showcasing products to help residents mitigate and adapt to an already warmer climate. In Barrow, the Alaska Eskimo Whaling Commission spent a large part of its annual convention last year discussing, among other things, the perils of hunting bowhead whales from increasingly thinner ice. Some Alaskan natives, mindful that "traditional knowledge" is often considered merely anecdotal and lacking in scientific rigor, have set up a Web site (nativeknowledge.org), on which you can see two thousand people sharing much the same anecdotes: turtles appearing for the first time on Kodiak Island, birds starving on St. Lawrence Island, thunder first heard on Little Diomede Island, coastal storms undercutting houses at Shishmaref, snowmobiles falling through the ice in Nenana. . . . Already the central Arctic is warming 10 times as fast as the rest of the planet, outpacing even our attempts to describe it. In Canada's Northwest Territories, Inuit Eskimos saw their first robin last summer, though there's no word in Inuit for "robin."

Those who remain skeptical that the Arctic is undergoing a period of rapid climate change point out that the region has always gone through cycles of warming and cooling, sometimes in just decades or even years. Perhaps, according to this argument, natural fluctuations cause the water to warm and cool, and the ice to thin and thicken as atmospheric pressures, water currents and wind patterns change. But the latest findings coming out of the Arctic suggesting a longer-term trend are hard to dismiss, even if their ramifications may not be felt elsewhere for many years. Recently unclassified submarine data, for example, show that the ocean's covering of snow and ice has thinned in some places by up to four feet, or 40 percent, since the 1960's. And satellite data indicate that the ice's reach has receded at a rate of 3 percent per decade since the 1970's. Recently, warmer, saltier water from the North Atlantic has crept farther into the Arctic basin than ever seen before. And that water is about 2.7 degrees Fahrenheit warmer than it was only a decade ago, causing further melting. Current models predict that global temperatures will rise by between 0.9 and 3.6 degrees Fahrenheit by 2050. And the rate of increase could be three to five times higher in the Arctic.

For their part, the black guillemots of Cooper Island cannot foretell the future. But just this past year George did devise an ingenious way for his birds to narrate the past. In the feathers of any guillemot are a host of chemical compounds that reveal aspects of its physiology, and one such compound—a naturally occurring carbon isotope called delta 13C—gives, in effect, a snapshot of what that bird has eaten from the carbon-based food chain in the past six months, just as a human autopsy can reveal the deceased's last meal. Since each region of the Arctic also has a different carbon signature—the sub-Arctic Bering Sea, for example, is biologically highly productive and therefore possesses more of the isotope than the more northerly Beaufort and Chukchi Seas—a particular feather's delta 13C content can identify not just *what* that bird may have eaten, but also *where.*

Knowing that delta 13C is permanently preserved in feathers; knowing, too, that many Barrow-area guillemots going back as far as the 1880's had been shot, stuffed and housed in museum collections throughout the country, George sought permission to analyze those birds' feathers in order to compare them with feathers taken from his Cooper Island colony. Permission was granted, and guillemot feathers going back 120 years arrived at his Seattle home from collections in Philadelphia, Fairbanks and several points in between. George immediately sent them off to a lab, and when he plotted his data on a map, the results were startling. While the delta 13C content of the 19th-century birds was quite high, suggesting they'd had to fly as far south as the Bering Sea in winter to find ice cracks in order to fish, the more recent feathers possessed far less of the isotope, indicating that the birds had been able to winter some 500 miles to the north, in the Beaufort and Chukchi Seas. It suggested that over the past 120 years increasingly warmer temperatures were causing the pack ice to recede, causing cracks to open up farther and farther north. In effect, George had taken his 27-year study and back-cast it to show that guillemots were tracking more than a century of warming.

"Skeptics can always find fault with the instrumentation used to take temperatures back in the 1880's," he says. "But with the carbon isotope, it shows a huge decrease in delta 13C from 1880 to the present, which only makes sense if the birds were wintering farther and farther to the north. It's incredibly powerful. It's more than a hundred years. And it is," he says in a moment of gravity, "the only interpretation of this data."

VI: In Which George Confronts His Own Mortality

Cooper Island, June 18, 2 o'clock in the afternoon. In the huge amphitheater of the sky, several weather systems are playing simultaneously—rain to the south, cumulus clouds to the north—but for the moment the feature attraction is directly overhead: azure skies, not a cloud above, and the sun warming the air to an astonishing 35 degrees. With no wind for the first time in over two weeks (and no trees to rustle in a breeze), the island's birds have complete dominance of the sound stage: honk of geese, warble of snow buntings and—the soundtrack of the Arctic—the upward yodeling of long-tailed ducks as they fly in perfect V-formation above our heads. My, what a beautiful day!

The good weather is of more than passing interest to us right now. Four days ago, just minutes before he was scheduled to airlift me off the island and resupply George and Tamara with food, Gary Quarles radioed that he wouldn't be coming out after all. With the sky thick with fog, visibility was dangerously low, he said. And besides, Barrow Search and Rescue was short two pilots and had put him on 24-hour call. He'd get out to Cooper as soon as possible, he promised, weather permitting. But that was four days ago, and though we keep gazing up, looking for signs of our deliverance, there has been no indication of a chopper in the sky. For George, this is how it always goes. One year, he ran out of food on Cooper and radioed to Barrow, arranging for resupply, but the dispatcher went on vacation without forwarding George's message, and two weeks passed before someone happened to walk by the radio and heard his plaintive voice: "This is Cooper Island, *can you read me?*" Now, sitting outside the weatherport in the uncommon sunshine, he shrugs and says, "Nothing to do but hope the weather stays clear and look forward to some very positive news. I'd say you're getting the full Cooper now."

In the course of his 27-year study, George has not missed a single summer on Cooper Island, but there was a time in the mid-90's when he thought of giving it up. Back in Seattle, in what he refers to as "the dark years," his marriage was coming apart, he needed to spend more time with his school-age son, Karl, and he began to doubt the value of all this work. "Like it or not," he says, "I've had a number of relationships go to hell because I always leave on June 1, saying, 'I'm going to the moon—goodbye.' And the financial commitment is not insignificant. And so, up to 1995, there was a feeling of—'Wait a minute. What's going on here? Is this just another study with utility to only a subset of ornithologists? What have I done with my life?' " But when, rather suddenly, his bird study began to intersect with the larger story of climate change, George recommitted to Cooper Island and his colony of guillemots in ways he'd never imagined.

Among his colleagues in Barrow, George is a local hero for the tenacity he has shown out on Cooper Island. And his graph displaying the advancement of egg-laying among his colony of guillemots has been given a special place of honor on the wall of the Climate Monitoring and Diagnostics Laboratory in Barrow, next to the graphs of several multimillion-dollar government studies. But aside from completing his dissertation in 1998 and having an article accepted by the academic journal *Arctic*, he hasn't published his results or sought a wider audience. And when I ask him why that is, he looks down, he removes and cleans his glasses with his shirt, and when he looks up again, he speaks in a slow, deliberate voice. "It makes me feel really bad that I haven't gotten this out earlier," he begins. "And so it's hard for me to talk about. I think that whatever characteristics cause people to do long-term studies are somehow linked to their not wanting or needing to be published. But I don't want to make excuses. I'm 55. My father died when he was 54. I don't want to say that I outlived my dad and then fritter away the next 20 years. Or die and have someone say of me, 'He had a data set that could have really added to the debate.' Now," he goes on, "there's almost an obligation. Especially with George Bush in office, and people saying, 'Is climate change real?' You still get these people who say, 'Do you really think it's happening?' and I'm, like, '*What is it you don't understand?!*' It needs to get out, and it needs to get out soon. People say that it's happening naturally, and why should we worry? But the world may not have the stability we think it has. This," he says, gesturing around the island, "is evidence that stasis isn't operating."

George does not involve himself with the various strategies—conservation, reducing fossil-fuel emissions, reforestation programs—that may if not reverse then at least mitigate what the vast majority of scientists now believe to be a world-wide warming trend driven in part by human activity. As George sees it, his job is to question and observe until he fully understands the workings of his own particular planet—this strip of sand and gravel 25 miles off the coast of North America. But to witness all the changes that have come to Cooper Island and its birds over the past 30 years is to wonder when those changes will work their way up the food chain to us, despite civilization's capacity to buffer us from the day-to-day pressures of natural selection that formed the species as we know it. To be stranded on Cooper Island is to be reminded of the larger sphere on which we are all confined, along with all the changes we may have wrought. George shakes his head and looks off. "Aside from the nuclear threat," he says, "there hasn't been much in science that has the potential to affect a larger percentage of the population's everyday life."

And so, George is stepping up the pace. Having all but abandoned his original ornithological inquiry in favor of an all-out assault on Arctic warming, he plans to put a portable weather station on Cooper to get more precise correlations between the climate and his birds. He has talked to a scientist with a robotic airplane about flying out to Cooper to photograph the island from a bird's-eye view. He plans to begin sampling fat in chicks and adult guillemots to see if their nutrition can be related to ice conditions. And he has applied for a berth on a government icebreaker cruise next year to study the guillemots' winter habitat out on the ice.

Finally, at long last, Dave Ramey's voice comes over the radio, informing us that Quarles is on his way. And within half an hour, we can see his helicopter cruising toward us over the ice of Elson Lagoon. After two and a half weeks spent on Cooper Island time, my departure seems to occur in fast-forward. The helicopter touches down by the weatherport in a swirl of dust and gravel. With Quarles waving me in, I bid a hasty farewell to George and Tamara. And after I throw in my packs and jump on board, we lift off in the deafening roar of the rotar-chop.

As we climb into the sky, Cooper Island begins to recede and lose detail. Out on the ocean, a line of pressure ridges—huge, colliding slabs of blue-green sea ice—rises up like the skyline of a distant city. Above them, a thin dark line of cloud—what the Inupiat call "water sky"—reveals the invisible presence of an ice crack out on the horizon. Give or take a passing icebreaker, or a native hunter looking for seals from an ice floe, the next group of human beings is probably in Svalbard, Norway, on the other side of the globe.

By the time we reach our cruising altitude of 1,000 feet, Cooper Island is just a patch of gravel in the vastness of the frozen ocean. From up here there's a temptation to employ the tropes of a hundred nature writers, to think of Cooper—of the Arctic in its entirety—as some distant, untouched environment, following only the rhythms of nature that the Alaskan poet John Haines once wrote of: "A place where the clocks are stilled and the sun still holds some of its ancient power; another kind of rhythm dominates existence: *When the ice goes out, when the fish come, when the geese and ducks begin to gather.*" But already those rhythms have been disrupted. The birds come back to the island earlier and earlier each year, the ice pulls offshore and the birds starve. And if the seas do rise, even Cooper Island, at nine feet above sea level, will go under. It may take years—it may not happen in his lifetime—but George plans to get himself a raft, just in case.

When I look below me, I can barely find the weatherport and the three yellow domes. But then I spot George out by the north beach, knees bent, telephoto scope up to his eye. Clinging to the frozen rim of the inhabited world, with his levitating tents and his wavering radio antenna, his shotgun and his limp polar-bear fence, he looks to the skies and waits.

About the Contributors

NATALIE ANGIER, whose science writing for *The New York Times* won her the 1991 Pulitzer Prize, started her career as a founding staff member of *Discover* magazine, where her beat was biology. In 1990 she joined the *Times*, where she has covered genetics, evolutionary biology, medicine, and other subjects. Her work has appeared in a number of major publications, and she is the author of three books: *Natural Obsessions,* about the world of cancer research (recently reissued in a new paperback edition); *The Beauty of the Beastly;* and the national bestseller *Woman: An Intimate Geography,* published originally in 1999 and now available in paperback. Currently she is working on a new book, *The Canon: What Scientists Wish That Everyone Knew About Science,* and serving as the editor for Houghton Mifflin's *The Best American Science and Nature Writing 2002.* She is also the recipient of the American Association for the Advancement of Science–Westinghouse Award for excellence in science journalism and the Lewis Thomas Award for distinguished writing in the life sciences. She lives in Takoma Park, Maryland, with her husband, Rick Weiss, a science reporter for *The Washington Post,* and their daughter, Katherine Ida Weiss Angier.

She writes: "I started working on this story on September 12, 2001, driven, as we all were, by an overwhelming desire to do something, to counter an unspeakable outrage through the small, specific, and distinctly human act of using one's voice. I wanted to say, I needed to say, You will not succeed, you bastards. You can level two of the tallest skyscrapers in the world, and all the thousands of people and their universes within. But as long as there are lives, and sun, and energy left behind, you inevitably will lose; for life has a way of repeating itself, of loving the extravagant life that it creates from itself, and of building back what you seek in your lazy, sputtering self-indulgent hostility to tear down. Anabolism trumps catabolism every time. And the generosity and beauty that we see in nature, that is expressed most exquisitely as human altruism at work in the face of horror, are variations on the simple words that life loves best: I am."

LISA BELKIN has been a contributing writer to *The New York Times Magazine* since 1995, specializing in medical and social issues. She also writes the "Life's Work" column for the *Times*'s *Workplace* section, about the intersection (and collision) of life and work. Since joining the *Times* in 1982 as a news clerk in the Washington bureau, she has covered family/style news, business, and culture, as well as being a national correspondent based in Houston and reporting on the New York City hospital system. She is also the author of two books: *First Do No Harm,* about a hospital's ethics committee; and *Show Me a Hero,* about a neighborhood torn apart by a judge's desegregation order. Among the numerous honors she has received are the Front Page Award from the Newswoman's Club of New York, the John Barlow Martin Award for Public Interest Magazine Journalism, the American Medical Writer's Association Book of the Year, and the online journal *Salon*'s Top Ten Non-Fiction Books of the Year. Born in New York City, Ms. Belkin graduated, magna cum laude, from Princeton University in 1982. She is married to Dr. Bruce D. Gelb. They have two sons.

She adds this update to "The Made-to-Order Savior": "In the year since I first met Henry and Molly and their families, much has changed. Molly has a head full of thick curls now, and she is back at school, without a mask. After school she is hard to keep up with—riding horses, playing soccer, practicing ballet, being a Brownie scout. She still has a feeding tube so that her mother can get enough nutrition into her, but otherwise she leads what her parents consider a normal life. And she still has a special relationship with her brother, Adam, but gets mad at him when he bites the head off her Barbies.

"Henry's life is also worlds removed from where it was when my article first appeared. Back then I worried that he would not live to see his story published. Instead, for reasons his doctors don't completely understand, he got better. He was left with a nasty skin condition, which is probably caused by chronic graft-versus-host disease. But compared to the misery he's been through, the itching and inflammation are only a minor burden. In September of 2001 he did something his parents feared he would never do—he started kindergarten and he's had the best attendance record in the class. A month later his baby brother, Joe Strongin Goldberg, was born. So far he hasn't gnawed on any Barbies.

"Yet while nearly everything has changed for these families, one thing remains the same. Henry and Molly are not cured. They still have Fanconi anemia, with all the future risks that brings. 'Please don't declare him healthy,' Laurie warned when I told her I'd be writing this epilogue. 'Who knows what will happen next?'"

David Berlinski was born in New York City in 1942 and educated at the Bronx High School of Science, Columbia College, and Princeton University, from which he received his Ph.D. He taught philosophy and logic at Stanford University during the 1960s, and during the 1970s worked as a management consultant with McKinsey and Company and as a senior quantitative analyst for the City of New York. During the late 1970s, Berlinski served as a professor of mathematics at the Université de Paris at Jussieu, and thereafter he held research positions at the Institute for Applied Systems Analysis in Austria and the Institut des Hautes Etudes Scientifiques in France. He has taught mathematics and philosophy at altogether too many American universities. His books include *On Systems Analysis; Black Mischief: Language, Life, Language and Logic; A Tour of the Calculus; The Advent of the Algorithm;* and *Newton's Gift.* He is also the author of three novels. He now lives in Paris.

"The sciences are plainly in an uncertain state," he says by way of introduction to "What Brings a World into Being?" "There is in this nothing amiss. We are born to grope and fumble—a destiny admirably achieved. It is not the details that are missing. The details have been gushing from a thousand telescopes, laboratories, learned journals, and Internet sites. What we do not yet know is what those details mean. Enter thus the magical concepts, all glitter and flash—cosmic inflation, multiple universes, Darwinian evolution, self-organization, chaos, catastrophe, systems complexity. And information, the moment's verbal charm. The genome? Pure information. The mind? Ditto. The very cosmos? Ditto yet again.

"Now information is informally a handy way of suggesting what facts convey or sentiments affirm. But information is also a profound and subtle technical concept, the mathematical theory an ornament of contemporary thought.

"Trouble arises when weight is placed on the mathematical concept that it cannot bear. The diamond bright theory *is* diamond bright, but diamonds do what diamonds can do, and that is often very little. So, too, information. Renunciations are now in force: they form the burden of my essay. Information is not a great creative force; it does not bring things into being; it has no causal powers. And out in the badlands beyond the *mathematical* theory, it explains very little."

JOSEPH D'AGNESE has written articles for *This Old House, The New York Times, Saveur, Garden Design,* and *New Jersey Monthly.* In 1999, he became a contributing editor to *Discover* magazine. He began his career as a writer and editor for *DynaMath,* a children's mathematics magazine published by Scholastic. His two picture books for kids—on the lives of mathematicians Carl Friedrich Gauss and Leonardo Fibonacci—are forthcoming from Henry Holt. D'Agnese lives in Hoboken, New Jersey.

"Some scientific disciplines get all the attention," he reports, "and others go ignored. The summer this story on tissue engineering ran, President Bush sharply limited embryonic stem cell research, and cloning, with its attendant fears of soulless, marching replicants, continued to be a dirty word. Strangely, tissue engineering—an allied field which seeks to create human tissue and organs from the sick patient's own cells, and which is more successful, more advanced, and less controversial—got no play. While working on the story, I was surprised to learn that tissue engineers don't really understand why a couple of cells sprinkled on a plastic mold multiply into a living tissue or organ. Once they're clinging to a three-dimensional form, the cells somehow manage to commune with their siblings and work together to form a functioning whole.

"That process is embodied by the Vacantis, four brothers who are the field's reigning family. Having myself grown up in a family of three brothers, I was intrigued by the notion that a scientist could work with his brothers and actually get something accomplished. The Vacantis certainly do. Since we published, the UMass team has isolated living human sporelike cells from the seemingly dead flakes of skin found in common household dust. The evolutionary implications are fascinating: When we breathe in dust, are we exchanging genetic material with other living things?"

JULIAN DIBBELL (jdibbell@yahoo.com) is the author of *My Tiny Life: Crime and Passion in a Virtual World* and has been writing about the culture and politics of digital technology for over a decade. His work has appeared in *Feed, Wired, The Village Voice, Time, Harper's Magazine, Spin,* and many other publications, including over a dozen anthologies. He lives in South Bend, Indiana.

"When I wrote it," he explains, "'Pirate Utopia' was about an arcane subspecies of secret communication known as steganography. It was not, I regret to say, about Osama bin Laden, though I would love to be able to claim that his prominent place in this February 2001 essay represents a canny premonition on my part. In fact it was more like an accident. At the time, bin Laden was still something of an arcane subject himself, and the fact that he had recently come under suspicion of hiding secret messages in publicly available Internet porn struck me not as cause for alarm but as fodder for a Leno joke or two—and also an amusing way of introducing my readers to the topic at hand. Had I known what bin Laden's name would mean to the world by the autumn of 2001, I would certainly have wiped that ironic smirk off my lead. And I think I'd have taken the terrorist's connection to the topic at hand a little more seriously.

"Never mind that there turns out never to have been much proof that anyone, let alone al Qaeda, uses digital steganography to communicate over the Internet. And never mind that even after September 11, new versions of the charge—from the White House's fear of clandestine communiqués in al Qaeda broadcasts to one supermarket tabloid's discovery that BIN LADEN'S TIMEX SENDS CODED MESSAGES!—remain hard to put much faith in.

"The fact is, 'Pirate Utopia' is as much about the political fantasies surrounding steganography as it is about the technological facts. It's about technology not as tool but as icon: as dream image of political possibility. And while it's true that, when I wrote the essay, the post-hippy anarcho-utopianism of which steganography once seemed so redolent was the possibility that interested me most, I did recognize others. Already, as the essay notes, the technology had been adapted to a corporate fantasy of total control in the realm of intellectual property. And even in my smirking lead I saw how effectively it might stand for a nightmare of civil life suffused with the hidden chaos of terrorism.

"It fills me with nostalgia to say, however, that I could never have guessed how real that nightmare would become."

CHRISTOPHER DICKEY is the Paris Bureau Chief and Middle East Editor of *Newsweek* magazine. He's also the author of several books, including a memoir, *Summer of Deliverance,* which was picked by *The New York Times* as one of the notable books of 1998. He is probably better known for writing about terrorists than scientists. His 1997 novel, *Innocent Blood,* was a psychological study of the kind of American boy-next-door who might bring Islamic terror home to the United States. Among Dickey's recent honors was an Overseas Press Club Award, shared with other *Newsweek* staffers, for coverage of Osama bin Laden's networks before and after September 11.

So what's all this got to do with a GFP bunny? "Not a whole lot. It was just a story that interested me," he says. "Eduardo Kac may have spun a tale about his rabbit Alba, but the concept behind genetically modified art is fascinating. Kac staked his claim on this supposedly luminous rodent to explore the idea of 'the other,' and his ambition is to create beings that are about as utterly other as he can imagine. Yet there's also a strangely cozy quality to his lifework. Alba was supposed to become a family pet. I guess what Kac wants, like many scientists and artists (and terrorists, for that matter) is to play God. But in the event, he proved all too human. As for Alba, I occasionally fantasize about green-glowing paté, but as far as I know she still hasn't entered the food chain."

TIM FOLGER is a contributing editor for *Discover* magazine. He now lives in northwestern New Mexico, where, between freelance assignments, he is trying to finish a book about the Navajo code used by the Marine Corps in the Second World War. He is also the series editor of *The Best American Science and Nature Writing,* published by Houghton Mifflin. In a previous life, before returning to college to study physics, and long before becoming a journalist, he worked for a few years on ships in the North Sea and off the New England coast.

He writes: "It seems extraordinary to me that nearly a century after the creation of quantum mechanics, physicists still don't agree on what the theory tells us about the universe we—and maybe all those other versions of us—live in. I am comforted by the thought that if David Deutsch is right, then this book will generate an infinite amount of royalties across the multiverse."

DARCY FREY is the author of *The Last Shot: City Streets, Basketball Dreams.* His writing has appeared in *The New York Times Magazine, Harper's Magazine, Rolling Stone,* and many other publications. He is the winner of a

National Magazine Award and the Livingston Award, and his work has been collected in several anthologies, including *The Best American Essays 1994*. He lives in New York City.

"Before meeting George Divoky," he explains, "I spent several weeks in the frigid winter darkness of Barrow, Alaska, interviewing one environmental scientist after another, trying to find a way to put a human face on the issue of global warming. Everyone kept telling me I had to talk to George. When I finally did meet up with him, in a Barrow coffee shop, I realized right away that I'd found the field's most intriguing character. Global warming is such a maddeningly intangible, ungrounded subject, but the thing about George is that for three months of the year, he literally lives on the ground, in an unheated tent on a remote barrier island off the northern coast of Alaska. Three months later, when George returned to his island, I went with him—and spent three freezing weeks stranded in the Arctic Ocean, working with George, unsure whether my flight home would ever arrive."

ATUL GAWANDE, a surgical resident in Boston, has been a staff writer on medicine and science for *The New Yorker* since 1998. He is also a research fellow at the Harvard School of Public Health and has served as a senior health policy adviser in the Clinton administration. His first book, *Complications: A Surgeon's Notes on an Imperfect Science,* was published in 2002 by Metropolitan Books.

"A minister I know was the one who got me thinking about blushing," he says. "He is a chronic blusher, turning wine-red and self-conscious at the slightest provocation—even, as everyone in his congregation knew, while giving sermons. Yet for decades he still got up at the altar week after week, though often he hated it. He made himself, he told me. It is this perplexing intertwining of physiology and emotion and the will that makes blushing so intriguing. And the fact that surgeons have begun doing operations to take the physical phenomenon of it away raised the possibility of understanding perhaps a little more about how these things are intertwined in all of us. Now, as for me, do I blush? Of course. Pretty easily, in fact. But with my dark skin, almost no one knows."

JEROME GROOPMAN holds the Dina and Raphael Recanati Chair of Medicine at the Harvard Medical School and is Chief of Experimental Medicine at the Beth Israel Deaconess Medical Center. He serves on many scientific editorial boards and has published more than 150 scientific articles. His research has focused on the basic mechanisms of cancer and AIDS and has led to the development of successful therapies. His basic laboratory research involves understanding how blood cells grow and communicate ("signal transduction"), and how viruses cause immune deficiency and cancer. Dr. Groopman also has established a large and innovative program in clinical research and clinical care at the Beth Israel Deaconess Medical Center, an institution that provides specialized medical services to people with AIDS and cancer. He has also been active in community education projects, fostering AIDS awareness among teenagers and young adults. He has authored several editorials on policy issues in *The New Republic, The Washington Post, The Wall Street Journal,* and *The New York Times.* His first popular book, *The Measure of Our Days,* published in October 1997 by Viking, explores the spiritual lives of patients with serious illness and the opportunities for fulfillment they sometimes find. In 1998, he became a staff writer in medicine and biology for *The New Yorker.* His second book, *Second Opinions: Stories of Intuition and Choice in the Changing World of Medicine,* was published in February 2000.

He gives this background to "The Thirty Years' War": "In December 1999, I attended a scientific meeting on blood diseases and cancer. As often occurs at such gatherings, experimental data were communicated not only from the podium in lecture halls but among groups of scientists huddled in the corridors. It was during such a corridor conversation that I first learned of a new therapy for chronic myelogenous leukemia, called Gleevec. My grandmother had died of that disease, and as a practicing oncologist I had treated afflicted patients with limited success. Now, finally, a rational therapy that targeted the deranged machinery of the leukemic cell had been developed. Initial clinical results indicated it had startling salutary effects.

"The idea for an article germinated in my mind, to explore how this remarkable new drug came to be and what it said about the American cancer enterprise launched nearly three decades before. I began to read newspaper articles and Congressional reports that prefigured Nixon's war on cancer, and I was shocked to learn how miscast they were. Members of the scientific community had led the public to believe that sufficient knowledge existed to conquer cancer and that with enough will and resources, the malady would be eradicated in short order. Would the elegant biology and profound clinical results with Gleevec lead to the same type of global overpromising?

"My aims in writing 'The Thirty Years' War' then broadened. I sought to distinguish hype from hope, to critically assess the national cancer effort, and to craft a clear perspective on the accomplishments and failures of the past, on the present state of knowledge in cancer medicine, and on reasonable expectations for the future."

In the course of her career, SARAH BLAFFER HRDY has done anthropological fieldwork in a number of countries. She is the author of *The Black-man of Zinacantan; The Langurs of Abu: Female and Male Strategies of Reproduction; The Woman That Never Evolved;* and *Mother Nature: A History of Mothers, Infants, and Natural Selection,* which was selected by both *Publishers Weekly* and *Library Journal* as one of the Best Books of 1999, and which won the Howells Prize for Outstanding Contribution to Biological Anthropology. She has been elected to the National Academy of Sciences, the American Academy of Arts and Sciences, and the California Academy of Sciences.

She writes: "My mother, like most college-educated women of her generation, was under the impression that if she rushed over to pick me up every time I cried, she would be spoiling me, conditioning me to cry more. According to long-standing psychological wisdom, babies were assumed to be 'blank slates' ready to be molded by their caretakers. It was only in the second half of the twentieth century that this view started to change.

"The main impetus for this change was British psychiatrist John Bowlby. He argued that all primate infants, including human ones, are born powerfully motivated by the 'set goal' of remaining in contact with the mother. Human infants need a 'warm, intimate, and continuous' relationship with the mother to fully develop.

"Bowlby's ideas have been widely accepted. But in the last quarter century, sociobiologists and anthropologists like myself have begun to rethink some of Bowlby's starting assumptions. In particular, Bowlby was thinking in terms of the nuclear family in which the father worked and the mother stayed home with children. But human family arrangements in the past—like families today—were more flexible and opportunistic than that model would lead us to expect, and caretaking arrangements may have been more variable than the ape model of a mother in constant contact with her infants that Bowlby had taken for granted.

"Hence, it was no accident that such questions attracted a working mother like me. By profession, I am a scientist, but deep down, my motivations are primarily personal. I am a wife, the mother of three children, looking forward one day to being an involved grandmother. But I also desired time away from children in order to work. The world around me is changing at an ever more rapid pace, bringing with it new modes of child rearing. How are these changes likely to affect our children, as well as their descendants in the future? Just what is the bottom line on the emotional needs of human infants?"

Women hackers are rare. Among women over fifty, CAROLYN MEINEL may be the only one in the world. Her mailing list of some twenty-four thousand hackers and high-traffic Web site (www.happyhacker.org) testify to her popularity among "white hat" hackers. She is the author or coauthor of three popular books (including *The Happy Hacker: A Guide to Mostly Harmless Hacking*), six technical books, many research papers in refereed journals, and over two hundred magazine articles. In addition, she has been an editor of Jim Baen's *Destinies* magazine, senior contributing correspondent to *Washington Technology,* and a regular contributor to *Technology Review.* Her adventures as a political activist for space technology and missile defense have been the topic of many newspaper, magazine, and TV news stories. Among her achievements are being the cofounder and former president of the L-5 Society, now known as the National Space Society, where she also was the editor of its magazine, now known as *Ad Astra;* serving as the executive director of the American Space Frontiers Political Action Committee; and coordinating the defeat of the 1979 United Nations Agreement Governing Activities of States on the Moon and Other Celestial Objects. She has also worked for the Defense Advanced Research Projects Agency and has even appeared as a character in Ed Regis's book *Great Mambo Chicken and the Transhuman Condition.* Since 1988 she has run a consulting company, M/B Research (www.techbroker.com), providing competitive intelligence for high-technology companies and consulting in computer security. She lives in Cedar Crest, New Mexico.

Of "Code Red for the Web," she explains: "It all started at the Def Con hackers' convention, Las Vegas, August 1995. Def Con was like the cantina scene in *Star Wars.* All sorts of characters thronged together, from buttoned-down salarymen to tattooed and pierced freaks. Since I'm a bit of a hacker, I soon found a niche in this cantina.

"Something seemed a bit odd about Def Con, however. I had expected demonstrations and groups hanging around computers trying out legal, harmless, and insanely fun hacks. Instead, there was a lot of glorifying of just plain crime. And—strangest of all—the place was crawling with federal agents with arrest powers. I tried to imagine a convention of people charged with protecting banks that would attract crowds who looked, walked, and quacked like bank robbers. That's pretty much the impression Def Con gave me. Just what was the Def Con convention series designed to accomplish? I resolved to dig into this question.

"At first there were a number of other reporters trying to understand whatever the heck was going on. However, anyone who angers computer criminals soon learns how good they are at stealing e-mail, defacing Web sites with insults and dirty pictures, and running up bogus credit card bills against people they oppose. Meanwhile, federal investigators hunger for what reporters know about cybercrime. Tell all, or go to jail for contempt of court. So it's not surprising when reporters tiptoe away from the hard questions.

"Sure enough, when I crossed the line, hackers did their worst against me. Big deal, I know how to protect myself in cyberspace. What finally got me frightened was the FBI, which began by warning me to quit writing about hackers and escalated to threatening to arrest me. I finally made an issue of their harassment in a press conference. I haven't heard from a federal agent since. Hackers and the FBI did, however, make it easier to research this story. By raising my profile, whistle-blowers came to me with bits and pieces of how the federal government appears to have built a rogue hacker army with the objective of waging deniable cyberwar. 'Code Red for the Web' shows just the tip of the iceberg of this story."

OLIVER MORTON writes mostly about scientific knowledge and its technical and cultural effects. He is a contributing editor at *Wired* magazine and in the mid 1990s was editor in chief of *Wired UK*, its European sister publication. Before that he worked at *The Economist*, spending much of his time there as science and technology editor. His work has appeared in *Nature, Science, The New Yorker, Newsweek International, Discover, Prospect*, and a range of other newspapers and magazines up to and including *The Hollywood Reporter* (but only once, alas . . .). He is also a member of the SETI policy subcommittee of the International Academy of Astronautics and a consultant to the Near Earth Object working group of the International Astronomical Union. His new book, *Mapping Mars* (Picador, September 2002), contains a lot of geologists, a dead musk ox, and a happy hour, among many other things.

He offers this background to "Shadow Science": "I first heard about the ideas that became the Kepler mission in 1992, at a meeting in San Juan Capistrano where planetary scientists were discussing all sorts of possible missions that might be undertaken under the auspices of the then-about-to-begin Discovery program. It was a bit like a vast Hollywood pitch meeting, with seventy different ideas for spacecraft put forward for discussion, acclaim, criticism, and the chance to get a budget as big as the biggest blockbuster movie's. The pitches didn't quite have the zing they do in the film business—not so much 'It's *Buffy* meets *Braveheart,*' more 'It's Pioneer Venus meets the Cassini infrared imaging spectrometer'—but for a journalist fascinated by planetary science, it was a terrific couple of days.

"I can't remember for sure whether I found the pitch for Bill Borucki's transit telescope attractive at the time, but I know that within a year or so I had become convinced that it was a neat idea, even if I had no way to judge how practical it was. I started to keep an eye on its progress. As the field of extra-solar planet-spotting leapt to life in the mid 1990s, I watched Borucki refine Kepler again and again, ceaselessly making it a more attractive project but never getting it funded. By the time I came to write about him for *Wired,* I was almost as interested in Borucki's persistence as in his ideas.

"Science journalism has a tendency to seek out the mavericks trying to overturn the mainstream, the would-be revolutionaries kicking at sacred cows; we're always sniffing after paradigm shifts. Borucki, it seems to me, epitomizes a more common and possibly more productive way to stand out from the crowd. He's not opposed to the mainstream, or championing ideas others have rejected; he just happens to have become consumed by an idea that the mainstream has neglected. In the 1980s he saw a poorly marked path that led in the same general direction as the road his peers were following, and he became determined to take that less-traveled path, convinced it was a shortcut.

"And now, finally, his conviction has spread to the people with control over the green light. In December 2001, NASA approved two new Discovery missions: Dawn, which will explore the asteroid belt, and Kepler. The mission will now be launched in 2006 (tight budgets squeezed out the possibility of a 2005 launch). And soon after that, if everything works as well as he's trying to ensure it will, and if the earth is not some strange sort of cosmic exception, Bill Borucki will start to see the shadows of other worlds."

MARY ROGAN is an award-winning journalist and essayist for both Canadian and American publications. In 2000 she won the President's Medal, which is awarded to the best magazine article in Canada. Born and raised in New York, Ms. Rogan currently lives in Toronto.

On the subject of "Penninger," she says: "There were times when I found it overwhelming to be around Josef. The scope of what he's undertaking is staggering. But always, what emerges, is his extraordinary humanity. That, even more than his genius, is Josef's real gift."

SALLY SATEL is a W. H. Brady Fellow at the American Enterprise Institute, a staff psychiatrist at the Oasis Clinic in Washington, D.C., and a lecturer at the Yale University School of Medicine. She was an assistant professor of psychiatry at Yale University from 1988 to 1995 and was a Robert Wood Johnson Health Policy Fellow with the Senate Labor and Human Resources Committee. Dr. Satel has testified before the House Ways and Means Committee, the House Government Reform and Oversight Committee, and the Senate Special Committee on Aging. She has published articles in several professional journals, including *Journal of Clinical Psychiatry, American Journal of Psychiatry,* and *Journal of the American Medical Association,* and in popular publications such as the *Wall Street Journal, The New York Times, National Review, The New Republic,* and *Slate.* Dr. Satel is the author of *PC, M.D.: How Political Correctness Is Corrupting Medicine* (Basic Books, 2001) and *Drug Treatment: The Case for Coercion* (AEI Press, 1999).

"I was prompted to write 'Medicine's Race Problem'," she explains, "because of the heated reaction to an article published in the *New England Journal of Medicine* in the spring of 2001. In short, the article examined race-related differences in response to a medication for heart failure. There was nothing remarkable about the article. Its data were properly analyzed, and its rationale was familiar: that it is important to test the effects of cardiovascular drugs in African Americans and whites because of a well-established tendency for the two groups to respond differently to the same medications.

"To my amazement, the article provoked a hostile reaction. One of *NEJM*'s editors wrote an accompanying editorial angrily accusing the researchers of 'medical racial profiling,' and subsequent newspaper accounts deemed the research offensive, even racist, because it implied genetic differences among races.

"Not only was I surprised by the vehemence of the response, I was struck by an emerging double standard. It was permissible, on the one hand, for research to be directed toward discovering social reasons for the poorer health of some minorities—indeed, the federal government, schools of public health, and health care philanthropies pour hundreds of millions of dollars into this effort. However, if inquiry into health disparities between groups yielded genetic differences or seemingly innate physiological differences between races, then some professionals decried the inquiries and even sought to have funding stopped. The article is an account of this tension within the medical research profession, and its implications."

LAUREN SLATER is a psychologist and the author of three books: *Welcome to My Country, Prozac Diary,* and *Lying.* Her articles and essays have appeared in numerous publications and have been twice anthologized in *The Best American Essays;* she has also been a National Magazine Award Finalist. She is working on a new book, *Great Psychological Experiments of the 20th Century,* to be published by W. W. Norton in 2003.

She offers these thoughts on "Dr. Daedalus": "I've long wondered about the ethics involved in various 'enhancement technologies,' be it Prozac for a mood boost or Ritalin for an energy boost, so when I came across the even more extreme idea of wings for human beings, I was enchanted and repelled and wanted to explore the competing claims those reactions had on me. Thus was born this essay."

MICHAEL SPECTER has, since 1998, been a staff writer at *The New Yorker,* where he writes mostly about science and technology. Before joining the magazine he worked at *The New York Times,* where he was a senior correspondent based in Rome, and prior to that posting, he was Moscow bureau chief. In both those jobs he frequently wrote about health care, scientific developments, and public health issues such as AIDS in Africa or the aging population of Europe. He started writing about science while he was a reporter at *The Washington Post,* where, from 1986 to 1990, he served as a medical and science reporter on the national staff. He lives in New York City with his wife, Alessandra Stanley, a reporter for *The New York Times,* and their daughter.

He writes: "I became interested in the issue of neurogenesis because it is rare that a firmly held scientific doctrine is completely overturned. For years, it has been assumed by all that humans develop no new brain cells after their infancy. Yet, over the past several years, that has proven untrue. In fact, we seem to generate new brain cells throughout our lives. It remains unclear if those new cells could one day help cure degenerative diseases. The debate about that continues; several recent studies suggest that new cells are not constantly created in the most complex and important parts of the brain. Yet other work suggests the opposite. No doubt the debate will continue for the next several years, as proof is gathered, examined, and either accepted or rejected by the scientific community."

MARGARET TALBOT is a contributing writer at *The New York Times Magazine* and a senior fellow at the New America Foundation. She is the recipient of a 1999 Whiting Writer's Award for her essay writing, examples of which have appeared in *The New Republic, The Atlantic Monthly, The Anchor Essay Annual,* and *The Art of the Essay,* as well as in *The New York Times Magazine.* She lives in Washington, D.C., with her husband, Arthur Allen, and their children, Isaac and Lucy.

Of what brought her to "A Desire to Duplicate," she reports: "When my editor, Daniel Zalewski, asked me if I wanted to look into the Raelians' announcement that they were trying to clone a dead infant, I was immediately intrigued. Who could resist the opportunity to hang around with a bunch of French-Canadian UFO cultists? I knew that, at the very least, I wouldn't have to scramble for vivid details. But as I reported the story, I began to see their plan mainly in terms of its metaphoric significance. Whether the Raelians ever succeeded in cloning a human or not, someone would. And in the meantime, it seemed to me, they had tapped into a deep longing shared by some people far more conventional and respectable than they—a longing for a union of science and science fiction, for an almost religious understanding of biotech and its possibilities. Something about why the Raelians wanted reproductive cloning—and why other people I spoke with, from fertility doctors to bereaved mothers, wanted it—made me think about the complicated and unforeseen hopes and desires that scientific progress rouses. I hope the article conveys some of that, as well as a sense of what's wrong with reproductive cloning."

GARY TAUBES is a freelance writer whose reporting on science, technology, and public health issues has appeared in numerous national publications over the past twenty years. He studied physics at Harvard, engineering at Stanford, and journalism at Columbia and is a three-time recipient of the National Association of Science Writer's Science-in-Society Journalism Award. He is also the author of two books, *Nobel Dreams* and *Bad Science.* Taubes began reporting on controversial science and pathological science (defined by the Nobelist Irving Langmuir as "the science of things that aren't so") in 1985 and has since made a habit of it.

"This article is the third in an unplanned series of articles on dubious science in public health and nutrition that I wrote for the journal *Science,*" he says. "In effect, what I did was take the lessons I had learned over a decade of reporting on the extraordinary difficulty of doing good science and of discovering reliable knowledge, and I then applied them to the field of public health. What I discovered was that much of what we now consider nutritional dogma—cutting back on the salt or fat in your diet, for instance, will reduce your risk of heart disease—could not actually be supported by the existing data. This led me to spend as much research effort documenting the history and politics of these issues as I did the science itself. I then wrote the articles to convey the entire picture—history, science, politics, and sociology—rather than the timely 'what's happening now' angle, which is virtually meaningless without context and yet is the common fodder of the daily and weekly press."

NICHOLAS WADE, a science reporter at *The New York Times,* was educated in England at Eton and at King's College, Cambridge. He worked on two scientific weekly magazines, *Nature* in London and *Science* in Washington, before joining *The New York Times* in 1981 as an editorial writer. He was the newspaper's science editor from 1990 to 1996. He has written several books, including *The Nobel Duel,* a case study of scientific competition; *Betrayers of the Truth* (with William J. Broad), an analysis of fraud and self-deception in science; and most recently *Life Script,* an account of the human genome project and its implications.

"Bjørn Lomborg's book *The Skeptical Environmentalist* struck me immediately as a serious contrarian view of an important issue," he says, "and therefore deserving of a newspaper's attention. Besides which, his personal story is irresistible—a Greenpeacenik who set out to prove Julian Simon wrong and ended up being converted to many of Simon's skeptical views about environmentalist claims of imminent disaster.

"The reaction to Lomborg's work has so far been on the shrill side. Some scientists and environmentalists have tried to paint him as a person of no weight, a non-scholar whose arguments they are therefore not obliged to take seriously. *Scientific American* published four articles attacking him but, according to Lomborg, refused to allow him a response.

"But Lomborg's criticisms of the environmental litany are trenchant, as his argument that the persistent overstatements of the environmentalist camp, however well meant, are liable to cause a vast misallocation of resources. Lomborg is not going away. With his smiling young face and Greenpeace credentials, he is hard to dismiss as a conservative crank in bed with industrial polluters. Indeed he has now been appointed director of Denmark's new Institute for Environmental Evaluation."

STEVEN WEINBERG is a member of the Physics and Astronomy Departments of the University of Texas at Austin. His research has been honored with numerous prizes and awards, including in 1979 the Nobel Prize in

Physics and in 1991 the National Medal of Science, as well as election to both the U.S. National Academy of Sciences and Britain's Royal Society, and over a dozen honorary doctoral degrees. He is the author of over two hundred scientific articles on elementary particle physics, cosmology, and other subjects and has also written for such periodicals as *The New York Review of Books, The Times Literary Supplement, The Atlantic Monthly, Time,* and *The New Republic.* His books include *Gravitation and Cosmology: Principles and Applications of the General Theory of Relativity* (1972); *The First Three Minutes* (1977); *Discovery of Subatomic Particles* (1983); *Elementary Particles and the Laws of Physics* (with Richard P. Feynman) (1987); *Dreams of a Final Theory: The Search for the Fundamental Laws of Nature* (1993); a trilogy, *The Quantum Theory of Fields* (1995, 1996, 2000); and *Facing Up: Science and Its Cultural Adversaries* (2001). His writing on science for the general reader has been honored with the Gemant Award of the American Institute of Physics and the Lewis Thomas Prize for the Scientist as Poet of Rockefeller University. Dr. Weinberg was educated at Cornell, Copenhagen, and Princeton and taught at Columbia, Berkeley, MIT, and Harvard before coming to Texas in 1982.

He writes: "Amherst College held a public lecture series on 'Science and the Limits of Explanation' in the autumn of 2000. I was glad to accept their invitation to speak in this series, as for some years I had been bothered by hearing the remark that science describes but does not explain. As things worked out, I was rewarded with a responsive audience and a beautiful October day in New England. This is the written version of my talk, as it was published the following May in *The New York Review of Books.*"

"Dr. Daedalus" by Lauren Slater. Copyright © 2001 by *Harper's Magazine.* All rights reserved. Reproduced from the July issue by special permission. "Crimson Tide" by Atul Gawande. Copyright © 2001 by Atul Gawande. All rights reserved. Reprinted by kind permission of the author. First published in *The New Yorker.* This essay appears, in slightly different form, in *Complications: A Surgeon's Notes on an Imperfect Science* by Atul Gawande, published in 2002 by Metropolitan Books. "The Made-to-Order Savior" by Lisa Belkin. Copyright © 2001 by Lisa Belkin. All rights reserved. Reprinted by kind permission of the author. First published in *The New York Times Magazine.* "A Desire to Duplicate" by Margaret Talbot. Copyright © 2001 by Margaret Talbot. First published in *The New York Times Magazine.* Reprinted with the permission of the Wylie Agency, Inc. "Medicine's Race Problem" by Sally Satel. Copyright © 2001 by the Board of Trustees of the Leland Stanford Junior University. Reprinted with permission from *Policy Review,* a publication of the Hoover Institution, Stanford University. All rights reserved. "The Thirty Years' War" by Jerome Groopman. Copyright © 2001 by Jerome Groopman. Reprinted by permission of William Morris Agency, Inc., on behalf of the author. First published in *The New Yorker.* "The Soft Science of Dietary Fat" by Gary Taubes. Reprinted with permission from *Science,* Vol. 291, pp. 2536–2545. Copyright © 2001 American Association for the Advancement of Science. "Brothers with Heart" by Joseph D'Agnese. Copyright © 2001 by Joseph D'Agnese. All rights reserved. Reprinted by kind permission of the author. First published in *Discover.* "I Love My Glow Bunny" by Christopher Dickey. Copyright © 2001 by Christopher Dickey. All rights reserved. Reprinted by kind permission of the author. First published in *Wired.* "Rethinking the Brain" by Michael Specter. Copyright © 2001 by Michael Specter. Reprinted by permission. First published in *The New Yorker.* "Penninger" by Mary Rogan. Copyright © 2001 by Mary Rogan. Reprinted by permission. First published in *Esquire.* "Mothers and Others" by Sarah Blaffer Hrdy. Copyright © 2001 by Sarah Blaffer Hrdy. Published in the May 2001 *Natural History* magazine; adapted and abridged from "The Past and Present of the Human Family," one of the Tanner Lectures on Human Values, with permission to reprint courtesy of the University of Utah Press, Salt Lake City, and the Trustees of the Tanner Lectures on Human Values. "Of Altruism, Heroism and Nature's Gifts in the Face of Terror" by Natalie Angier. Copyright © 2001 by The New York Times Co. Reprinted by permission. "Pirate Utopia" by Julian Dibbell. Copyright © 2001 by Julian Dibbell. All rights reserved. Reprinted by kind permission of the author. First published in *Feed.* "Code Red for the Web" by Carolyn Meinel. Copyright © 2001 by Carolyn Meinel. All rights reserved. Reprinted by kind permission of the author. First published in *Scientific American.* "What Brings a World into Being?" by David Berlinski. Copyright © 2001 by *Commentary.* Reprinted from *Commentary,* April 2001, by permission; all rights reserved. "Quantum Shmantum" by Tim Folger. Copyright © 2001 by Tim Folger. All rights reserved. Reprinted by kind permission of the author. First published in *Discover.* "Shadow Science" by Oliver Morton. Copyright © 2001 by Oliver Morton. First published in *Wired.* Reprinted with the permission of The Wylie Agency, Inc. "Can Science Explain Everything? Anything?" by Steven Weinberg. Copyright © 2001 by Steven Weinberg. Originally published in *The New York Review of Books.* "The Eco-Optimist" by Nicholas Wade. (Originally titled, "From an Unlikely Corner, Eco-Optimism."). Copyright © 2001 by The New York Times Co. Reprinted by permission. "George Divoky's Planet" by Darcy Frey. Copyright © 2002 by Darcy Frey. All rights reserved. Reprinted by kind permission of the author. First published in *The New York Times Magazine.*

A Note from the Series Editor

Finding good science writing is not easy. In 2001, it got harder. Two splendid magazines, rich sources for this anthology's first two editions, shut down. *The Sciences*, published by the New York Academy of Science, was a thoughtful, intelligent, and visually beautiful journal that, without ever condescending to its readers, managed to convey the intellectual adventure of science. *Lingua Franca*, which spiritedly covered the world of academia, featured many fine science-related articles. In addition, readers of this edition will notice that Julian Dibbell's "Pirate Utopia" originally appeared on the Web site feed.com, which has since ceased operation. Feed often ran smart, insightful essays on science and technology. All these publications will be missed.

Submissions for next year's volume can be sent to Jesse Cohen, c/o Editor, The Best American Science Writing 2003, Ecco/HarperCollins, 10 East 53rd Street, New York, New York 10022. Please include a brief cover letter; manuscripts will not be returned. Submissions made electronically are also welcomed and can be e-mailed to< jesseicohen@netscape.net>.

Other Books in THE BEST AMERICAN SCIENCE WRITING Series:

THE BEST AMERICAN SCIENCE WRITING 2000

Edited by James Gleick
ISBN 0-06-019734-X (hardcover)
ISBN 0-06-095736-0 (paperback)
ISBN 0-694-52399-2 (audio)

This first volume in an annual series carries the imprimatur of Pulitzer Prize nominee James Gleick, one of our foremost chroniclers of scientific social history.

This stellar collection includes the writings of James Gleick, George Johnson, Jonathan Weiner, Sheryl Gay Stolberg, Deborah M. Gordon, Francis Halzen, Timothy Ferris, Stephen S. Hall, Floyd Skloot, Denis G. Pelli, Douglas R. Hofstadter, *The Onion*, Don Asher, Natalie Angier, Stephen Jay Gould, Susan McCarthy, Peter Galison, and Steven Weinberg.

"***The Best American Science Writing 2000*** **is richly informative, wide-ranging, and intellectually provocative. It conveys the ongoing struggle of scientists to understand the physical world, and themselves.**"—Alan Lightman

THE BEST AMERICAN SCIENCE WRITING 2001

Edited by Timothy Ferris
ISBN 0-06-621164-6 (hardcover)
ISBN 0-06-093648-7 (paperback)

Pulitzer Prize and National Book Award nominee Timothy Ferris, one of the foremost writers about astronomy, edited this second volume of outstanding science writing. Like the 2000 edition, *The Best American Science Writing 2001* covers the full range of scientific inquiry—from biochemistry, physics, and astronomy to genetics, evolutionary theory, and cognition.

The contributors include: John Updike, Michael S. Turner, Natalie Angier, Joel Achenbach, Erik Asphaug, John Archibald Wheeler, Stephen S. Hall, Richard Preston, Peter Boyer, John Terborgh, James Schwartz, Ernst Mayr, Greg Critser, Andrew Sullivan, Malcolm Gladwell, Helen Epstein, Debbie Bookchin and Jim Schumacher, Stephen Jay Gould, Tracy Kidder, Jacques Leslie, Robert L. Park, Alan Lightman, and Freeman Dyson.